Genetic Improvement of Tropical Crops

Hugo Campos • Peter D.S. Caligari

Genetic Improvement of Tropical Crops

Springer

Hugo Campos
International Potato Center
Lima, Peru

Peter D.S. Caligari
Verdant Bioscience,
Galang, North Sumatra, Indonesia

ISBN 978-3-319-59817-8 ISBN 978-3-319-59819-2 (eBook)
DOI 10.1007/978-3-319-59819-2

Library of Congress Control Number: 2017954958

Printed on acid-free paper

This Springer imprint is published by Springer Nature
The registered company is Springer International Publishing AG
The registered company address is: Gewerbestrasse 11, 6330 Cham, Switzerland

Foreword

Agri-food systems in the tropics will have to evolve rapidly over this century to keep pace with expanding and more diverse food demands from a rapidly growing more urban population (especially in sub-Saharan Africa), putting pressure on the natural resource base, against the backdrop of climate change and increased biotic and abiotic stresses. Sustained or increased genetic gain will be essential to meet these challenges. It is fortunate as we look to these challenges that we are at a time of unprecedented expansion of genomic resources, biotech methods, bioinformatics, and statistical approaches which can support genetic improvement. Hence, the book opens with important contributions on genomic selection and a statistics update including new approaches for handling messy field research data through linear mixed models. The central part of the book clearly lays out the status of genetic improvement among some of the most important tropical crops and key elements needed for future development. This book will appeal to a wide range of audiences as a synthesis of our state of knowledge in this area of such critical importance.

The crop chapters are interesting both for some of the underlying shared constraints and messages and also for the diversity of the contexts and the variable progress made across the crops. A majority of the chapters focus on what are usually considered staple food crops (bananas, cassava, maize, rice, and sweetpotato), where international agricultural research, much of it at the CGIAR, has played a key role, reflected in the authorships of these chapters. For these crops, sustained public investment is likely to be a perquisite for broad-based agricultural investment with an array of public-private partnerships needed for fast dissemination of new varieties. Two chapters cover industrial crops (sugarcane and oil palm) where crop improvement is predominantly a private sector endeavor. The editors of the book are to be commended for bringing all this diversity together in a single volume.

Although ample treatment is given of genomic selection and of opportunities for genetic modification, most of the chapters make the argument that conventional plant breeding schemes will continue to be the main driver of genetic gain, although guided by increasingly better genomic information and statistical analyses. However, the different crops are at quite different points along a spectrum of understanding and managing genetic variability which lies at the heart of crop improvement.

Tropical maize which has been produced on the progress made with hybrids in temperate maize and rice lie on one side of the spectrum, whereas cassava and sweetpotato on the other. So maize breeding at CIMMYT has mostly shifted to developing inbred lines, with open-pollinated varieties only provided for less productive environments. Interestingly, although private companies in the tropics mostly generate revenue from selling hybrids, they are prepared also to provide open-pollinated varieties. Rice has benefited from sustained crop improvement and better genomic information to get ahead of the curve such that SNP markers are available for the major yield enhancing functional genes associated with a significant part of the yield increase achieved thus far. Significant investment into cassava and sweetpotato breeding has been more recent, and heterozygosity makes breeding intrinsically more challenging for these crops. Cassava breeders from CIAT argue that until inbred lines of cassava are available for hybrid breeding, then cassava will not be able to achieve further significant yield gain, while sweetpotato breeders at CIP are pursuing a novel approach of developing split breeding populations for crossing to exploit hybrid vigor. Sugarcane, bananas, and oil palms as long-cycle crops face additional challenges for breeding, but banana because of the need to get back to a seedless and sterile variety perhaps faces some of the toughest challenges for breeding of all the crops in this fascinating book.

The updated breeding results presented here clearly attest to the value of investment in public plant breeding efforts to accelerate genetic gain through improved varieties as central to the rural transformation needed to improve the quality of life of rapidly growing populations in the tropics.

With the exception of maize, rice, and oil palms, these are clonally propagated crops which pose particular problems for seed systems to handle bulky and perishable planting materials. So not surprisingly, these chapters pay especial attention to developing seed systems with some important lessons learned across this crop group.

So, I sincerely hope the readers will enjoy reading this book as much as I have and that it contributes to and stimulates learning among all those involved in one way and another in supporting crop improvement of this absolutely vital set of tropical crops which will be one of the most pressing endeavors of humanity over the rest of this century.

Graham Thiele
CGIAR Research Program on Roots
Tubers and Bananas led by the International Potato Center
Lima, Peru

Preface

The human population is growing at a significant rate and is destined, although predictions vary, to reach 9.5 billion by 2050. However, such a global figure disguises more dramatic growth population current facts and trends affecting the tropics:

- By 2050, about 50% of the world's population will call the tropics "home", and the third largest country in terms of population, after India and China, will not be the United States any longer. Instead, it will be a tropical country, Nigeria.
- Seven out of nine countries where over 50% of the population growth is expected between now and 2050 (India, Nigeria, Democratic Republic of the Congo, Ethiopia, United Republic of Tanzania, Indonesia, and Uganda) are tropical countries.
- More than two-thirds of the world's population living in extreme poverty live in the tropics.
- By 2050, the tropics will host most of the world's people and two-thirds of its children.

Furthermore, the tropics are facing urbanization rates rising faster than that being experienced globally. Therefore, there is an increasing need to provide to its inhabitants not only food security but also nutrient security. At the same time, they are facing accelerated environmental degradation, and the predicted impacts of global climate change will affect. These countries more severely harder than most others in terms of their ability to provide their own food while also striving to increase their food exports and income. It is very clear that the challenge, of just sustaining, not to mention to significantly increasing the production of affordable, nutrient-rich staple crops in tropical countries, is daunting.

Unfortunately, for historic, agronomic, and political reasons, most of the attention, especially for research and development into agriculture and food production, and particularly crop genetic improvement, has been on a few major crop species. These are ones that have been cultivated on a historical basis in temperate regions of the world, mainly Europe, North America, and Central Asia. Notwithstanding that continuing effort, there is a concerning plateau in several major temperate crop species in terms of their response to artificial selection. Moreover, yield stagnation has been reported in some of the world's most intensive cropping systems such as rice in East Asia, maize in South Europe, and wheat in Northwest Europe.

What can be done, in the face of the trends, facts, and predictions described above, to increase the food and nutrient security of the growing mass of people calling the tropics home? It should be remembered that this also has a major impact on the economic prospects of many tropical countries where agriculture remains as a major economic and social force in terms of income and employment generation.

One of the more effective ways to increase food production in the tropics is not only to secure the availability of locally produced, affordable, quality food and contribute to food security at the household level but also to increase food production in a sustainable manner to generate family income and, at a more aggregate level, build export revenues. A major component of in this will be through the development and adoption of new improved varieties through genetic enhancement. In general terms, such varieties will provide farmers with higher yielding ability and with higher efficiencies in exploiting lower chemical inputs alongside enhanced tolerance to abiotic stresses, such as drought and heat, and biotic ones, such as pest and diseases. Increasingly, concurrently with the above, crop genetic improvement of tropical crops is being used to increase the content of compounds associated with the well-being and health status of people, particularly expectant mothers and children under 5 years old.

This book is our humble effort, aided by our talented colleagues as authors, to fill in the dearth of information and insight about the genetic improvement of crops adapted to tropical conditions, thus providing a fresh, updated yet rigorous perspective of the status and prospects for the genetic improvement of a diverse array of tropical crops. In order to enrich and expand their knowledge, while conveying more value to its readers, by design, the book provides breadth through:

- Addressing crops propagated through seed and crops propagated by diverse vegetative means. A conspicuous difference between temperate and tropical agriculture is the disproportionately high number of tropical staple crops which are propagated through diverse plant organs other than via botanical seeds.
- The selected group of authors assembled reflects the increasing share of global plant breeding endeavors carried out by industry and includes the perspective of private experts in plant genetic improvement.
- Crops mainly used for household or traditional food production, but also others which are grown to be processed by agroindustry, such as oil palm and sugarcane, are discussed.

The book is organized into two main parts. Its first part, enabling technologies, covers two aspects which are relevant across crops, namely, how to maximize the use of genetic information through current molecular approaches and how to use statistics as a tool to sustain increased genetic gains and breeding efficiency. Also covered are the possibilities of a molecular breeding approach of recent application in crop plants, genomic selection, which effectively removes many of the constraints hampering a meaningful impact in terms of genetic gains and selection efficiency that former molecular breeding tools encountered.

The second part of the book provides an updated view of seed-propagated crops, such as rice and maize, as well as crops propagated through vegetative means such as sweetpotato, cassava, banana, and sugarcane. Each chapter addresses the main breeding objectives, markets served, current breeding approaches, biotechnology, genetic progress observed, and in addition a glimpse into the future for each of these selected and important tropical crops.

While thinking about, planning, and compiling this book, we were also acutely aware of the diminishing numbers of professionals, academics, and students who are following or developing careers in agriculturally related subjects worldwide. This is in fact nowhere more obvious than in genetic improvement and plant breeding. If one considers the information above and the rising population, then put it together with an ever-decreasing number of students registering for relevant courses, the prospect is frightening and more so in the tropics because of the relatively lower number of universities, research organizations, and funding opportunities to develop the next wave of passionate experts in the genetic improvement of crops.

This book will not rescue the precarious state of plant breeding, but we trust it will at least form a basis for continued effort to improve tropical crops. We hope it may just stimulate a few more researchers to consider allying themselves with those breeders who are making valiant efforts to improve the various crops discussed, more students to pursue graduate studies in tropical crops, and funding organizations to consider increasing their support for the genetic improvement and other aspects of tropical crops. If such accomplishments take place, our work and that of the chapter contributors would be more than fully justified.

Lima, Peru — Hugo Campos
Talca, Chile — Peter D.S. Caligari

Acknowledgments

The path to writing this book began about 18 months ago, when after finishing a previous writing assignment we wondered how to contribute to revamp the rather neglected – in terms of research and development funding as well as critical mass – genetic improvement of crop species growing in the tropics. It has been a wonderful voyage, not devoid of fun nor exempted from unexpected events, and, above all, hugely rewarding. Now that the same path is leading us to the end of this journey, we wish to thank the many talented people who made this book possible.

First, thanks must go to all authors contributing chapters, who selflessly devoted part of their most priceless asset, time, to discuss among themselves and then write draft chapters, which they then polished into what would subsequently become the final chapters assembled in this volume. We equally thank the organizations that they are affiliated with for allowing them to take on this writing task. Nevertheless, we accept that any of the errors remaining are solely ours and our inattentive reading through all this material – a task from which we learned a lot.

We would also like to sincerely thank the Springer team, namely, Roberta Gazzarolle for being the first willing to trust us and support this project and then Luciane Christante de Mello and Susan Westendorf for expert editorial support and encouragement, and for gently keeping us in line with our declared deadlines.

Hugo Campos would like to express his gratitude to many mentors and colleagues and particularly to Patricio Barriga (RIP) at the Universidad Austral de Chile for introducing him, many years ago, to the fascinating word of genetics and crop genetic improvement, and to the International Potato Center for the support to finish this book. He also wishes to wholeheartedly thank his wife, Orietta, and two grown-up children, Ignacio and Noelia, for their constant love, understanding, encouragement, and for tolerating the arrival of yet another writing assignment.

Peter Caligari would also like to thank the many colleagues and especially his mentor and supervisor Professor Sir Kenneth Mather (RIP), in Birmingham University, who helped him understand the wonderful intricacies of genetics at all levels and its profound importance in so many aspects of the world about us. Like Hugo, his family as ever, have been patient and tolerant while he withdrew into "the world of his computer" – he thanks them for their love, patience, and understanding.

This book is dedicated to our families and their future.

Contents

Contributors

Delphine Amah International Institute of Tropical Agriculture, Ibadan, Nigeria

Maria I. Andrade International Potato Center, Maputo, Mozambique

Yassier Anwar Verdant Bioscience, Timbang Deli Estate, Galang, North Sumatra, Indonesia

Felicity C. Atkin Sugar Research Australia, Gordonvale, Australia

Syamsul Bahri Verdant Bioscience, Timbang Deli Estate, Galang, North Sumatra, Indonesia

Allan Brown International Institute of Tropical Agriculture, Arusha, Tanzania

Peter D.S. Caligari Verdant Bioscience, Timbang Deli Estate, Galang, North Sumatra, Indonesia

Hugo Campos International Potato Center, Lima, Peru

Edward E. Carey International Potato Center, CSIR-CRI, Kumasi, Ghana

Hernan Ceballos Centro Internacional de Agricultura Tropical, Cali, Colombia

Jennifer Ciomas Verdant Bioscience, Timbang Deli Estate, Galang, North Sumatra, Indonesia

Gregory O. Edmeades Independent Consultant, Cambridge, New Zealand

Brian P. Forster Verdant Bioscience, Timbang Deli Estate, Galang, North Sumatra, Indonesia

Antonio Franco Garcia Escola Superior de Agricultura Luiz de Queiroz (ESALQ), Universidade de São Paulo (USP), Piracicaba, SP, Brazil

Salvador A. Gezan University of Florida, Gainesville, FL, USA

Wolfgang J. Grüneberg International Potato Center, Lima, Peru

Clair H. Hershey Centro Internacional de Agricultura Tropical, Cali, Colombia

Kshirod K. Jena Plant Breeding Division, International Rice Research Institute, Metro Manila, Philippines

Deborah Karamura Bioversity International, Kampala, Uganda

Eddy S. Kelanaputra Verdant Bioscience, Timbang Deli Estate, Galang, North Sumatra, Indonesia

Jerome Kuriba National Agricultural Research Organization, Kampala, Uganda

Jan W. Low International Potato Center, Nairobi, Kenya

Hassan Mduma International Institute of Tropical Agriculture, Arusha, Tanzania

Robert O.M. Mwanga International Potato Center, Kampala, Uganda

Eero A.J. Nissila Plant Breeding Division, International Rice Research Institute, Metro Manila, Philippines

Newfields Consulting Ltd, Punkalaidun, Finland

Fazrin Nur Verdant Bioscience, Timbang Deli Estate, Galang, North Sumatra, Indonesia

Moses Nyine International Institute of Tropical Agriculture, Kampala, Uganda

Rodomiro Ortiz Swedish University of Agricultural Sciences, Uppsala, Sweden

B.M. Prasanna International Maize and Wheat Improvement Center, ICRAF House, Gigiri, Kenya

Safrina Rahmah Verdant Bioscience, Timbang Deli Estate, Galang, North Sumatra, Indonesia

Heru Rusfiandi Verdant Bioscience, Timbang Deli Estate, Galang, North Sumatra, Indonesia

Umi Setiawati Verdant Bioscience, Timbang Deli Estate, Galang, North Sumatra, Indonesia

Baihaqi Sitepu Verdant Bioscience, Timbang Deli Estate, Galang, North Sumatra, Indonesia

Joanne K. Stringer Sugar Research Australia, Indooroopilly, Australia

Rony Swennen International Institute of Tropical Agriculture, Arusha, Tanzania

David Telengera National Agricultural Research Laboratories, Kampala, Uganda

Walter Trevisan Independent Consultant, Dekalb, IL, USA

Robooni Tumuhimbise National Agricultural Research Organization, Kampala, Uganda

Brigitte Uwimana International Institute of Tropical Agriculture, Kampala, Uganda

Luís Felipe Ventorim Ferrão Escola Superior de Agricultura Luiz de Queiroz (ESALQ), Universidad de São Paulo – ESALQ, Piracicaba, SP, Brazil

G. Craig Yencho Department of Horticulture, North Carolina State University, Raleigh, NC, USA

Marvellous Zhou Department of Plant Breeding, South African Sugarcane Research Institute, Mount Edgecombe, South Africa

Part I
Enabling Technologies

Chapter 1
Statistical Approaches in Plant Breeding: Maximising the Use of the Genetic Information

Joanne K. Stringer, Felicity C. Atkin, and Salvador A. Gezan

1.1 Introduction

Breeding programmes deal with large number of activities, including evaluation of hundreds or thousands of genotypes and selection of the best individuals to comprise the next generation of individuals or to be released as new cultivars. Genetic testing is an expensive task that constitutes the largest activity performed in any breeding programme. Phenotyping of genotypes is particularly demanding on small breeding programmes, such is the case of most tropical crops, and for this reason, all activities that aim to maximise (or optimise) the use and quality of the information generated from genetic tests are critical. The basis for this evaluation and selection originates from data and information generated from field and greenhouse experiments, so these need to be carefully planned and analysed.

Genetic tests can be optimised through three different ways: (1) design of experiments, (2) implementation and measurement of trials and (3) statistical analysis. Appropriate selection of the experimental design, their implementation and then their statistical analyses can yield considerable benefits resulting in greater precision of estimates of genetic parameters leading to increased genetic gains from successful selections, and better operational decisions that depend on information obtained from genetic tests, such as heritability, genotype-by-environment interactions, trait-to-trait correlations, etc. Such optimisation can be classified into 'a priori'

J.K. Stringer (✉)
Sugar Research Australia, Indooroopilly, Australia
e-mail: jstringer@sugarresearch.com.au

F.C. Atkin
Sugar Research Australia, Indooroopilly, Gordonvale, Australia
e-mail: fatkin@sugarresearch.com.au

S.A. Gezan
University of Florida, Gainesville, FL, USA
e-mail: sgezan@ufl.edu

H. Campos, P.D.S. Caligari, *Genetic Improvement of Tropical Crops*,
DOI 10.1007/978-3-319-59819-2_1

and 'a posteriori': the former related to actions that are implemented before the experiment is established (i.e. at the design stage), while 'a posteriori' are those actions that are critical to implement once the experiment is established and often relate to tools to be used in for statistical analysis.

There is a plethora of classical and modern literature on the ideal characteristics of a wide array of experimental designs. However, no single design will suit all experimental objectives and environmental conditions found in field tests around the world. Hence, the choice of the 'best' design must be made carefully. Statistical and computational tools can be used to generate experimental layouts with great efficiency, where, as always, the principles of replication, randomisation and blocking are critical (for more details about these principles, see Welham et al. (2014)).

Randomised complete block (RCB) designs are the most frequently used in plant breeding. Blocking is important to minimise variability, and this design is effective when within-replicate (or block) variability is relatively small. Where there is large site heterogeneity, or when there are many genotypes to be evaluated, other experimental designs can be more efficient. For example, incomplete block (IB) designs allow for a better control of site heterogeneity by specifying smaller compartments that include a (planned) subset of the genotypes to be tested. IB designs are often generated by implementing an alpha design, a particular class of IB design where the number of genotypes (or entries) is a multiple of block size (John and Williams 1995). Another efficient alternative is the use of row-column (RC) designs that consider both row and columns within a replicate as complete or incomplete blocks. Both of these designs provide greater control of site heterogeneity and can be generated using an array of public and commercial software. For more details about the use and analysis of these designs in the context of plant breeding, we recommend Williams et al. (2002). Other efficient design options include the use of restricted randomisation such as latinisation, nested structures and spatial designs (Whitaker et al. 2002), which can increase the efficiency of the experiments.

For early generation variety trials where large numbers of genotypes are often tested, there may be insufficient planting material to replicate all genotypes. One of the most widely used designs is the use of grid plots where checks (or control genotypes) are repeated several times arranged in a block or incomplete block, depending on the experimental design implemented. Test genotypes are unreplicated and allocated at random to the remaining plots. Examples of this are the various augmented block designs developed by Federer (1956) and Federer and Raghavarao (1975). In an alternative approach, Cullis et al. (2006) proposed the use of partially replicated (p-rep) designs in which a subset of the test genotypes are replicated two or more times, and these are arranged in a resolvable spatial design. Then, the unreplicated test genotypes are randomly allocated to the remaining plots. For a fixed amount of resources, Cullis et al. (2006) found that p-rep designs result in a greater genetic gain than augmented designs.

The second optimisation of genetic testing focuses on the implementation of and measurement within a field design. Here, it is important to observe carefully all operational aspects of field testing, including documentation, labelling, site preparation and crop maintenance. One aspect that is critical here refers to preparing the site in such a way that environmental heterogeneity is minimised. This applies to all soil selection and preparation before planting and its management while the trial is active. In addition, to ensure the best quality of the phenotypic data originating from these trials, adequate definitions of response variables and clarity and consistency on measurement protocols are critical. Any actions that decrease experimental noise will increase the precision of the estimation of genetic parameters and, therefore, increase heritability estimates.

It is also important to collect the most accurate and reliable data that will be used to make decisions on which genotypes are rejected, advanced or ultimately commercially released. For example, Australian sugarcane breeders evaluate genotypes on the basis of their relative economic genetic value for traits of commercial importance – how much value would a genotype add to industry profitability if grown commercially (Wei et al. 2006).

Having collected the data, the genetic tests can be optimised through statistical analysis. This has been an area that has had several important advances over the last few decades. Of special interest for plant breeding is the use of linear mixed models (LMM) that combine estimation procedures such as residual maximum likelihood (REML) to estimate variance components and to predict random effects (or best linear unbiased predictions, BLUP). LMMs are an extension to the traditional linear models (LM) that allow for more flexible assumptions such as correlations among experimental units (e.g. temporal correlation) and among effects (e.g. by considering the numerator relationship matrix of genetic effects or BLUP) and heterogeneity of variances (e.g. different error variances for each block or site).

Modern analysis of complex and unbalanced data to obtain parameters, such as site-to-site and trait-to-trait genetic correlations, is possible by fitting LMMs that estimate variance components. Spatial analysis (Gilmour et al. 1997) of field experiments is a useful tool that incorporates the co-ordinates of the experimental units (plots or plants) into the LMM to account for physical proximity by modelling the error structure (i.e. correlations among observations), something that can be extended easily to also model competition among neighbouring plants. This is also particularly important with augmented and p-rep designs where spatial analysis allows for extracting better genetic information from the unreplicated test genotypes.

The greatest benefit of LMMs is that it is possible to combine data from many sources, with different levels of unbalance, into a complex model that will maximise the use of this information to estimate genetic parameters. For example, multi-environment trials (MET) use information from several trials, where not all genotypes are present in all sites, and for each site, there might be different numbers of replicates and precision and therefore heritabilities. These trials are evaluated together into a single LMM to estimate overall breeding values and genetic correlations among sites.

Many statistical tools can be implemented 'a posteriori' given a field dataset. One of these is post hoc blocking, where, for a given experimental layout (say a RCB design), a new blocking structure is superimposed on top of the original, and a new linear model is fitted as if the superimposed blocking structure belonged to the original design (Gezan et al. 2006). This tool increases the precision of estimates of heritability and of the predicted genetic values at little extra cost by only marginally increasing the complexity of the analysis.

In the next sections, some of these modern statistical approaches, such as interplot competition and spatial analysis, will be first defined and then illustrated in more detail.

1.2 Accounting for Interplot Competition

In early-stage selection trials, most plant improvement programmes face the challenge of finding a few incrementally superior individuals from among a large number of lines produced by cross-pollination (Stringer et al. 2011). Due to limitations on planting material and space for field testing, genotypes are often planted in trials in small, partly replicated, single-row plots. Such trials are subject to variation arising from spatial variability and interplot competition, which makes the identification of elite genotypes problematic. Unless accounted for, spatial variability and interplot competition may seriously affect the estimates of genetic merit and, hence, reduce genetic progress.

Interplot competition (also known as interference) arises when a treatment or response on one experimental or measurement unit may affect the response on neighbouring units (Martin and Eccleston 2004) and is caused by both genetic and environmental sources (Magnussen 1989). It is difficult to quantify and there is no universal method to account for the competitive interactions among genotypes. As resource limitations generally preclude the use of multi-row plots to account for interplot competition, statistical approaches have been developed to adjust for competition in the design and analysis of field trials.

One alternative is to use appropriate experimental layouts, such as the neighbour-balanced (NB) designs suggested by Williams (1952), Street and Street (1987) and Azaïs et al. (1993). However, these designs are not practical where large numbers of genotypes are to be screened, due to the number of replicates required to achieve balance between neighbouring genotypes (Kempton 1982).

In regard to statistical analyses, Besag and Kempton (1986) presented two approaches to estimate interplot competition. Building on earlier work by Kempton (1982), they developed the *phenotypic interference model*, which is a simultaneous autoregressive approach where competition is assumed to be directly related to yields of neighbouring plots. This has been applied successfully to a wide range of crops including sugar beet (Kempton 1982; Durban et al. 2001), potatoes (Connolley et al. 1993), swedes (Bradshaw 1989) and trees (Resende et al. 2005).

The second model developed by Besag and Kempton (1986) is the *treatment* or *genotypic interference model* and was originally proposed by Pearce (1957). In this model, competition effects are associated with genotype differences in characteristics such as plant height, tillering ability, date to maturity and canopy size (Kempton and Lockwood 1984; Talbot et al. 1995). Here, competition effects are associated with the average genotypic value of the nearest neighbouring genotypes rather than the phenotypic response (Stringer et al. 2011). In addition, each treatment is assumed to have a direct effect and a neighbour effect on adjacent plots.

1.3 Incorporating Spatial Variation

In early-stage field trials, which are typically large, growing conditions may be quite variable across the trial area, leading to the phenomenon known as spatial variability. One of the oldest techniques available to minimise the effect of this variability is the method of check (or control) plots (Wiancko 1914) in which replicated plots are distributed over the trial site as checks and are used as a benchmark to assess the yields of test plots. It is assumed that the checks and test varieties show the same general pattern of response to soil fertility over a trial as the test varieties. If this is not true, then the method of check plots will actually increase the error of assessment (Kempton 1984a; Besag and Kempton 1986). An alternative approach, which may be more useful for dealing with small-scale variation, is spatial or nearest neighbour (NN) analysis where a plot parameter is adjusted by using information from immediate neighbours. Although Papadakis (1937) proposed the earliest NN method, it lacked efficiency.

Spatial methods were largely neglected by statisticians until Wilkinson et al. (1983) developed the *smooth trend plus independent error model* on which most spatial models have since been based (Stringer et al. 2012). Since then, there have been many alternative approaches, including the one-dimensional models of Gleeson and Cullis (1987) and the two-dimensional approaches of Cullis and Gleeson (1991). Spatial analysis has been successfully applied to early generation trials by Cullis et al. (1992), who found that the response to selection for the spatial method was greater than for check plot method proposed by Wiancko (1914). In all of these models, the covariance structure of the plot errors was modelled as a single component. These techniques were later extended by Gilmour et al. (1997), who demonstrated that modelling plot errors alone as a single process may not be appropriate in most cases, requiring the spatial variation to be partitioned into three components. This approach is currently used to analyse over 1000 cereal variety trials in Australia annually (Stringer et al. 2012) and has resulted in increased accuracy and precision in the estimates of genotype effects in a wide range of crops (Apiolaza et al. 2000; Dutkowski et al. 2002; Gilmour et al. 1997; Grondona et al. 1996; Qiao et al. 2000; Sarker et al. 2001; Silva et al. 2001; Singh et al. 2003).

The methods developed by Gilmour et al. (1997), and later refined by Stefanova et al. (2009), partition spatial variation into three additive components. Atkin (2012) defines these as:

- *Local trend*, reflecting small smooth changes due to parameters such as fertility, soil moisture and light
- Nonstationary *global trend* which is usually aligned with the columns and rows of a field trial and associated with large-scale changes across the trial, for example, large-scale moisture or fertility gradients

- *Extraneous variation*, which usually arises from management practices or experimental procedures that have recurrent patterns, such as spraying operations or serpentine harvesting (harvesting columns of rows in alternating directions)

Previous approaches at removing global trend involved first or second differencing of the data (Gleeson and Cullis 1987), but this often overcomplicated the model. Gilmour et al. (1997) recommend directly fitting nonstationary global trends through the use of polynomial or spline functions (Verbyla et al. 1999) to the row and column co-ordinates. The modelling approach developed by Gilmour et al. (1997) is a sequential approach and commences by including design factors such as replicates that reflect the trial design. Next, modelling local trend is undertaken using a first-order separable autoregressive process in the row and column directions. During the modelling procedure, diagnostic tools, such as the sample variogram and trellis plots, play a large role in determining what effects should be included in a model. This is always followed by formal assessment by using the Wald test for fixed effects and REML likelihood ratio test for random effects.

1.4 Modelling Competition and Spatial Variation

As indicated earlier, interplot competition arises when a treatment or response on one unit affects the response on neighbouring units (Martin and Eccleston 2004). For example, in sugarcane, estimates of cane yield are affected more by interplot competition than are estimates of sugar content when genotypes are evaluated in single-row plots (Fig. 1.1), because plants in adjoining plots compete for resources such as water, fertiliser and sunlight (Jackson and McRae 2001; McRae and Jackson 1998; Skinner 1961; Stringer and Cullis 2002). This often results in a negative correlation between neighbouring plots, biasing estimates of cane yield.

Although there are many approaches in the literature that individually model spatial variability or interplot competition, there are only a few studies that jointly account for both sources of bias. Durbán Reguera (1998) and Durban et al. (2001) presented one such approach. They used cubic smoothing splines to model spatial global trend together with the phenotypic interference model for competition (Stringer et al. 2011). Genotype effects were considered fixed and adjusted profile likelihood was used for parameter estimation (McCullagh and Tibshirani 1990). This model was limited by not considering genotypes to be random nor incorporating a spatial process to model local trend. In a small simulation study based on the Rothamsted downy mildew data, Durbán Reguera (1998) found that the profile likelihood gave biased estimates of the variance components and in some cases the competition parameter was also biased. However, when using McCullagh and Tibshirini's adjustment to the profile likelihood, bias in the parameters of interest was small. Matassa (2003) developed a method combining both models from Besag and Kempton (1986) for interplot competition together with the methods of Gilmour

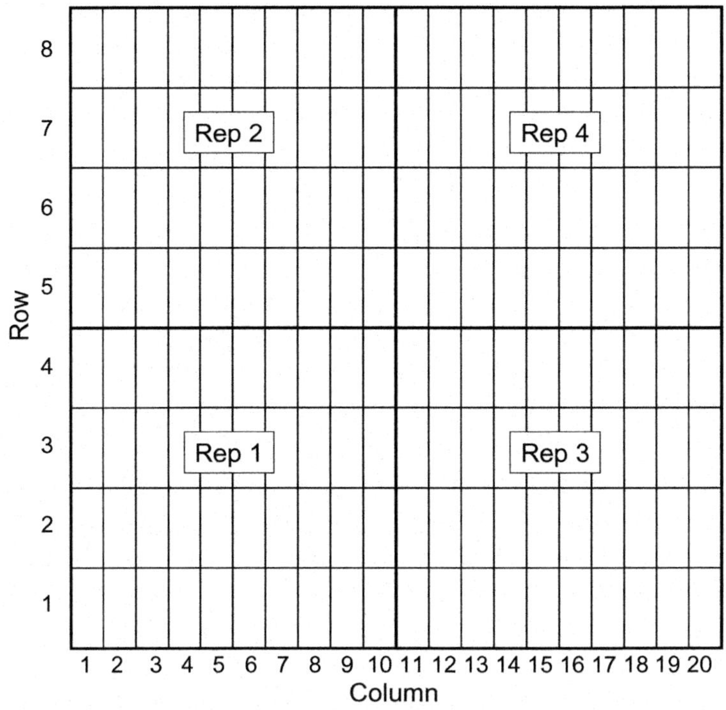

Fig. 1.1 A typical sugarcane family trial layout in a rectangular array of plots

et al. (1997) for spatial variability. Matassa's approach was similar to Durbán Reguera (1998) and Durban et al. (2001) in that genotype effects were fixed. However, Matassa (2003) used marginal likelihood and profile likelihood for parameter estimation. On comparing the estimation procedures in a simulation study, Matassa (2003) found that the preferred method depended on what terms were included in the design matrix and also on the sign of the trend parameter.

Stringer et al. (2011) developed an alternative approach to jointly model spatial variability and interplot competition. They partitioned spatial variability into global trend and extraneous variation (Gilmour et al. 1997) and allowed for both genotypic (Besag and Kempton 1986) and residual level competition. Genotype effects were considered to be random, as recommended by Smith et al. (2001), and REML was used for parameter estimation. Stringer et al. (2011) presented two simultaneous autoregressive processes to model competition at the residual level. They recommended an equal-roots second-order autoregressive model for trials where competition is dominant and an equal-roots third-order autoregressive model where both competition and spatial variability exist.

In sorghum breeding trials in Australia, parental lines are evaluated in single-row plots where both interplot competition and spatial variability are present (Hunt et al. 2013). Hunt et al. (2013) extended the methods of Stringer et al.

(2011) used for sugarcane clonal trials, by incorporating pedigree information into a LMM. This allowed Hunt et al. (2013) to partition total genetic effects into additive and nonadditive components for parent evaluation in the presence of both competition and spatial effects. The methods developed by Stringer et al. (2011) and Hunt et al. (2013) are routinely applied to sugarcane clonal and full-sib family (produced from biparental cross-pollination) trials from Queensland, Australia. In such trials, clones and families are evaluated in single-row plots and large spatial trends and interplot competition are regularly present. The presence of spatial variation and interplot competition effects in sugarcane family trials will probably have a similar effect on estimating additive genetic effects of sugarcane parents as with estimating total genetic or clonal effects and lead to biased estimates of breeding values (BV) for cane yield (in the presence of both spatial variation and interplot competition) and sugar content (in the presence of spatial variation only). In turn, this will bias the ranking of parents and so impact the outcomes of parental selection.

The two statistical models explained below (a basic RCB design model with and without modelling spatial variation) can be used to estimate additive genetic effects of sugarcane parents from family trials for parent selection (Atkin 2012) by applying some of the techniques used by Stringer et al. (2011) and Hunt et al. (2013).

1.4.1 Base Model: RCB Without Modelling Spatial Variation

Consider an experiment consisting of p trials that contains a total of m genotypes (families). Each trial is laid out in a rectangular array of r rows and c columns ($n = r \times c$) (Fig. 1.1). Where the data are ordered as rows within columns, the mixed linear model for $\mathbf{y}^{(n\times 1)}$ combined across trials is

$$\mathbf{y} = \mathbf{X}\mathbf{b} + \mathbf{Z}_g\mathbf{g} + \mathbf{Z}_u\mathbf{u} + \mathbf{e}$$

where $\mathbf{b}^{(b\times 1)}$ is a vector of fixed effects with the associated design matrix $\mathbf{X}^{(n\times b)}$; $\mathbf{g}^{(mp\times 1)}$ contains the random genotype and genotype by environment effects of m entities in each of p trials with indicator matrix $\mathbf{Z}_g^{(n\times mp)}$; $\mathbf{u}^{(d\times 1)}$ contains the random replicate effects with associated design matrix $\mathbf{Z}_u^{(n\times d)}$; and $\mathbf{e}^{(n\times 1)}$ is a vector of plot error effects combined across trials. Vector $\mathbf{b}$ contains only an overall mean effect for each trial or more complex design structures.

Here, vector $\mathbf{g}$ is the random genotypic effect of unique parents (for each trial), where a sugarcane parent can be used as either a male or a female, or both. Using a *biparental* model (or a reduced animal model) (Mrode 2005; Quaas and Pollak 1980), vector $\mathbf{g}$ is then further partitioned into additive and nonadditive genetic effects as per Costa e Silva et al. (2004). The prediction of BVs described here is also applicable to the next model described below: the spatial model.

1.4.2 Spatial Model: RCB Plus Modelling Spatial Variation

The above model can be extended to include the partitioning of spatial variability, where vector **b** contains an overall mean for each trial, as well as trial-specific modelling due to global trend (Stefanova et al. 2009). Global trend is accommodated in the model by using design factors such as linear row and/or linear column effects or by fitting spline functions to the row and column co-ordinates (Verbyla et al. 1999). Vector **u** includes effects associated with the modelling of extraneous variation due to experimental procedures and blocking design factors specific to each trial or sub-trial (in cases where a trial comprised of two or more sub-trials). For each trial, vector **e** is further partitioned into a vector that represents a spatially dependent process and a vector of residual errors (Gilmour et al. 1997).

Local spatial trend is modelled using a first-order separable autoregressive (AR) process in the row (AR(1)) and column directions (AR(1)), as recommended by Cullis et al.(1998), Gilmour et al. (1997) and Grondona et al. (1996). After fitting the local trend, diagnostic tools such as the sample variogram and trellis plots (Gilmour et al. 1997) can be used to determine if global spatial trend and/or extraneous variation needed to be included in the model. An example of a theoretical variogram for an AR(1) × AR(1) process is given in Fig. 1.2. This variogram has a

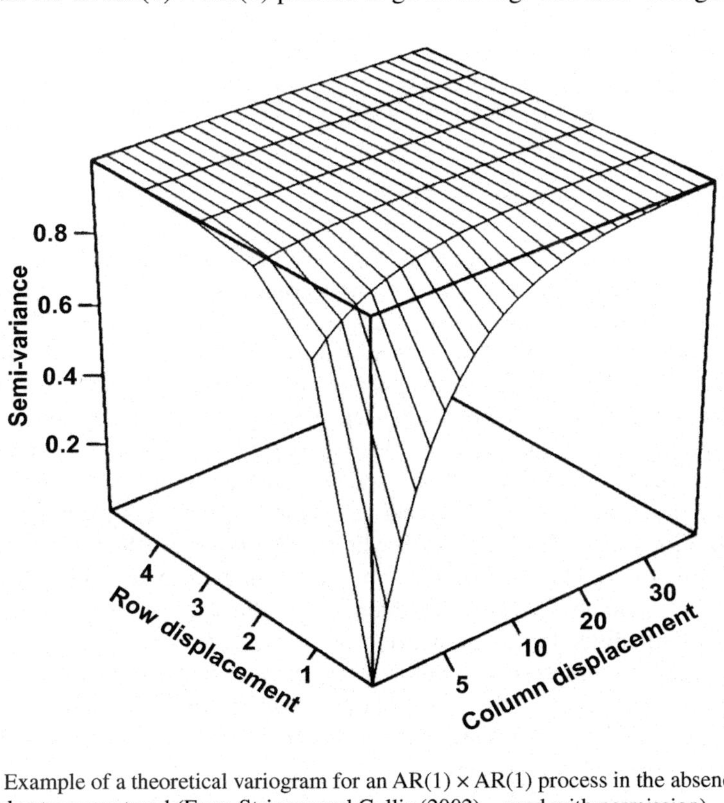

Fig. 1.2 Example of a theoretical variogram for an AR(1) × AR(1) process in the absence of both global and extraneous trend (From Stringer and Cullis (2002) – used with permission)

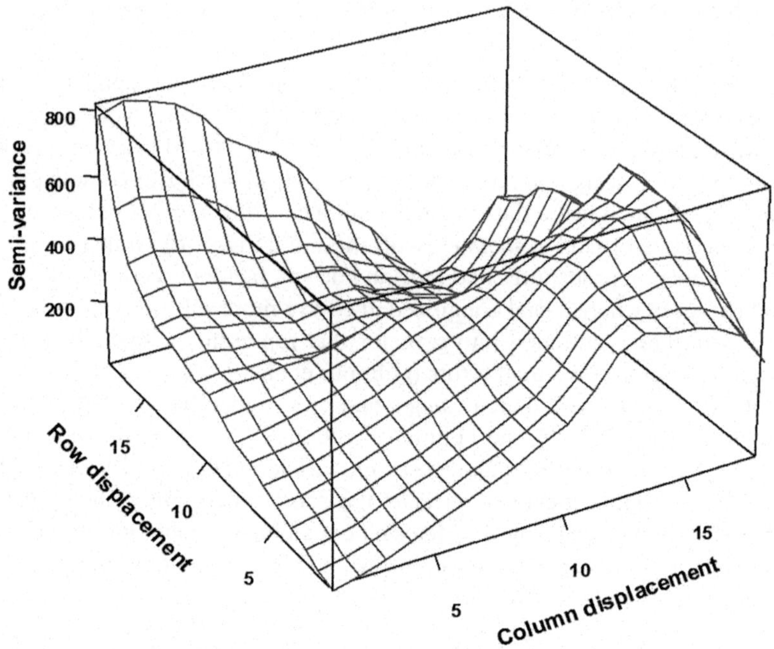

Fig. 1.3 Example of a sample variogram for an AR(1) × AR(1) process indicating the presence of a global trend in the row and column direction (From Stringer and Cullis (2002) – used with permission)

smooth appearance and an exponential increase in the row and column directions reaching a plateau giving it a 'tabletop' appearance. Departure from this smooth appearance indicates the presence of extraneous variation; similarly, if the sample variogram fails to reach a plateau in the row and/or column direction, this indicates the presence of a global trend that needs to be incorporated into the LMM (Stringer and Cullis 2002). An example of the presence of a global trend is given in Fig. 1.3 for which a linear row and column effect would then be fitted. The inclusion of these fixed effects is based on visual inspection of the sample variogram followed by a formal assessment using the Wald test (Agresti 1990). An example of extraneous variation is given in Fig. 1.4 for which a random row effect would be fitted. The inclusion of random effects is also based on visual inspection for the sample variogram, followed by the use of the likelihood ratio test to ascertain if the change in REML log-likelihood for random effects is significant.

These types of analyses are routinely performed in the Australian sugarcane breeding programme using the statistical package ASReml (Gilmour et al. 2006) and could be extended to other tropical crops where competition and spatial effects are often experienced in field trials.

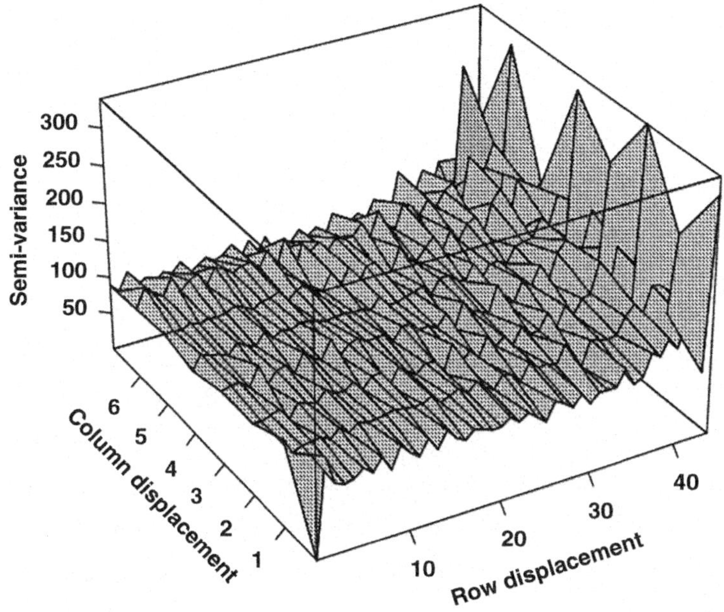

Fig. 1.4 Example of a sample variogram for an AR(1) × AR(1) process indicating the presence of extraneous variation (From Stringer and Cullis (2002) – used with permission)

1.5 Further Approaches That Incorporate Genotype-by-Environment Interactions

Mixed model analyses of data from multi-environment trials (METs) can be used to partition the total variation into sources such as trial, genotype and genotype-by-environment (G×E) interactions. Although this provides an estimate of the magnitude of G×E, it does not provide any insight into the nature of G×E effects (Kempton 1984b). Multiplicative methods are particularly useful at describing G×E interactions and have been widely used in a fixed-effects setting. The earliest of these was the *regression on mean model,* where either the phenotypic values or interaction is regressed on environmental indices. This was first suggested by Yates and Cochran (1938) and enhanced by Finlay and Wilkinson (1963); however, this approach assumes genotypes respond linearly to environmental change (Flores et al. 1998). Freeman (1973) suggested the use of multiplicative methods in genetic analyses, and of these, the additive main effects and multiplicative interaction (AMMI) has been used very widely (Gauch and Zobel 1988). AMMI combines the additive analysis of variance for main effects with the multiplicative principal component analysis for the interaction. However, AMMI requires data to be balanced and, hence, it can be too restrictive for the analysis of MET data for most crops (Smith et al. 2001).

A key issue often neglected in G×E studies is the need to model plot-level residuals (Smith et al. 2001). Individual trials routinely exhibit spatial variability (Atkin 2012) as a correlation in residuals among neighbouring plots, and it is common for residual variances to differ among trials. Estimates of genotype main effects and G×E interactions may be biased if this is not accounted for (Cullis et al. 1998). An approach which overcomes these limitations was developed by Smith et al. (2001), in which spatial variability within a trial is partitioned into local and global trends and extraneous variation using the methods of Gilmour et al. (1997) (described previously), and the heterogeneity of residual variance among trials is accounted for. This is called the *factor analytic model* (FA) and implies that genetic effects are correlated between trials; hence, it allows the genetic variation at each environment to differ and allows for different covariances between pairs of environments (Smith et al. 2001). This flexibility requires a large number of variance components to be estimated. However, the particular FA model proposed by Smith et al. (2001) uses the algorithms of Thompson et al. (2003) by providing a parsimonious fit for parameter estimation. Therefore, the multiplicative mixed model with random genotype effects, i.e. the FA model (as used by Chapman et al. 2004), is currently considered the most appropriate approach for the analysis of G×E interactions in breeding programmes for many crops including sugarcane.

1.6 Final Remarks

The array of statistical tools available in quantitative genetics for the analysis of messy and complex data that originates from breeding trials is diverse, noisy and continuously evolving. The emergence of powerful statistical software that can deal with this data has allowed breeders to extract more information from each experiment, including aspects such as competition and spatial correlations. In addition, the availability of large quantities of molecular data to be incorporated into the linear mixed models, for example, by calculating observed relationships among genotypes based on molecular markers (VanRaden 2008), has widened the options to improve and optimise the design and analysis of genetic experiments. Here we have presented some tools, but these modern tools will constitute, in the near future, a daily part of all statistical analysis performed by many breeding programmes across the world.

References

Agresti A (1990) Categorical data analysis. Wiley, New York

Apiolaza LA, Gilmour AR, Garrick DJ (2000) Variance modelling of longitudinal height data from a *Pinus radiata* progeny test. Can J For Res 30:645–654

Atkin FC (2012) Using family data to maximise gentic gain from parent selection in a sugarcane breeding program. PhD Thesis, The University of Queensland

Azaïs JM, Bailey RA, Monod H (1993) A catalogue of efficient neighbour-designs with border plots. Biometrics 49:1252–1261
Besag J, Kempton R (1986) Statistical analysis of field experiments using neighbouring plots. Biometrics 42:231–251
Bradshaw JE (1989) Interplot competition in yield trials of swedes (*Brassica napus* ssp. *rapifera* L.) Euphytica 42:135–140
Chapman SC, Jackson PA, Rattey A (2004) Genotype by environment interactions across the Australian sugarcane industry. Proc Aust Soc Sugar Cane Technol 26:8 pp
Connolley T, Currie ID, Bradshaw JE, McNicol JW (1993) Inter-plot competition in yields of potatoes (*Solanum tuberosum* L.). Ann Appl Biol 123:367–377
Costa e Silva J, Borralho NMG, Potts BM (2004) Additive and non-additive genetic parameters from clonally replicated and seedling progenies of *Eucalyptus globulus*. Theoret Appl Genet 108:1113–1119
Cullis BR, Gleeson AC (1991) Spatial analysis of field experiments: an extension to two dimensions. Biometrics 47:1449–1460
Cullis BR, Gleeson AC, Thomson FM (1992) The response to selection of different procedures for the analysis of early generation variety trials. J Agric Sci 118:141–148
Cullis BR, Gogel BJ, Verbyla AP, Thompson R (1998) Spatial analysis of multi-environment early generation variety trials. Biometrics 54:1–18
Cullis BR, Smith AB, Coombes NE (2006) On the design of early generation variety trials with correlated data. J Agric Biol Environ Stat 11:381–393
Durbán Reguera ML (1998) Modelling spatial trends and local competition effects using semi-parametric additive models. PhD Thesis, Department of Actuarial Mathematics and Statistics, Heriot-Watt University
Durban M, Currie ID, Kempton RA (2001) Adjusting for fertility and competition in variety trials. J Agric Sci 136:129–140
Dutkowski GW, Silva JCE, Gilmour AR, Lopez GA (2002) Spatial analysis methods for forest genetic trials. Can J For Res 32:2201–2214
Federer WT (1956) Augmented (or Hoonuiaku) designs. Hawaii Plant Rec 55:191–208
Federer WT, Raghavarao D (1975) On augmented designs. Biometrics 31:29–35
Finlay KW, Wilkinson GN (1963) The analysis of adaptation in a plant breeding programme. Aust J Agric Res 14:742–754
Flores F, Moreno MT, Cubero JI (1998) A comparison of univariate and multivariate methods to analyze G × E interaction. Field Crops Res 56:271–286
Freeman GH (1973) Statistical methods for the analysis of genotype-environment interactions. Heredity 31:339–354
Gauch HG Jr, Zobel RW (1988) Predictive and postdictive success of statistical analyses of yield trials. Theoret Appl Genet 76:1–10
Gezan SA, Huber DA, White TL (2006) Post-hoc blocking to improve breeding value prediction. Can J For Res 36:2141–2147
Gilmour AR, Cullis BR, Verbyla AP (1997) Accounting for natural and extraneous variation in the analysis of field experiments. J Agric Biol Environ Stat 2:269–293
Gilmour AR, Gogel BJ, Cullis BR, Thompson R (2006) ASReml user guide, 2.0 edn. VSN International Ltd, Hemel Hempstead
Gleeson AC, Cullis BR (1987) Residual maximum likelihood (REML) estimation of a neighbour model for field experiments. Biometrics 43:277–288
Grondona MO, Crossa J, Fox PN, Pfeiffer WH (1996) Analysis of variety yield trials using two-dimensional separable ARIMA processes. Biometrics 52:763–770
Hunt CH, Smith AB, Jordan DR, Cullis BR (2013) Predicting additive and non-additive genetic effects from trials where traits are affected by interplot competition. J Agric Biol Environ Stat 18:53–63
Jackson P, McRae TA (2001) Selection of sugarcane clones in small plots: effects of plot size and selection criteria. Crop Sci 41:315–322

John JA, Williams ER (1995) Cyclic and computer generated designs, Monographs on Statistics and Applied Probability, vol 38. Chapman & Hall, London, 255 pp
Kempton RA (1982) Adjustment for competition between varieties in plant breeding trials. J Agric Sci 98:599–611
Kempton RA (1984a) The design and analysis of unreplicated field trials. Vortrage Pflanzenzuchtung 7:219–242
Kempton RA (1984b) The use of biplots in interpreting variety by environment interactions. J Agric Sci 103:123–135
Kempton RA, Lockwood G (1984) Inter-plot competition in variety trials of field beans (*Vicia faba* L.). J Agric Sci 103:293–302
Magnussen S (1989) Inter-plant interactions and their influence on within and among plot variances. Scand J Forest Res 4:369–377
Martin RJ, Eccleston JA (2004) Variance-balanced designs under interference for dependent observations. J Stat Plan Inference 119:207–223
Matassa V (2003) Optimisation of experimental design and analysis for sugarcane variety trials. PhD Thesis, The University of Queensland
McCullagh P, Tibshirani R (1990) A simple method for the adjustment of profile likelihoods. J R Stat Soc Series B Stat Methodol 52:325–344
McRae TA, Jackson PA (1998) Competition effects in selection trials. Proc Aust Soc Sugar Cane Technol 20:154–161
Mrode RA (2005) Linear models for the prediction of animal breeding values, 2nd edn. CABI Publishing, Wallingford, 344 pp
Papadakis JS (1937) Méthode statistique pour des expériences sur champ. Bull Inst Amél Plante à Salonique 23:30 pp
Pearce SC (1957) Experimenting with organisms as blocks. Biometrika 44:141–149
Qiao CG, Basford KE, DeLacy IH, Cooper M (2000) Evaluation of experimental designs and analyses in wheat breeding trials. Theoret Appl Genet 100:9–16
Quaas RL, Pollak EJ (1980) Mixed model methodology for farm and ranch beef cattle testing programs. J Anim Sci 51:1277–1287
Resende MDV, Stringer J, Cullis B, Thompson R (2005) Joint modelling of competition and spatial variability in forest field trials. Braz J Math Stat 23:7–22
Sarker A, Singh M, Erskine W (2001) Efficiency of spatial methods in yield trials in lentil (*Lens culinaris* ssp. *culinaris*). J Agric Sci 137:427–438
Silva JCE, Dutkowski GW, Gilmour AR (2001) Analysis of early tree height in forest genetic trials is enhanced by including a spatially correlated residual. Can J For Res 31:1887–1893
Singh M, Malhotra RS, Ceccarelli S, Sarker A, Grando S, Erskine W (2003) Spatial variability models to improve dryland field trials. Exp Agric 39:151–160
Skinner JC (1961) Sugar cane selection experiments 2. Competition between varieties. Technical Communications. BSES, Brisbane
Smith A, Cullis B, Thompson R (2001) Analyzing variety by environment data using multiplicative mixed models and adjustments for spatial field trend. Biometrics 57:1138–1147
Stefanova KT, Smith AB, Cullis BR (2009) Enhanced diagnostics for the spatial analysis of field trials. J Agric Biol Environ Stat 14:392–410
Street AP, Street DJ (1987) Combinatorics of experimental design. Oxford University Press, Oxford, 400 pp
Stringer JK, Cullis BR (2002) Application of spatial analysis techniques to adjust for fertility trends and identify interplot competition in early stage sugarcane selection trials. Aust J Agric Res 53:911–918
Stringer JK, Cullis BR, Thompson R (2011) Joint modeling of spatial variability and within-row interplot competition to increase the efficiency of plant improvement. J Agric Biol Environ Stat 16:269–281
Stringer JK, Smith AB, Cullis BR (2012) Spatial analysis of agricultural field experiments. In: Hinkelmann K (ed) Design and analysis of experiments, Special designs and applications, vol 3. Wiley, Hoboken, pp 109–136

Talbot M, Milner AD, Nutkins MAE, Law JR (1995) Effect of interference between plots on yield performance in crop variety trials. J Agric Sci 124:335–342
Thompson R, Cullis B, Smith A, Gilmour A (2003) A sparse implementation of the average information algorithm for factor analytic and reduced rank variance models. Aust NZ J Stat 45:445–459
VanRaden PM (2008) Efficient methods to compute genomic predictions. J Dairy Sci 91:4414–4423
Verbyla AP, Cullis BR, Kenward MG, Welham SJ (1999) The analysis of designed experiments and longitudinal data by using smoothing splines. Appl Stat 48:269–311
Wei X, Jackson P, Cox M, Stringer J (2006) Maximising economic benefits to the whole sugarcane industry from the BSES-CSIRO sugarcane improvement program. Proc Aust Soc Sugar Cane Technol 28:181–186
Welham SJ, Gezan SA, Clark SJ, Mead A (2014) Statistical methods in biology: design and analysis of experiments and regression. Chapman and Hall/CRC, Boca Raton. 608 pp
Whitaker D, Williams ER, John JA (2002) CycDesigN: a package for the computer generation of experimental designs. Version 2.0. CSIRO, Canberra. 33 pp
Wiancko AT (1914) Use and management of check plats in soil fertility investigations. J Am Soc Agron 6:122–124
Wilkinson GN, Eckert SR, Hancock TW, Mayo O (1983) Nearest neighbour (NN) analysis of field experiments. J R Statist Soc B 45:151–211
Williams EJ (1952) Experimental designs for serially correlated observations. Biometrika 39:152–167
Williams ER, Matheson AC, Harwood CE (2002) Experimental design and analysis for tree improvement, 2nd edn. CSIRO Publishing, Collingwood, 214 pp
Yates F, Cochran WG (1938) The analysis of groups of experiments. J Agric Sci 28:556–580

Chapter 2
Genomic Selection: State of the Art

Luís Felipe Ventorim Ferrão, Rodomiro Ortiz, and Antonio Augusto Franco Garcia

2.1 Introduction

Plant breeding underpins successful crop production and involves the modification of genotypes to improve yield, field performance, host plant resistance to pests, and end use quality. Traditionally, genetic progress has been achieved by phenotypic evaluations in field trials. It is undeniable that important advances were obtained in the last decades. It is important, however, to take into account the time required to achieve these gains. In practice, approaches based in phenotypic metrics are coupled with long testing phases resulting in slow genetic gain per unit of time.

Since the 1980s, with the advent of molecular markers and the perception of its advantages, new opportunities were opened for its use in breeding programs. The central purpose is to assist (or support) the selection using DNA information. Named as marker-assisted selection (MAS), the application was motivated by the opportunity to reduce cost and time and, consequently, increase the expected genetic gain (Lande and Thompson 1990). Additionally, the use of markers was seen as an important alternative to increase the understanding of the genetic architecture of a quantitative trait, which has always been unclear and intriguing.

Among the MAS methods, the first to be widely accepted in the animal and plant breeding was termed quantitative trait loci (QTL) mapping. Quantitative traits refer to phenotypes that are controlled by two or more genes (i.e., multigenes) and affected by environmental factors, thus resulting in continuous variation in a

L.F.V. Ferrão • A.A.F. Garcia (✉)
Escola Superior de Agricultura Luiz de Queiroz (ESALQ),
Universidade de São Paulo (USP), Piracicaba, SP, Brazil
e-mail: lfelipe.ferrao@gmail.com; augusto.garcia@usp.br

R. Ortiz
Swedish University of Agricultural Sciences, Uppsala, Sweden
e-mail: rodomiro.ortiz@slu.edu

H. Campos, P.D.S. Caligari, *Genetic Improvement of Tropical Crops*,
DOI 10.1007/978-3-319-59819-2_2

population (Mackay et al. 2009). QTL are regions in the genome that harbor genes that govern a quantitative trait of interest (Doerge 2002). The concept that underlies QTL analysis is to split the mapping task into two components: (1) identifying QTL and (2) estimating their effect (Jannink et al. 2010).

Despite the importance of elucidating the genetic bases of quantitative loci, the QTL mapping approach has drawbacks that prevent its routine application in breeding programs (Bernardo 2008). The linkage disequilibrium induced in experimental populations, for instance, restricts the relevance of results to the families (or population) under study (Heffner et al. 2009). Additionally, QTL mapping has a better performance for traits controlled by major genes, which is an unusual scenario for traits with agronomic importance (Goddard and Hayes 2007). Supported by these inconveniences, Meuwissen et al. (2001) proposed a promised methodology that was popularized later as genomic selection (GS). It is opportune a brief description about the facts that drive the development of the GS methodology.

As previously mentioned, QTL mapping failed in its practical application (Dekkers 2004; Bernardo and Yu 2007; Xu 2008). In addition, high-throughput genotyping was boosted by next-generation sequencing techniques (NGS) that significantly reduced the cost per marker (Poland and Rife 2012). The availability of cheap and abundant molecular markers changed, in different aspects, the form in which DNA information could be inserted in genetic studies. Firstly, genotyping was automated and, in some cases, outsourced, permitting a routine and feasible application. Secondly, a vast number of genome-wide single nucleotide polymorphism (SNP) markers were discovered in many species (He et al. 2014). Lastly, computational and statistical methods converged to handle the effective analysis of the vast amount of molecular data. All of these contributed to the development of a new method of marker-assisted selection, with greater success.

Hence, if, on the one hand, traditional QTL analysis is based on the detection, mapping, and use of QTL with large effect on a trait selection, on the other hand, GS works by simultaneously selecting hundreds or thousands of markers covering the genome so that the majority of quantitative trait loci are in linkage disequilibrium (LD) with such markers (Meuwissen et al. 2001). Formally, the core of GS is the absence of any statistical test to declare if a marker has a statistically significant effect. Even effects that might be too small will be used to compute the genomic estimated breeding value. In addition, when markers covering the whole genome are used, LD is assumed between QTL and markers across all families resulting in wider applications, even for traits with low heritability (Goddard and Hayes 2007).

It has been predicted for over two decades that molecular information have the potential to redirecting resources and activities in breeding programs (Meuwissen et al. 2001; Goddard and Hayes 2007; Crossa et al. 2010; Jannink et al. 2010; Nakaya and Isobe 2012). GS has emerged as the method closest to achieving this goal. In this chapter, theory and practice will be discussed to detail how the methodology may reshape breeding programs and facilitate selection gains.

2.2 Practical and Theoretical Requirements for Genomic Selection Implementation

The previously mentioned features make GS a product of this millennium with real prospects for success. To this end, some practical and theoretical requirements are necessary for an effective implementation. In practical terms, genotyping and the definition of training and testing data sets constitute important aspects. In theoretical terms, biological and genetic concepts will be reflected by the final GS performance. This following section intends to present some details about the practical and theoretical factors which underlie GS implementation.

2.2.1 *Practical Implementation*

For a consolidated breeding program, with breeding schemes well defined that are consistently supported by good germplasm and experimentation, practical usage of genomic prediction can be considered straightforward. In general, it depends on critical decisions about which materials should be predicted and, in particular, financial and physical resources to be available for genotyping and phenotyping. These requirements are formally summarized by the subdivision of the program in three data sets, generically named "populations." The population term, in GS context, should be interpreted as a set of genotypes, where the predictive models will be trained, validated, and applied. These concepts have a close relationship with terms commonly used in the statistical learning area, especially topics on resampling and cross-validation (James et al. 2013).

The first data set is the training population (TRN). This set is also known as the reference population or discovery data set (Goddard and Hayes 2007; Nakaya and Isobe 2012; Desta and Ortiz 2014). In this step, a predictive model is defined, and the allelic effects are estimated. The individuals belonging to TRN (accesses, lines, clones, double haploid, families, etc.) must be genotyped and phenotyped for the traits of interest. A common challenge is the definition of which individuals should compose this reference population. There is not a standard way to answer this question. In theory, this population is composed by promising materials, on which the breeder has particular interest to apply selection methods and, hence, obtain new cultivars. As will be described in the next topic, this specification will have important consequences on the predictive ability of GS.

Next, a second data set called the validation or testing population (TST) should be defined (Goddard and Hayes 2007; Nakaya and Isobe 2012; Desta and Ortiz 2014). In general, this population is slightly smaller than the TRN and also includes individuals that must be genotyped and phenotyped. The role of the TST, simply stated, is to check the efficiency predictive equation defined in the previous step. The genome-estimated breeding values (GEBVs) are obtained using the marker effect estimated in the TNR and correlated with the true phenotypic values (Desta and Ortiz 2014). This result is called predictive accuracy (Ould Estaghvirou et al. 2013) and has been commonly reported as the standard metric to evaluate GS efficiency. Its magnitude will provide an important measure of GS ability to predict phenotypes, based solely on genotypic data.

The last data set is commonly called the breeding population (Goddard and Hayes 2007; Nakaya and Isobe 2012; Desta and Ortiz 2014). This is the population where GS will be directly applied, so it is the major focus of the breeding programs. Given the satisfactory accuracy value obtained in the last step, the molecular markers become the unit of evaluation in the breeding program. The effects estimated in the TST and validated in the TRN will be used, therefore, to predict new phenotypes. At this moment, selection will be guided solely by marker information (Lorenz et al. 2011). For this reason, selection can be performed at early stages (e.g., seedlings inside greenhouses), thus resulting saving of time and field evaluations (assuming that costs of genotyping are smaller).

Figure 2.1 shows the importance of these populations. As illustrated, all of them are connected, and the effects estimated in the first step will be used in all subsequent steps. In this sense, the use of an appropriate genomic model is a critical step (an in-depth discussion is provided in the Statistical Method section). Although populations are presented as physically separated, a single population may serve all the three functions.

Genotyping and phenotyping are important aspects to consider for practical implementation. The final state of a trait will be the cumulative result of a number of causal interactions between the genetic makeup of the genotype and the environment in which the plant developed (Malosetti et al. 2013). Therefore, it is common that genetic and nongenetic sources are decomposed and studied, making the experimental design and agricultural practices two fundamental aspects to be considered during the data set definition. Certainly, GS success is closely dependent of the environment in which the phenotypes are measured and on the presence of genotype-by-environment (GxE) interaction.

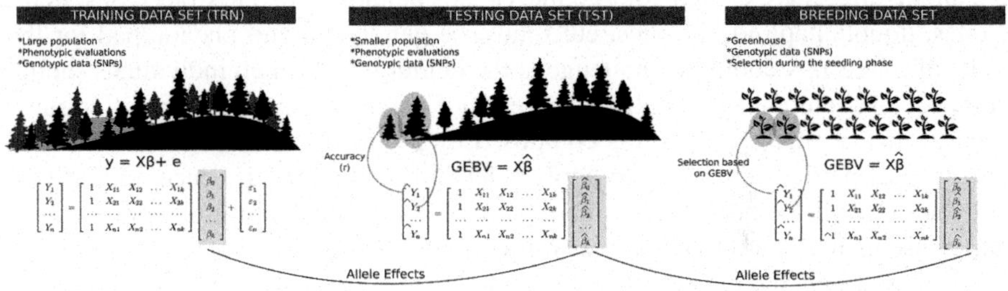

Fig. 2.1 Genomic selection (GS) implementation. Allele effects (β) estimated in the training data set (*TRN*) are used in all subsequent steps. In the testing data set (*TST*), these effects are used to predict the genomic estimated breeding values (*GEBV*), which are correlated with the true phenotypic values. This correlation value is termed as predictive accuracy (r), and it is an important indicator of efficiency. Breeding data set is the GS target. Prediction based on molecular information is performed, and genotypes are selected in early stages (seedlings), using the alleles effects as selection criteria

Regarding the genotyping, as already mentioned, procedures have been advanced by the rapid progress of NGS methodologies. Genotyping by sequencing (GBS) is a product of this rapid advance and combined the possibility to simultaneously perform marker discovery (SNPs) and genotyping across the population of interest (Elshire et al. 2011). Unlike the traditional genotyping methods where these two steps are performed separately, GBS is a one-step approach which makes the technique truly rapid, flexible, and perfectly suited for GS studies (Poland and Rife 2012). Although markers based on solid arrays (chips) or PCR may be used in GS studies, the number of reports using GBS and its variants is significantly higher. However, the number of genotyping methods techniques is under constant development, and we will certainly see more progress in this area in the next years.

2.2.2 *Theoretical Aspects Related to Predictive Capacity*

The predictive ability will be dependent on the genetic and nongenetic factors under analysis. Having a reasonable understanding of theoretical aspects that underlie these factors helps to guide GS implementation and, hence, improve the predictions. A central concept, closely linked with the theoretical definition of GS, is the linkage disequilibrium (LD). Also known as allelic association, LD is the "nonrandom association of alleles at different loci" (Flint-Garcia et al. 2003). The correlation between polymorphisms is caused by their shared history of mutation and recombination. The terms linkage and LD are often confused. Although LD and linkage are related concepts, they are intrinsically different. Linkage refers to the correlated inheritance of loci through the physical connection on a chromosome, whereas LD refers to the correlation between alleles in a population (Ott et al. 2011). Generally, all of the sources that affect Hardy Weinberg (HW) equilibrium could potentially have an influence on LD patterns (Flint-Garcia et al. 2003). In the GS context, LD concept plays a key role, as the distance along which LD persists will determine the number and density of markers and experimental design needed to perform an association analysis (Flint-Garcia et al. 2003; Mackay and Powell 2007).

Several studies have been proposed to elucidate other factors which affect predictive ability. If a large number of QTL contribute to trait variation, the following equation described by Daetwyler et al. (2013) is appropriate to predict the expected accuracy: $\sqrt{N_p h^2 \left[N_p h^2 + M_e \right]^{-1}}$, where N_p is the number of individuals in the TRN,h^2 is the heritability of the trait, and M_e is the number of independent chromosome segments.

A critical parameter is M_e, since it is inversely proportional to the accuracy. The success of GS is directly associated with the genetic distance between the reference population (TRN), where the model is trained, and the breeding population, where the estimated marker effects are used as unit of selection. This equation formalizes the idea, considering that one always expects a reduction in the predictive ability when the genetic distance increases. As a practical consequence, it is expected to have lower predictive accuracy when generations are far apart, for instance. Thus, understanding the genetic background where the model will be trained for the selection target (the breeding population) is essential to success.

The population size is another factor in the formula. Its importance is clear for two reasons, as pointed out by de Los Campos et al. (2013). First, the accuracy of estimated marker effects increases with sample size, because bias and variance of estimates of marker effects decrease with increasing sample size. Second, an increase in sample size may also increase the extent of the genetic relationship between TRN and TST data sets, which was previously described as an important factor. Population size has been highly variable in GS studies. In a revision on the subject, Nakaya and Isobe (2012) showed that, for cereals such as maize, barley, and wheat, an average size of 258 individuals has been used in the TRN data set. On the other hand, this value is larger in forest studies, where, on average, 673 individuals constitute the TRN. Studies in plants have been shown that smaller TRN sizes are required, relative to studies in animal. The authors point out two factors for this: (1) the narrow genetic diversity in plant populations, which is mainly caused by self-crossing reproduction, and (2) the quality of phenotypic evaluations, as good experimental design is more common in plant than in animal breeding.

Heritability is the biological factor highlighted in the formula. Heritability is defined as the proportion of phenotypic variance among individuals in a population that is due to heritable genetic effects. It is, therefore, expected to increase accuracy for traits governed by genetic factors and with less environmental effects. The direct relationship between accuracy and heritability is supported by simulation (Daetwyler et al. 2013).

2.3 Statistical Methods Applied to Genomic Predictions

GS studies involve the prediction of breeding values using DNA information (Fig. 2.1). For this, the inference of marker effects and their connection to phenotypes is considered the final stage. Given its importance, this section was designated to describe the use of linear models to predict breeding values, highlighting differences between philosophies of analysis in statistical learning. As a final topic, we discuss GS models that have been commonly used in plant breeding.

2.3.1 Linear Models and a Gentle Introduction to Statistical Learning

Prediction begins with the specification of a model involving effects and other parameters that try to describe an observed phenomenon. In GS context, a statistical model is proposed to associate phenotypic observations with variations at the DNA level. A large number of models may be defined to link these variables.

A particularly useful class are linear models, where various effects are added and assumed to cause the observed values (Garrick et al. 2014). A linear relationship is considered the simplest attempt to describe the dependency between variables. For this reason, it is often the starting point to model some phenomena. Supported by a consolidated theory, this class of models has a statistical value and genetic interpretation that are useful for biometric research.

The attempt to develop an accurate model, which can be used to predict some important metric, is called statistical learning (James et al. 2013). In many GS implementations, linear regression models are used to this end to describe the genetic values. Linear regression, simply stated, is a method that summarizes how the average values of an outcome variable vary over subpopulations defined by linear functions of predictor variables (Gelman and Hill 2007). In the context of genomic prediction context, phenotypes are the response (or dependent) variable, and they are regressed on the markers (predictors or independent variable) using a regression function. As pointed out by de Los Campos et al. (2013), this regression function should be viewed as an approximation to the true unknown genetic values, which can be interpreted as a complex function involving the genotype of the ith individual at a large number of genes, as well as its interactions with environmental conditions.

In statistical notation, a regression model may be represented by:

$$y_i = \beta_0 + \sum_{j=1}^{p} \beta_j x_{ij} + \epsilon_i$$

This means that the phenotypic observation of the individual i is made up of the sum of the following components: β_0 is an intercept, x_{ij} is the genotype of the ith individual ($i = 1,...,n$) at the jth marker ($j = 1, \ldots, p$), and β_j is the regression coefficient, corresponding to marker effects. The ϵ_i term represents random variables capturing nongenetic effects, which can emerge due to imperfect linkage disequilibrium between markers and QTL or model misspecification.

As noted, the phenotypic response is influenced by more than one predictor variable, as expected for quantitative traits. The postulated model may be an idealized oversimplification of the complex real-word situation, but in such cases, empirical models provide useful approximations of the relationship among variables (Rencher and Schaalje 2008). Intuitively, the regression model boils down to a mathematical construct used to represent what we believe may represent the mechanism that generates the observations at hand.

In matrix notation, the same model may be represented as $y = \mathbf{XB} + \mathbf{e}$, where $\boldsymbol{y}$ is a $n \times 1$ phenotypic vector, $\mathbf{X}$ is a $n \times p$ marker genotype matrix, $\mathbf{B}$ is a $p \times 1$ vector of marker effects, and $\mathbf{e}$ is a $n \times 1$ vector of residual effects. Again, the response vector is made up of the value of the linear predictor plus the vector of residuals. The linear predictor consists of the product of the marker genotype (matrix $\mathbf{X}$) and the estimated marker effects $(\hat{B})$. Thus, the linear predictor $(X\hat{B})$ is another vector containing the expected value of the response, given the covariates, for each individual i.

In essence, statistical learning refers to a set of approaches for estimating the regression coefficients (James et al. 2013). There are two main reasons one may wish to estimate these coefficients: prediction and inference. GS studies, in general, are focused in predictions. A set of inputs **X** (molecular markers) are readily available, and the output y (phenotypic values) should be predicted. In this setting, the way in which the coefficients are estimated is often treated as a black box, in the sense that one is not typically concerned with the exact form of **B** but whether it yields accurate predictions for the phenotypic values. Strictly speaking, this is the main difference between prediction and inference. In inference studies, we wish to estimate these coefficients and know their exact form, in order to understand which predictors are associated with the response.

In regression analysis, inference on the regression coefficient (marker effects) can be performed using different approaches. For example, one approach might be to derive a function for the marker effects that maximizes the correlation between predicted values and their unobserved true values. Alternatively, another approach might be to minimize the prediction error variance, which is the expected value of the squared difference between predicted values and their unobserved true values (Garrick et al. 2014). This last criterion is referred to as ordinary least squares (Rencher and Schaalje 2008) and is widely used in regression analysis. Intuitively, the idea is sensible: given that we are trying to predict an outcome using other variables, we want to do so in such a way as to minimize the error of our prediction (Gelman and Hill 2007).

A direct relation between regression and quantitative genetic concepts can be formulated. Considering regression models with one predictor, under an additive model and two alleles at a locus, the estimated regression coefficient may be interpreted in terms of the average effect of an allelic substitution, which quantifies the variation of the phenotypic values when an allele is replaced by its alternative (Falconer and Mackay 1996). As a biological consequence, two copies of the second allele have twice as much effect as one copy, and no copies have zero effect. The underlying assumption here is that the marker will only affect the trait if it is in linkage disequilibrium with an unobserved QTL.

Some points deserve attention during regression analysis with multiple predictors. First is the interpretation of the regression coefficients. The interpretation for any given coefficient is, in part, contingent on the other variables in the model (Rencher and Schaalje 2008). Typical advice is to interpret each coefficient "with all the other predictors held constant." Secondly, the dimensionality problems occur when the number predictor vastly exceeds the number of records. In this case, the use of usual theory to infer marker effects is not adequate. In traditional QTL studies, this inconvenience was avoided because predictors were added on regression models if they significantly improved the fit of existing models. As a statistical consequence, the data dimensionality was maintained, and least square estimators could be used without further problems. However, GS models suggest using all available molecular markers as covariates in a unique linear model. This leads to a situation where some kind of penalization is required in order to maintain the data dimensionality.

Dimensionality is a topic commonly discussed in statistical learning and deserves some comments. Predictive accuracy was previously mentioned as the gold standard metric in GS studies. Although the mathematical proof is beyond the scope of this chapter, it is possible to show that two statistical components are directly associated with this task. In order to minimize the error, it is necessary to select a statistical learning method that simultaneously achieves low variance and low bias. High variance refers to more flexible models, meaning that any small change in the original data set causes considerable change in regression coefficient estimative. In GS context, it means that marker effects have more variation between training sets. On the other hand, bias refers to the difference between an estimator's expectation and the true value of the parameter being estimated. In another words, it is the error introduced by approximating a real-life problem considering simpler models. As a general rule, more flexible models result in higher variance and lower bias. The balance between both metrics determines the predictive ability, and, for this, the term bias-variance trade-off is commonly used (James et al. 2013).

Next, we discuss how data dimensionality and bias-variance trade-off are considered under the frequentist and Bayesian framework. Genetic assumptions used in each approach are also discussed.

2.3.2 *Modeling Philosophy: Frequentist × Bayesian Approach*

Two philosophies to estimate genomic breeding values have been widely discussed in the literature. See, for example, Heslot et al. (2012); Kärkkäinen and Sillanpää (2012); Gianola (2013); de Los Campos et al. (2013).

The frequentist approach, in general, uses markers for estimating the realized relationships, directly computing the breeding value in a mixed model context. On the other hand, a Bayesian framework focuses on the inference of marker effects, and the genetic value of an individual is obtained by the sum of these estimated effects. Regardless of the modeling philosophy, the central problem for both approaches is how to deal with the number of markers (p) which vastly exceeds the number of individuals (n) or, in a statistical learning context, how to deal with the bias-variance trade-off.

In traditional analysis (in matrix notation), the least square estimator of **B** (regression coefficients) treats **X** as a fixed matrix and satisfies the system of equations: $\mathbf{X}'\mathbf{X}\mathbf{B} = \mathbf{X}'\mathbf{y}$, where **B** may not be a unique solution. If $n << p$, $\mathbf{X}'\mathbf{X}$ is singular having a zero determinant, the **B** estimator is not unique, and the variance is infinite. Thus, an infinite number of solutions can be obtained, and these estimators cannot be used either as an inferential or as a predictive machine (Gianola 2013). One way of tackling the data dimensionality issue is by considering the introduction of constraining (or shrinking) on the size of the estimated coefficients. This approach, referred as regularization (James et al. 2013), can often substantially reduce the variance at the cost of increasing the bias.

Frequentist and Bayesian approaches have different perspectives on how this penalty should be considered. Frequentists derive an estimator by adding a penalty to the loss function (e.g., penalized maximum likelihood) (Kärkkäinen and Sillanpää

2012; James et al. 2013). In the Bayesian context, regularization is inserted directly into the model formulation by specifying an appropriate prior density for the regression coefficients (Gianola 2013; de Los Campos et al. 2013). As a consequence, Bayesians consider the assumptions of model sparseness as a part of the model formulation (prior density), while in the frequentist view it is assumed part of the estimator (Kärkkäinen and Sillanpää 2012).

In what follows, the most useful GS models and their genetic assumptions will be presented. Frequentist and Bayesian methods are divided and discussed for clarity purposes. Here, the idea is not to advocate for one over the other, rather to discuss some relevant points that differentiate them.

2.3.2.1 Frequentist Approaches

Oversaturated models are addressed in a frequentist framework by adding a penalty during the parameter estimation, which significantly reduces their variance. See, for example, Whittaker et al. (2000); James et al. (2013); de Los Campos et al. (2013). In the machine-learning literature, this is attained via ad hoc penalty functions that produce regularization (Gianola 2013). In penalized regressions or shrinkage methods, estimators are derived as solutions to an optimization problem that balances model goodness of fit to the training data and model complexity. Several penalized estimation procedures have been proposed, and they differ on the choice of penalty function (de Los Campos et al. 2013).

One of the first methods proposed for genomic prediction was Ridge Regression (RR) (Whittaker et al. 2000), which is equivalent to best linear unbiased prediction (BLUP) in the context of mixed models (Habier et al. 2007). Another special penalized regression method is known as least absolute angle and selection operator (LASSO) (Tibshirani 1996). There is an assumed penalty function which underlies the difference between these methods: LASSO makes the regression coefficients shrink more strongly than RR. Additionally, the penalty induced by LASSO may involve zeroing out some coefficients and shrinkage estimates of the remaining effects; therefore, LASSO combines shrinkage and an indirect variable selection (hence the "selection operator" in its name). Regardless of the penalty function used, penalized regressions will result in biased estimators of the marker effects. However, the small bias induced is paid off with reduced variance for the parameters (de Los Campos et al. 2013).

As pointed out in the last topic, the least squares approach estimates the regression coefficients using the value that minimizes the residual sum of squares (RSS). Following the previous statistical notation:

$$\mathrm{RSS} = \sum_{i=1}^{n}\left(y_i - \beta_0 - \sum_{j=1}^{p}\beta_j x_{ij} \right)^2$$

Ridge Regression is similar to least squares, except that the coefficients are estimated by minimizing the RSS plus by a penalty function

$$\sum_{i=1}^{n}\left(y_i-\beta_0-\sum_{j=1}^{p}\beta_j x_{ij}\right)^2+\lambda\beta_j^2=\mathrm{RSS}+\lambda\beta_j^2$$

The second term in the Ridge Regression formula is called shrinkage penalty and is responsible for the regularization (Whittaker et al. 2000; James et al. 2013). The central question is deciding how stringent the regularization process should be. As the penalty grows, the degree of shrinkage becomes stronger, and eventually all of the marker effects are shrunk to zero. A model with all regression coefficients close to null values is not desirable, as it does have any power to prediction. However, when this is close to the null value, the penalty term has no effect, and RR will produce the least square estimates. Hence, an ideal scheme is one that can selectively shrink the regression coefficients; i.e., markers with small or no effects should be severely penalized, whereas those with larger effects should not be shrunk at all (Xu and Hu 2010).

The penalty assumed in the RR will shrink all of the coefficients toward zero, but none of them will be set equal to zero. As pointed out by James et al. (2013), this may not be a problem for prediction studies, but it can create some challenges during the interpretation, given the number of predictors is quite larger. In this sense, the LASSO approach is an alternative to ridge regression, and the coefficients are estimated with the following equation (Tibshirani 1996):

$$\sum_{i=1}^{n}\left(y_i-\beta_0-\sum_{j=1}^{p}\beta_j x_{ij}\right)^2+\lambda\left|\beta_j\right|=\mathrm{RSS}+\lambda\left|\beta_j\right|$$

LASSO and RR present a similar formulation. However, the LASSO penalty has the effect of forcing some of the coefficients to be exactly zero. For this reason, LASSO is known to perform not only regularization but also variable selection. Models with this feature are referred in the literature as sparse models, given the ability to subset variables (James et al. 2013). Additional details about regularization are presented by James et al. (2013) and Habier et al. (2009).

In the context of GS, an important relation is commonly addressed. There is a close connection between Ridge Regression and kinship-BLUP, a methodology where the breeding values are predicted based on their kinship. Best linear unbiased prediction (BLUP) was developed by Henderson (1949, 1950) in seminal articles applied to genetic and breeding. The main purpose was to estimate fixed effects and breeding values simultaneously. Important properties of BLUP were incorporated in its name: Best means it maximizes the correlation between true and predicted breeding values or minimizes prediction error variance; Linear because predictors are linear functions of observations; Unbiased is a desired statistical propriety related to the estimation of realized values for a random variable; and Prediction involves predicting the true breeding value. BLUP has found widespread usage in the genetic evaluation of domestic animals because of its desirable statistical properties (Mrode and Thompson 2005).

The traditional BLUP approach relies on pedigree information to define the covariance between relatives. Formally, the vector of random effects (e.g., breeding values) is assumed to be multivariate normal, where the variance parameter is indexed by the numerator relationship matrix (called **A** matrix). The connection between traditional BLUP and GS studies is the computation of this covariance using DNA information: genomic relationship matrix (called **G** matrix) (VanRaden 2008). The replacement of the **A** matrix by the **G** matrix constitute the theoretical bases of the GBLUP approach, the standard method used in GS studies.

Though conceptually similar, GBLUP and BLUP have distinct performance. The main difference is the possibility to compute a realized kinship matrix using molecular information, instead of using only expected values based on pedigree record. As pointed out by Mrode and Thompson (2005), "in pedigree populations, G discriminates among sibs, and other relatives, allowing us to say whether these sibs are more or less alike than expected, so we can capture information on Mendelian sampling." Heffner et al. (2009) point out four mechanisms responsible for the divergence between realized relationships from their expectations: random Mendelian segregation, segregation distortion, selection, and pedigree recording errors.

GBLUP has some other important features that make it widely used in GS (Mrode and Thompson 2005; VanRaden 2008; Crossa et al. 2014): (1) the accuracy of an individual's genomic estimated breeding value (GEBV) can be calculated in the same way as in pedigree-based BLUP, using software and concepts well known in breeding routines; (2) GBLUP information can be incorporated with pedigree information in a single-step method; and (3) in contrast with the penalized regression (RR and LASSO), the dimensions of the genetic effects are reduced from $p \times p$ (where p is the number of markers) to $n \times n$ (where n is the number of individuals), which is more efficient for computing purposes. An important assumption assumed in these methods is that markers are random effects with a common variance. Under a genetic perspective, this assumption may be unrealistic because markers may contribute differently to genetic variance. This is addressed by Bayesian models, discussed in the following topic.

2.3.2.2 Bayesian Approach

Before describing the most useful Bayesian models applied in GS studies, a brief description of central concepts in Bayesian inference is presented.

Simply stated, Bayesian inference determines what can be inferred about unknown parameters, given the observed data (Kruschke et al. 2012). From another perspective, Gelman et al. (2014) point out that: "by Bayesian data analysis, we mean practical methods for making inferences from data using probability models for quantities we observe and for quantities about which we wish to learn. The essential characteristic of Bayesian methods is their explicit use of probability for quantifying uncertainty in inferences based on statistical data analysis."

Formally, Bayesian analysis begins with the definition of a descriptive model, just as in classical statistics. Likewise, inference and prediction continue to be few of the major objectives. The great convenience, as described by Gelman et al. (2014), is the possibility to yield a complete distribution over the joint parameter space. So, the inference of a parameter is made in terms of probability statements, which has a commonsense interpretation (Kruschke et al. 2012).

All Bayesian inference is derived from a simple mathematical relation about conditional probabilities. When the rule is applied to parameters and data, Bayes' theorem can be conventionally written as:

$$p(\theta|y) = \frac{p(y|\theta) \times p(\theta)}{p(y)}$$

Termed as Bayes' rule, y is the observed data, and θ is a vector of parameters in the descriptive model. The posterior distribution, $p(\theta|y)$, specifies the relative credibility of every combination of parameters given the data. This is a probability distribution and, hence, provides the most complete inference that is mathematically possible about the parameters values. The term $p(y|\theta)$ is the likelihood and represents the probability that the data is generated by the model with parameter. The term $p(\theta)$ is called the prior distribution and represents the strength of our belief in θ without any observation. Finally, $p(y)$ is called the "evidence," or marginal likelihood, and is the probability of the data according to the model determined by summing across all possible parameters values weighted by the strength of belief in those parameters values. For details, see Kruschke (2011) and Gelman et al. (2014).

GS models are simple expansion of Bayes' rule assuming a hierarchical multiple linear regression as the explicit model (Meuwissen et al. 2001; Gianola et al. 2009). The term hierarchical (also named multilevel) is used when information is available on several different levels of observational units. Figure 2.2 is an adaptation from Kruschke et al. (2012) and is used for an intuitive explanation of the multiple layers (levels) assumed during a regression analysis.

The fact that GS models are just an expansion of these ideas is important. As pointed out by Kärkkäinen and Sillanpää (2012), it is a common challenge to understand the "widespread fashion of mixing the model and the parameter estimation in a way that it is hard to follow what is the model, the likelihood, and the priors and what is the estimator." Much of this critical view is directed to the collection of the Bayesian methods used for genomic predictions. The term "Bayesian alphabet" was coined by Gianola et al. (2009) to refer to the number of letters of the alphabet used to denote various Bayesian linear regression used in GS studies. These models are specified as Bayesian hierarchical regression and, in general, differ in the priors adopted for the regression coefficients, while sharing the same sampling model: a Gaussian distribution with a mean vector represented by a regression on the markers (SNP) and a residual variance. For a formal mathematical description, a notation similar to that presented by de Los Campos et al. (2013) is used here:

$$p(\mu,\beta,\sigma^2|y,\omega) \propto p(y|\mu,\beta,\sigma^2) \times p(\mu,\beta,\sigma^2|\omega)$$

$$p(\mu,\beta,\sigma^2|y,\omega) \propto p(y|\mu,\beta,\sigma^2) \times p(\mu) \times p(\beta|\omega) \times p(\sigma^2)$$

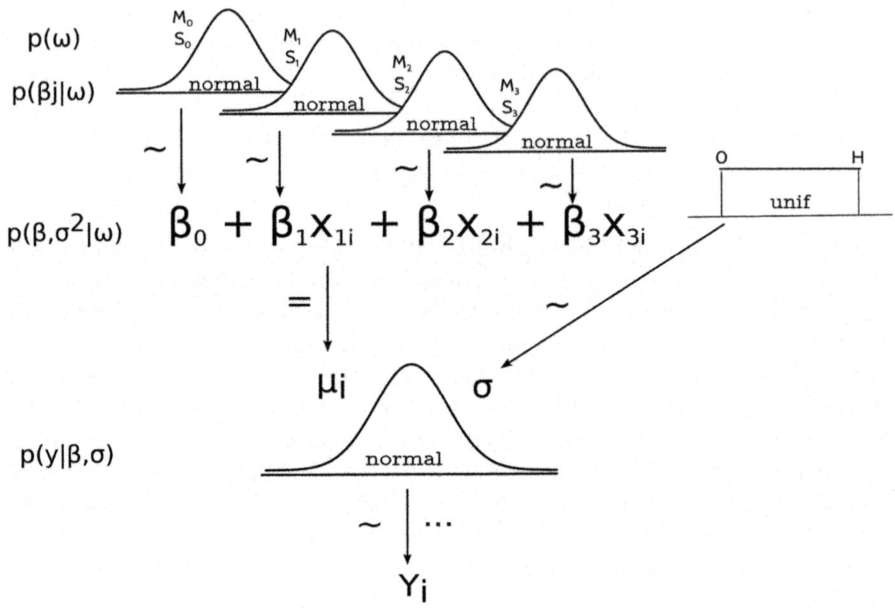

Fig. 2.2 Hierarchical Bayesian multiple linear regression. At the *bottom* of the diagram, the data *Yi* depends of the regression model and illustrates the likelihood function. The *arrow* has a "~" symbol to indicate that the data are normally distributed with a mean and a standard deviation. The ellipsis next to *arrows* denotes the repeated dependency across the observations. Moving up the diagram, the "=" signs indicate a deterministic dependency. The regression coefficients and the standard deviation are the parameters of the regression model. One layer above is represented the beliefs (prior) on these parameters. The prior distribution is a joint distribution across the five-dimensional parameter space, defined as the product of five independent distributions. The last layer is the hyperparameters and expresses our belief about the distribution of the regression coefficients and the standard deviation (Adapted from Kruschke et al. (2012))

where $p(\mu, \beta, \sigma^2|y, \omega)$ is the posteriori density of model to unknowns μ, β, σ^2 given the data (y) and hyperparameters (ω); $p\,(y|\mu, \beta, \sigma^2)$ is the likelihood of the data given the unknowns, which for continuous traits are commonly independent normal densities, with mean $X\beta$ and variance σ^2; and $p(\mu, \beta, \sigma^2|y, \omega)$, factorized in the equation, is the join prior density of model unknowns, including the intercept (μ) that is commonly assigned a flat prior; the regression coefficients (β), for which are commonly assigned IID informative priors; and the residual variance (σ^2), for which is commonly assigned a scaled inverse chi-square prior with degree of freedom *d.f* and scale parameter *S* (Gianola 2013; Pérez and de los Campos 2014).

The basic idea behind the model description was shown in Fig. 2.2. Some prior distributions may be changed, but the concept is the same; GS models should be interpreted as a variant of multiple regression models, described hierarchically. For example, the BayesA method, initially proposed by Meuwissen et al. (2001), may be formally described in layers, where at the first stage a normal multiple regression is assumed; at the second a normal conditional prior is assigned to each marker effect, all possessing a null mean but with a variance that is specific to each marker; and, lastly, it assigns the same scaled inverse chi-squared distribution for the hyperparameters.

As previously mentioned, the central point is how to deal with oversaturated models. In a Bayesian context the sparseness is included in the model by specifying an appropriate prior density for the regression coefficients. Supported by the infinitesimal model, which states that a quantitative trait is controlled by an infinite number of unlinked loci and each locus has an infinitely small effect (Fisher 1919), it is reasonable a priori belief that most of the predictors have only a negligible effect, while there are a few predictors with possibly large effect sizes. A prior density which represents these beliefs has a probability mass-centered near zero and distributed over nonzero values, with a reasonably high probability for large values (Kärkkäinen and Sillanpää 2012).

In-depth discussions about prior densities assigned to marker effects, as well as the hyperparameter definition, are presented by Kärkkäinen and Sillanpää (2012), Gianola (2013), and de Los Campos et al. (2013). Based on how much mass these densities have in the neighborhood of zero and how thick or flat the tails are, a general classification into three big categories was presented by de Los Campos et al. (2013).

Starting with the Gaussian prior, methods that assign this prior to the marker effects are referred to as Bayesian Ridge Regression (BRR) (Pérez et al. 2010). This mimics the RR approach (or the BLUP) when a specific penalty is assumed. RR-BLUP and BRR both perform shrinkage step that is homogeneous across markers. The second class of densities is called "thick-tailed priors." Two widely accepted methods which represent this class are BayesA (Meuwissen et al. 2001) and Bayesian LASSO (Park and Casella 2008). Relative to the Gaussian prior, these densities have higher and thicker tails. This induces shrinkage of marker effect estimates toward zero for smaller effects and less shrinkage for markers with larger effect estimates.

There is a third group of models ("point of mass at zero and slab priors"), which include BayesB (Meuwissen et al. 2001) and BayesC (Habier et al. 2011). For this class, the prior assumption is that marker effects have identical and independent mixture distributions, where each has a point mass at zero with probability Ï€ and a univariate-t distribution (BayesB) or a univariate-normal distribution (BayesC) with probability $1 - \pi$. When π=0, BayesB can be seen as a special case of BayesA; and the BayesC is identical to RR-BLUP. Alternative Bayesian models are discussed by Gianola (2013), which largely are expansions of the mentioned theory.

An important point under investigation, when different prior densities are tested, is the search for a better description of the genetic architecture (Gianola 2013; de Los Campos et al. 2013). For example, the BRR approach considers the marker effects as sampled from a normal distribution with fixed variance; hence, as a practical consequence, the effects are shrinking to the same degree assuming our beliefs that the trait is controlled by many loci with small effects. In contrast, the BayesB makes the assumption that most loci have no effect on the trait and thus more markers are left out of the prediction model; so, our preliminary hypothesis is that the trait is controlled by relatively few loci, whose effect vary in size.

The central question which underlies the choice for a Bayesian model is "What distribution should be used?" As previously noted, the answer is closely associated with the genetic architecture of a trait, which is commonly seldom known. Motivated by this, it has been a common practice in GS research to start by testing different models which represent the biological phenomenon.

2.3.2.3 Practical Lessons About Statistical Methods Used for Genomic Selection

Recently, a significant number of simulated and empirical studies were published comparing genomic prediction models (Heslot et al. 2012; Resende et al. 2012b; Crossa et al. 2013; Daetwyler et al. 2013). Among the additive models, Bayesian regressions and the GBLUP method have mainly been used in animal and plant breeding. GBLUP is attractive due to its straightforward implementation using existing mixed model software, relative simplicity, and limited computing time. Bayesian methods were not widely used until around 20 years ago, given the release of several packages that allowed Bayesian analysis to be performed easily and quickly on a standard desktop computer (Stephens and Balding 2009).

A summary of some models commonly used in GS studies is given in Table 2.1. Alternative methodologies are discussed by Heslot et al. (2012), de Los Campos et al. (2013), Zhou et al. (2013), Desta and Ortiz (2014), and Gianola (2013). The methods described were classified according to the approach used for the analysis, maintaining the same structure used in the last section. Table 2.1 highlights some attributes, such as genetic architecture, regularization, and variable selection. Genetic architecture is cited as a generic way to determine which models are able to weight markers of small and large effects. The regularization is a common feature for all GS methods; however some models are able to combine regularization and variable selection. A software commonly used and a classification of complexity are also presented. Here, complexity was defined as the number of parameters estimated during the inference. Bayesian models with variable selection and a mixture of distributions were classified as having high complexity (given the higher number of parameters to be estimated).

Currently, a great number of software/packages are freely available. As a general rule, "push a button" interfaces are not provided, and, hence, a minimal background in statistics and computation is required. Sampling methods (Monte Carlo Markov Chain), commonly used in Bayesian approaches, require more computational demand and, consequently, more time for performing the analysis.

Regarding the performance of these models in practice, simulation studies using frequentist and Bayesian methods have shown similar results for traits governed by many loci, which closely resemble the infinitesimal genetic model (de Los Campos et al. 2013; Daetwyler et al. 2013; Wang et al. 2015). A slight advantage of variable selection methods was observed for simulated traits where fewer loci contributed to genetic variation (Coster et al. 2010; Daetwyler et al. 2013). Empirical studies have been conducted to confirm the results of simulation studies (Moser et al. 2009; Heslot et al. 2012; Resende et al. 2012b; Ferrão et al. 2016a). When models are compared, in a large majority of the cases, small differences in predictive ability are observed.

Table 2.1 Genomic selection methods and their particularity commonly applied to plant breeding

Method	Philosophy	Attributes	Software	Complexity
GBLUP[a]	Frequentist	Regularization and homogeneous genetic architecture	rrBLUP[f] (R package), AsREML[g], GenStat[h], Wombat[i]	Low
RR-BLUP[b]	Frequentist	Regularization and homogeneous genetic architecture	rrBLUP[f](R package)	Low
LASSO[c]	Frequentist	Regularization, flexible genetic architecture, and selection of covariates	glmnet[j](R package)	Low
BayesA[d]	Bayesian	Regularization and flexible genetic architecture	BGLR[k] (R package), GenSel[l]	Moderate
BayesB[d]	Bayesian	Regularization, flexible genetic architecture, and selection of covariates	BGLR (R package)[k], GenSel[l]	High
BayesC[e]	Bayesian	Regularization, flexible genetic architecture, and selection of covariates	BGLR (R package)[k], GenSel[l]	High

[a]VanRaden (2008)
[b]Whittaker et al. (2000)
[c]Tibshirani (1996)
[d]Meuwissen et al. (2001)
[e]Habier et al. (2011)
[f]Endelman (2011)
[g]Butler et al. (2009)
[h]Payne et al. (2011)
[i]Meyer (2007)
[j]Simon et al. (2011)
[k]Pérez and de los Campos (2014)
[l]Fernando and Garrick (2009)

Some hypotheses have been proposed to explain these differences. One is related to intrinsic features of the data (e.g., the ration between number of markers and records, span of LD, genetic architecture, etc.) that hinder model regularization and, consequently, result in similar results. Another possible argument is discussed by Tempelman (2015) and involves statistical and computational challenges associated with the Bayesian inference. The author pointed out some general issues concerning the hyperparameter specification, MCMC diagnostics and the problem of data dimensionality.

The final message about practical lessons of the statistical methods is that no single method has emerged as a benchmark model for genomic predictions. Hence, evaluation and reflection about advantages and drawbacks of each one model should be considered as an imperative step during the GS implementation. However, given that so much effort would be taken for data recording, it seems reasonable to test a number of models before applying them in real-word situations.

2.4 Genomic Selection and Plant Breeding

The biometric models accounting for genomic prediction became mature after building on Henderson's mixed linear model equations for BLUP of breeding values using pedigree and phenotype data. Their accuracy remains an active area of research, but genomic selection has already led to increased rates of genetic gain, particularly for traits with low heritability. For example, dairy cattle has improved as a result of decreasing generation intervals and increasing significantly selection intensity. Simulations and empirical studies have demonstrated that GS has potential to accelerate the breeding cycle, maintain genetic diversity, and increase the genetic gain per unit of time in plant breeding (Bernardo and Yu 2007; Heffner et al. 2009; Heffner et al. 2010; Resende et al. 2012b). All of these factors have created a lot of excitement and high expectations in the plant breeding communities. Furthermore, Rajsic et al. (2016) provide a general average cost framework to quantify prediction accuracy of effects and varying cost ratios of phenotyping to genotyping for comparing the economic performance of GS vis-à-vis phenotypic selection. They found that GS appears promising for traits with heritability below 0.25 unless the phenotyping costs is higher than genotyping and the effective chromosome segment number 100 or more. The following section describes how genomic predictions can be integrated into breeding efforts and result in higher genetic gains.

2.4.1 *Expected Genetic Gain: The Breeder's Equation*

The expected genetic gain is an important metric to quantify the progress of a breeding program. For this reason, it is known as the breeder's equation. One version of this equation weights the expected genetic gain by the cycle size, as follows (Desta and Ortiz 2014): $\text{GP} = {ir_a\sigma_a^2}/{L}$, where i is the selection intensity, r_a is the selection accuracy, σ_a^2 is the square root of additive genetic variance, and L is the cycle length.

Using this equation, GS has potential to capitalize on all four of the components (Desta and Ortiz 2014). First is operating under the selective accuracy. The use of molecular markers can be leveraged to estimate a relationship matrix or applied directly into regression models and increase the selection gain. It is well established in the literature that conclusions based on molecular information tends to be more reliable. In a mixed model context, the kinship computed via DNA information is able to consider the realized relationship among individuals, instead of an expected value supported by pedigree records.

The second term is the cycle length. This term has a special importance in perennial crops, where the breeding cycle is longer. In order to advance generations and accelerate the gain per unit of time, genomic predictions can be performed during seedling phase. In addition to saving time, this reduces cost by avoiding the necessity to maintain populations for several years in the field. A good perspective on the relationship between cost and gain is presented by Heffner et al. (2010). The authors reported that, for many crops, the time for a breeding cycle using GS might represent one-third or less than that used by phenotypic selection.

The use of many cycles per year directly affects the selection intensity and genetic variance, the two remaining components in the equation. In general, the selection intensity is raised by the ability to evaluate a large breeding population and consider a big screening nursery. Consequently, new genes and combination of them, not present in the breeding program, may be incorporated in the evaluation process.

At this point it would be helpful to contextualize the relationship between expected genetic gains and a cost-benefit approach. Although GS has been announced as a potential tool to assist selection in breeding programs, there are a number of practical problems in conventional breeding programs that GS cannot eliminate or suppress. A reflection about them should be considered before deploying GS as a breeding tool. Heslot et al. (2015) point out some challenges: (1) the choice of germplasm on which to apply MAS, once the germplasm should represent the final objectives expected in GS applications; (2) trade-offs between family size and number of families created for MAS; (3) integration of information for multiple traits, balanced between phenotypic selection and MAS at a constant budget; (4) disconnection between the population used to train the models and the elite breeding germplasm (breeding population); and (5) logistical issues involved with the integration of molecular information in breeding programs.

Most of these points are related to resource allocation toward phenotyping, genotyping, the genetic architecture of traits under analysis, and population size, as described by Heslot et al. (2015). According to the authors, the use of markers to achieve breeding gains requires consideration of the genetic gain achieved by the breeding program with and without GS. Looking only at the expected genetic gain formula, it seems clear that GS in contrast to traditional breeding schemes will increase genetic gain. However a cost-benefit analysis will take into account that genotyping all candidates might require reducing the size of the breeding population and result in a negative impact on breeding gains. A practical example: on the base of a breeding program flow, as general rule, the number of candidates to be evaluated is higher, and they are not fully inbred, making the logistics of genotyping and prediction more complicated and expensive. For this reason, Heslot et al. (2015) point out "This trade-off is even stronger in a phenotypic breeding program, because large populations early in the cycle are combined with high selection intensity on highly heritable traits (high plot-basis heritability), which can be extremely efficient and relatively Inexpensive. It is probably beneficial to use markers to select on a low heritability trait, such as yield, early in the cycle; in most crops, yield cannot be measured accurately on segregating populations, single plants, or small plots. At the same time, most of the individuals in early generations can be discarded efficiently using inexpensive phenotyping."

Important insights about the cut-benefit trade-off have been reported in empirical studies. Meuwissen (2009), in animal breeding, computed the expected accuracy for reference populations considering different sizes and different heritabilities. In traits with low heritability, an accuracy of 0.20 can be obtained with a large reference population (2000–5000 animals). In contrast, Akanno et al. (2014) simulated a case of limited resources considering a small training population in animal breeding (1000 individuals). In this case, the population size was considered appropriate; however, it required multigeneration training populations and the reestimation of marker effects after two generations of selection. Although these simulations were proposed in an animal scenario, the results reinforce the ability of genomic prediction to improve the genetic gain. The level of this improvement is strongly associated with resource allocation.

2.4.2 Genomic Selection and Plant Breeding Schemes

The previous section describes how GS has the potential to raise expected genetic gain. In this regard, some empirical studies have supported GS superiority compared with traditional phenotypic methods. In tree breeding, for example, the selection efficiency per unit of time was estimated to be 53–112% higher than phenotypic selection, thus resulting in a time reduction of 50% in the breeding cycle (Resende et al. 2012a). Higher genetic gain, compared with phenotypic selection and other conventional MAS, was also reported in biparental wheat populations (Heffner et al. 2011). Likewise, GS appears to be more effective than pedigree-based phenotypic selection for improving genetic gains in grain yield under drought in tropical maize (Beyene et al. 2016). These are some examples of success that have encouraged GS application in practical breeding programs. However, open questions remain about how to implement these ideas in well-established plant breeding programs (Jonas and de Koning 2013). This is a reality for many crops, especially in non-private institutions.

Innovative studies have been performed by the International Maize and Wheat Improvement Center (CIMMYT), and a good perspective about the subject was presented by Crossa et al. (2014). As a general rule, genomic information has been useful for the investigation of unknown population structure, predicting on unrecorded pedigree structure, correcting incorrect pedigrees, and predicting the genetic value of Mendelian sampling terms (random sampling of the genome of each parent, which should be interpreted as a deviation of the average effects of additive genes an individual receives from both parents from the average effects of genes from the parents common to all offspring). In terms of the algorithms/models used for predictions, in CIMMYT trials, no prediction model fits all situations (Crossa et al. 2013; Perez-Rodriguez et al. 2013; Crossa et al. 2014). Further, it is noteworthy to consider the context where these studies were performed. In particular, the investigation of GxE interaction (Burgueño et al. 2012; Lopez-Cruz et al. 2015), predictions in structured populations, and the response of the GS across years and breeding cycle (Arief et al. 2015; Jarquín et al. 2016) have revealed new perspectives on the use of molecular information in plant breeding. In this sense, the studies applied to maize and wheat, developed at CIMMYT, have been used as benchmark for others crops with similar objectives.

An interesting viewpoint of practical recommendations was described by Bassi et al. (2015). In wheat, a comparative analysis showed equivalence in costs between phenotypic evaluations and GS. Authors reported how GS methods may reshape traditional breeding schemes, in order to increase the genetic gain. In short, GS and on-field evaluations are interleaved, and there are no significant changes in traditional schemes. Selection based on molecular markers can be performed using plants in a seedling phase (inside greenhouse), avoiding additional costs with experimental area and phenotyping and, consequently, shorting the length of each cycle. It is worth mentioning that GS and on-field evaluations are proposed as complementary methods, such that neither completely replaces the other. GS has several advantages, but it should be stressed that phenotypic evaluations will always be necessary.

In order to summarize how the aforementioned ideas could be incorporated in classical breeding schemes, an intuitive representation is shown in Fig. 2.3, where a schematic of breeding inbred lines is presented using doubled haploids (originally discussed by Heslot et al. (2015)). As pointed out, GS can be use at each stage of cultivar development. At the bottom of Fig. 2.3, we present our personal perspective on GS implication in intrapopulational recurrent selection schemes applied to coffee (*Coffea canephora*). A remarkable feature of the coffee breeding program is the long testing phase, since it is a perennial species with a long juvenile period. Historically, coffee experiments have been performed in multiple locations and harvests (years of production), which results in high cost and a long time to achieve the final product. Figure 2.3 highlights the ability to advance generations by implementing GS during the seedling phase, inside greenhouse. As an immediate consequence, the breeding cycle will be reduced and the selection intensity increased. In contrast to conventional methods of recurrent selection in *C. canephora*, this technique is to reduce the total time required to advance a generation by two-thirds (5–6 years).

The *C. canephora* scheme may be expanded for other tropical species. As a general rule, any breeding scheme is based on three steps: crossing, evaluation, and selection. Directional selection occurs when a breeder induces the phenotypic mean of a trait to move in the desired direction over one or more generations. To achieve this, breeders impose a selection threshold, such that an evaluation guarantees that individuals above this threshold are selected as the progenitor of the next generation. As a consequence, these individuals will intercross and compose a new breeding cycle (crossing step). The assumption behind each of these concepts is that the selected individuals provide genetic progress, which involves allelic transmission and increases the frequency of selected alleles in the breeding population.

Different metrics can be used to drive the selection step. GS accuracies support the use of GEBV, rather than phenotypic metrics, to guide selection in plant breeding. In this scenario, phenotyping plays a crucial role in the process of estimating and/or reestimating marker effects. New germplasm that may eventually feed breeding programs and improve the base population, under the GS regime, will be useful for composing a training population, which will increase the sample size and allows

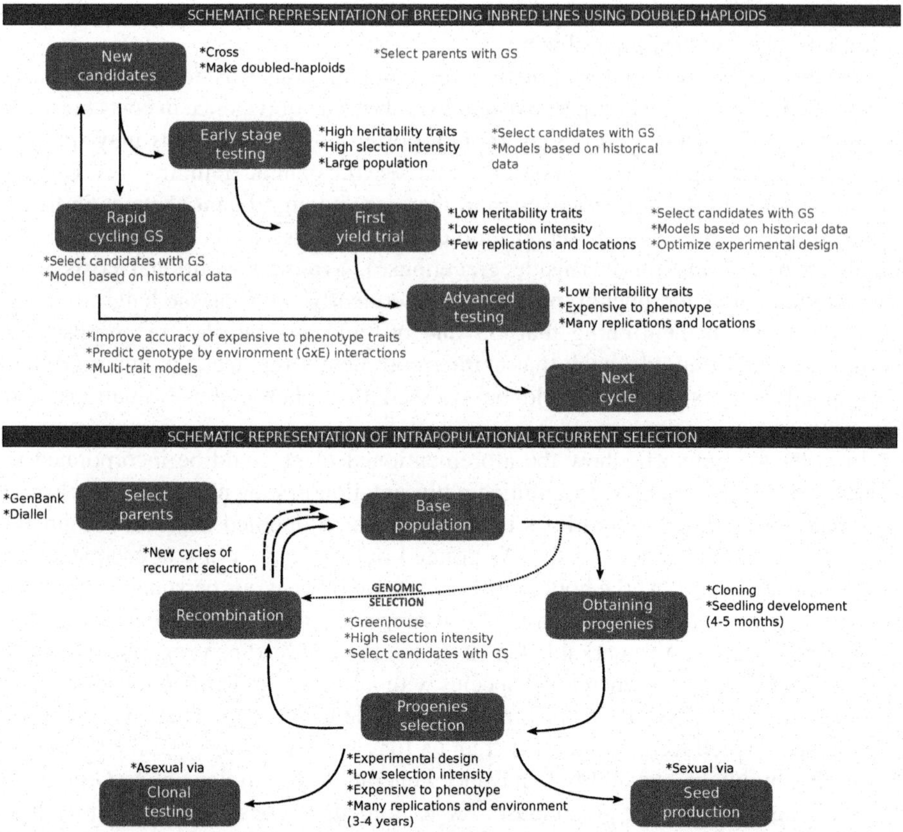

Fig. 2.3 Simple scheme of a breeding cycle with genomic selection (GS). For each stage, the figure presents side-by-side characteristics of classic breeding (in *black*) and potential applications of GS (in *gray*). The schematic of breeding inbred line, on the *top*, was originally described by Heslot et al. (2015), while the schematic of intrapopulational recurrent selection, on the *bottom*, is based on a *Coffea canephora* breeding program

new alleles be sampled. Field experimentation emerges as a crucial step at the end of the process when candidates that were selected by their GEBV should be tested in multiple environments.

It is noteworthy that marker effects may change as result of allele frequency changes or of epistatic interactions. Hence, model updating within breeding cycle should mitigate reduced gains caused by these mechanisms. In this context, there is an important routine of genotyping in the breeding program. For this end, maintaining a physical structure to genotyping may be expensive, and one solution is to outsource these services.

2.5 Challenges, Perspectives, and Trends

Previously, practical and theoretical aspects were discussed in order to elucidate GS application. It is clear that this new scenario not only reshapes the expectations of plant breeding but also brings a new context to investigate questions that raised researchers. This topic addresses challenges, perspectives, and trends that have been investigated in the plant breeding literature. The last subject in this section is a perspective on future directions in GS investigations.

2.5.1 A Multidisciplinary Solution to the Challenge of Big Data

GS is a multidisciplinary approach that involves interconnected areas, e.g., plant breeding, genetics, molecular biology, statistical genetics, and bioinformatics. The primary challenge during GS application is to connect all of these areas in an efficient framework. In practical terms this means collaborative work, where shared decisions among researchers, with different expertise, should guide the breeding program.

A challenge in this context is the "Big Data question" (James et al. 2013; Adams 2015). The term "Big Data" was designated for data sets that are so large or complex that traditional data processing applications are inadequate or inefficient. In GS studies, data sets with this magnitude are coming from new phenotyping technology, which are able to generate millions of measurements every day, and from "omics" projects that have been feeding huge public and private database with biological and molecular information. The necessity to store, process, and draw conclusion from such information is a challenge and necessity for modern breeders.

Although advances have been reported, it is noteworthy that the potential of GS does not invalidate or reduce the importance of two other areas in breeding programs: field evaluations and the continuity of traditional MAS research. In terms of field evaluation, the composition of good populations, the use of appropriated experimental designs, and the choice of promising parents continue to be important steps. These assignments are commonly activities of the so-called conventional breeder and, even in the presence of GS, remain as key point for success of plant breeding. In terms of traditional MAS approaches, we are reinforcing the importance to continue genetic mapping and QTL mapping researches. Genetic mapping studies have been important in modern genomic studies, especially during the genome assembly, which is useful for SNP prospection and subsequent genetic analysis. On the other hand, QTL mapping remains as the most appropriate approach for genetic architecture studies.

2.5.2 Genotype-by-Environment (GxE) Interaction

Recently, a large number of studies have been performed to address GxE interaction, and, therefore, different statistical models are reported in the literature. See, for example, Smith et al. (2005); Crossa (2012); Malosetti et al. (2013). GxE interaction occurs because different genotypes do not necessary response in the same way to equal conditions. An important point is the attempt to predict genotype performance over an environmental space.

In a mixed model context, genotypic performances across the environments have been modeled as correlated traits considering structured and unstructured covariance functions. A natural advantage is the flexible way in which these functions may be tested to describe the interactions and residual variance (Smith et al. 2005). Furthermore, when genetic effects are assumed as random, pedigree information can be incorporated, and more accurate breeding values may be computed via best linear unbiased prediction (BLUP). In coffee, for example, it was reported differences in the predictive accuracy of 10–17%, when comparing models that considered and ignored interaction effect (Ferrão et al. 2016b). Increases in the predictive ability by GxE modeling were reported by Burgueño et al. (2012), Lado et al. (2016), and Malosetti et al. (2016).

More recently, studies have advanced in order to incorporate modern information about environmental covariates (Jarquín et al. 2013; Heslot et al. 2014) and the explicit modeling of interaction between markers and environment (MxE) (Schulz-Streeck et al. 2013; Crossa et al. 2015; Lopez-Cruz et al. 2015). An important point in these studies is the possibility to decompose the effects into components that are constant across groups (environments or populations) and deviations that are group specific. From a quantitative genetics perspective, it is reasonable to expect that SNP effects may differ across populations and environments. In a breeding program, this may aid in the selection of generalist genotypes (good performance in all conditions, i.e., broad adaptation) or specialist genotypes (performance directed for a specific condition, i.e., narrow adaptation). In general terms, these insights are related with the classical breeding concepts about adaptability and stability.

2.5.3 GS in the Presence of Population Structure

Commonly, GS methods assume homogeneity of allele effects across individuals. However, this assumption ignores the fact that systematic differences in allele frequency and in patterns of linkage disequilibrium can induce group-specific marker effects (de Los Campos and Sorensen 2014). Although rarely discussed in GS context, population structure is a real prospect in plant breeding. These substructures are commonly caused by natural activities inside a breeding program, e.g., artificial selection, drift, and exchange of materials.

In genome-wide association (GWAS), it is known that population structure is an important source of spurious association between genetic variants and phenotypes. Principal components (PCs) methods are frequently used to account the population structure and "correct" for population stratification. Although important, such methods as the PCs induce a mean correction that does not account for heterogeneity of marker effects (de Los Campos et al. 2015b). Moreover, there are good reasons that support the hypothesis that, in heterogeneous populations, markers effects should be allowed to vary between groups. It is reasonable to capture this variation instead of treating it as potential confounder or ignoring it.

2.5.4 Epistasis and Dominance

GS models have been limited mostly to fit marker (or haplotypic) additive effects, either explicitly estimating the marker effects or implicitly through the so-called "genomic" relationship matrix (GBLUP method) (Vitezica et al. 2013). As previously cited, there is a natural trend to consider additive models as a starting point in GS investigations. Besides to capture a large portion of the genetic variation, additivity might be straightforward implemented. However, if most of the studies have addressed prediction taking into account only genes with additive effects, there is still a lack of reports dealing with the total genetic value, which include additive and nonadditive effects (Denis and Bouvet 2011).

Nonadditive variations result from interactions between alleles at the same locus (intra-locus) or interactions from different locus (inter-locus). Formally, intra-locus interactions are called dominance effects and can be defined as the difference between the genotypic value and the breeding value of a particular genotype (Falconer and Mackay 1996; Lynch and Walsh 1998). From the statistical point of view, dominance effects are interaction effects or within-locus interaction. On the other hand, interaction deviations or epistatic deviations refer to additional deviations when more than one locus are analyzed (Falconer and Mackay 1996; Lynch and Walsh 1998). Hence, the additivity assumed in GS studies may be derived from two sources: under a narrow view, refers to genes at one locus and means the absence of dominance, and in a broad view, refers to genes at different loci and means the absence of epistasis. In both cases, nonadditivity constitutes a major challenge for plant breeder (Holland 2001).

Considering dominance effects, recent studies have been shown superiority of models that took into account this source of variation. Dominance has theoretical and practical interest, because it is frequently used in crosses of animal breeds and plant lines. In tree breeding, for example, higher predictive accuracies were observed when dominance-additive variance ratio increases (Denis and Bouvet 2011). These results have been particularly interesting for tree improvement, where clonal cultivars can be produced. Considering animal and simulated data, Vitezica et al. (2013) point out advantages in recovering information when the dominance is modeled. In a similar direction, advantages to consider dominance effects are reported by Nishio and Satoh (2014) and Lopes et al. (2015).

There are well-defined cases of interactions at molecular level between gene products, but the real relationship between molecular interactions and complex phenotypes is often unclear. Considering classical quantitative genetics methods, the genetic component of variance are often poorly estimated providing the false impression that this source of variation is not important, as pointed out by Holland (2001). Lorenzana and Bernardo (2009) reported that including epistatic effects in prediction models will only improve accuracy if two conditions were considered: (1) if epistasis is present and (2) if it is accurately modeled. Currently, contrasting results have been reported adding some controversy about their importance in quantitative genetic analysis. Increased in predictive ability by the epistasis modeling is discussed by Hu et al. (2011), whereas Lorenzana and Bernardo (2009) have indicated that predictions were adversely affected. These results point out that importance of epistasis modeling can vary between species, type of crossing, and trait under analysis. It seems clear that, given the complexity of the subject, further research should be performed.

A critical viewpoint is presented by Lorenz et al. (2011): "if the predictive accuracy is lower when the epistasis is included, clearly epistasis was poorly modeled with the population sizes in this study." Another result that reinforce the epistasis importance is presented by Dudley and Johnson (2009), who concluded that epistatic effects are more important than additive effects in determination of oil, protein, and starch contents of maize. These results, and other reported in the literature, not only are remarkable for the importance of epistatic effects but also deserve attention for the necessity to better the description of nonadditive modeling in GS studies.

2.5.5 Polyploid Species

The GS application changes when polyploid species are considered. Challenges in this sense are not exclusive to GS but also include QTL, genetic mapping, and GWAS research. Important polyploid crops include sugarcane, wheat, potato, coffee, cotton, and some fruit species (e.g., apple and strawberry). Commonly, analytical frameworks assume a specific mode of inheritance and relation between alleles, based in diploid species, which does not fit in polyploid context (Dufresne et al. 2014). This difficulty is due to several complications evidenced in the polyploidy analysis, as follows: (1) larger number of genotypic classes, (2) poorly understood behavior of the chromosomes, (3) lack of molecular and statistical methods to precisely and efficiently estimate the genotypic classes, (4) ploidy level of the species, and (5) complexity of the interactions between alleles (Mollinari and Serang 2015).

Despite the significant number of polyploid tropical species and the increases of availability genomic data, there remain important gaps in the knowledge about polyploid genetics (Dufresne et al. 2014). A common practice adopted in polyploid analysis has been the interchange of knowledge and methods applied to diploid level. Although it is an approximation, it is a naive way of handling the problem, given the unrealistic and simplified assumptions that are assumed (Garcia et al. 2013).

Appropriate methods applied to genomic prediction in polyploid analysis are still in their infancy. The challenge begins before the modeling steps of genotype-phenotype relationship. Genotypic classification and SNP calling are not trivial tasks. A good perspective about the subject is presented by Garcia et al. (2013) and Mollinari and Serang (2015).

In polyploids, a locus may carry multiple doses of a particular nucleotide. Traditional molecular markers (e.g., AFLP and SSR) do not allow a straightforward estimation of this dosage at a given polymorphic locus. The development of modern genotyping technologies opened an important opportunity to evaluating the relative abundance of each allele (Mollinari and Serang 2015). Although progress were observed in tetrasomic polyploid species (e.g., potato species), more complex polyploid species, such as sugarcane and some forage crops, have not yet fully benefited from molecular marker information (Garcia et al. 2013). To circumvent these problems, the vast majority of genetic research in complex polyploids utilize only single-dose markers during the genetic analysis. So, all the modeling is performed considering the presence of polymorphisms in just one homologous chromosome per homology group. Among the limitations of this approach, it is noteworthy the impossibility to study the effects of allelic dosage, i.e., the effects of the number of copies of each allele at a particular locus in a polyploid genotype. Some studies have been shown that allelic dosage may be extremely important in gene expression in several polyploid species (Garcia et al. 2013; Mollinari and Serang 2015).

In order to advance in polyploid analysis, the measurement of relative abundances (dosage) of alleles is an important step. These estimated dosages may be modeled in association studies. The packages SuperMASSA (Serang et al. 2012) and fitTetra (Voorrips et al. 2011) are theoretical implementations of these ideas, however considering different approaches. It seems clear that subsequent steps involve the accommodation of these estimated allelic dosages into the predictive models. In addition, important gaps remain in our knowledge about the importance of additive and nonadditive effects during the genetic modeling, a critic subject in polyploids given their complex nature (e.g., multiple alleles and loci, mixed inheritance patterns, association between ploidy and mating system variation) (Dufresne et al. 2014). Nevertheless, studies in this direction are still modest and constitute a current challenge in future GS studies.

2.5.6 Genomic Selection 2.0: The Future Is Coming

Recently, GS studies have been not focusing only on predictive abilities but also in two other important features: identifying SNPs associated with the trait and understanding its genetic architecture (MacLeod et al. 2016). For this end, previous biological information and emphasis on estimated marker effects have been considered. Likewise, it has been a challenge in the incorporation of previous genetic evidence in new statistic methods. Some authors have named this current period as "Genomic Selection 2.0" (Hickey 2013; Boichard et al. 2016), representing the new era where information from sequencing data can be generated on millions of individuals and prior biologic results, such as causal mutation, may be considered in predictive models. The challenge in this scenario is determining which of these millions of variants are causal mutations, since the size of effects of such causal mutation is likely to be small.

Good perspectives are presented by Hickey (2013), who coined the term "Genomic Selection 2.0" (GS 2.0). According to the author, until the present, three types of GS investigations have been applied in breeding programs. The so-called GS 0.0 was the first method applied to genomic prediction and "assumed linkage disequilibrium between markers and causative mutations would drive prediction." One step forward, the GS 1.0 "primarily utilize linkage information via realized relationships with close relatives because training populations, although large by historical standards, are far from sufficiently large for linkage disequilibrium information to be very useful for making predictions about quantitative traits." The GS 2.0 is a label given to the current status, which involves large population sizes, millions of molecular data and automated phenotyping.

Considering animal breeding as our benchmark, the 1000 bull genomes project includes whole-genome sequences from 1682 cattle of 55 breeds, from which 67.3 million of genetic variants were identified – including 64.8 million SNP and 2.5 million of indels. The challenge is determining which genetic variants are causal mutations that underlying variations in complex traits. Mapped the causal mutations, these information may be included in genomic prediction investigations.

Broadly speaking, the problem of identifying relevant SNPs in high-dimensional data sets approximates GS methods with contemporaneous genome-wide association algorithm (GWAS). The primary rationale of GWAS investigations is to investigate the underlying biological phenomenon mapping variant genetics associated with important traits. Thus, it is reasonable to hypothesize that modern GS analysis may borrow particularity from GWAS method, i.e., identify important covariates and learn about underlying biological process and use them for prediction tasks.

In a statistical context, the scientific question behind these algorithms is naturally framed as a variable selection problem. Simply stated, "which variable (SNPs) under investigation are useful for prediction the outcome (phenotype)?" Mostly existing GWAS analysis is based on "single-SNP" approach (simply test each SNP, one at time, for association with the phenotype). An alternative is to consider the Bayesian variable selection regression (BVSR) approach, which resembles GS models except because the primary goal is to map SNPs with a biological signal, instead of to predict the genetic merit. In analytical terms, it stands out that Bayesian approach could access the predictive value of the SNP effects simply by computing the posterior probability (i.e., the posterior probability that its coefficient is not zero). Other natural advantages include the possibility to estimate heritability of complex traits, allowing for both polygenic and sparse models, and incorporating external genomic data into the priors, which can increase power and yield new biological insights (Guan and Stephens 2011).

There are many possible approaches to BVRS; see for a review O'Hara and Sillanpää (2009). Models based on a sparse ("spike and slab") prior for the coefficients of linear regression are few of the most widely used for GS purposes – previously discussed on the Statistical Methods topic. Regardless of the BVRS method, current studies have been focused in some important points: (1) traditional algorithms are based on computationally intensive MCMC methods; hence, advances have been driven to be able to perform analysis in a practical time frame considering large-scale problems; (2) typical genomic prediction studies do not produce easily interpretable measures of confidence that individual covariates have nonzero regression coefficients. A modern tendency may consider methods that are able to extract more information from signals (biological evidence) that exist in the data, instead of to be purely a predictive approach; (3) Bayesian framework has been continuously investigated for inferential and predictive purposes. In BVRS scenario, the hyperparameter definition has a crucial importance, given they are responsible for reflecting the sparsity of model and the typical size of nonzero regression coefficients. Both features are important on the inference about the genetic architecture and model complexity. Several rules to define hyperparameter have been suggested in the literature with a lack of consensus among different authors. In this sense, aggregate information from previous QTL mapping and GWAS studies have potential to drive these definitions. For instance, it is possible to consider assumptions that functional SNPs may tend to cluster near one another in the genome or make some SNPs that are better candidates for affecting the trait than other (O'Hara and Sillanpää 2009; Guan and Stephens 2011; Carbonetto and Stephens 2012).

From this perspective, MacLeod et al. (2016) reported the use of GS combined with mapping of causal variants. The BayesRC method incorporates prior biological evidence considering classes of variants to be enriched for causal mutation. In short, previous important evidence originated by variant annotation analysis or from a list of candidate genes may be used to enrich the data analysis. In a similar direction, but considering mapped QTLs as a priori known, Bernardo (2014) points out higher predictive ability when these information are considered as fixed effects in GS models. In wheat, Crossa et al. (2015) showed results considering the importance to identify markers with stable or specific effects across environments. Indirectly, this matter was addressed in the GxE Interaction section. However, it is noteworthy the focus study on the markers effects, rather than solely on predictive capacity. In rice, GS analysis has been proposed in conjunction with GWAS, in order to perform prediction and help on the genetic architecture comprehension (Spindel et al. 2015).

In terms of genotyping and bioinformatic steps, the example of the 1000 bull genomes project reflects the tendency to consider whole-genome sequences in thousands of individuals. Use of whole-genome sequences is supported by simulation studies, which show that SNP densities identified by this approach may provide 40% more accurate predictions than SNP identified by the available genotyping platforms (Meuwissen and Goddard 2010). In this scenario, costs per individual sequenced should be low, and, hence, an alternative may be to sequence all individuals at low coverage. By sequencing coverage, we mean the number of reads sequenced for a determinate site in the genome. In low-coverage data sets, among the drawbacks, a potential challenge in diploid species is the high probability that only one of the two chromosomes has been sampled at a site during the sequencing. As consequence, the genotype definition may be hampered (Nielsen et al. 2011). For this end, the development of computational and imputation methods that are able to deal with this scenario to, firstly, build consensus haplotypes and, then, impute full sequence information on the basis of these consensus haplotypes is important.

An area impacted by GS 2.0 is the phenotyping of quantitative traits that was relatively ignored until recently. In GS context, the perception that phenotype-marker association studies may only be useful considering reliable phenotyping have boosted the so-called high-throughput phenotyping platforms (HTPP). This new branch includes large set of available instruments (robotic and computing) to obtaining detailed measurements of plant characteristics (Cabrera-Bosquet et al. 2012). In general lines, HTPP allows to scan thousands of plants a day in an approach to science akin to high-throughput DNA sequencing (Finkel 2009). On one hand, accurate metrics for physiological process can be observed; but, in the other, the huge capacity of data acquisition required statistical and computational approaches able to summarize the information. Another perspective is the greater of flexibility to choose the traits to be considered in breeding programs. Many initiatives are under way to generate reference populations for traits that were long believed to be impossible to select.

At this point, it is helpful to review what we are attempting to achieve in this topic. Factually, GS research – like in all "omic" – have dynamic advances. As pointed out by Hickey (2013), GS 2.0 can be considered the state of the art. Our personal view is in accordance with the author, however, with additional remarks: (1) the importance to include biological information from multiple areas (e.g., chromatin structure analysis, candidate genes, causal mutation, and epigenetics); (2) continuous advances on the development of algorithms applied to imputation and haplotype construction, in special, for low-coverage sequencing; and (3) improvement of high-throughput phenotyping platforms, in order to achieve better phenotypic metrics and consider traits that are hard to be investigated by conventional methods.

2.6 Conclusions

Genomic prediction for selection is one of the most remarkable breeding proposals in recent years (Habier 2010; Hickey 2013). It may also address important connections between classical quantitative genetics and molecular biology, considering GS context (Gianola et al. 2009; de Los Campos et al. 2015a). Much of these optimistic discourses came from the success of GS in predictions of breeding values, when compared with the traditional MAS results. In fact, it is an undeniable huge potential. In animal breeding, in special, GS overlap the barrier of a simple promise and, currently, constitute in a reality (Hickey 2013). In plant breeding, the examples of success are still restricted for the so-called main crops, such as maize, rice, soybean, and wheat (Poland et al. 2012; Crossa et al. 2013; Jarquín et al. 2014; Crossa et al. 2014; Bassi et al. 2015; Spindel et al. 2015).

One point to be considered in tropical crops is that some species are not proper model systems. The term model systems is used in order to highlight some points that hinder or difficult GS application. Beyond the financial questions, some crops present long life cycles and traits are expressed in later stages of their cycle, some have high genetic load and inbreeding depression, for others large experimental areas are necessary, and for many there is the absence of a suitable genotyping platform. These are only some factors that can be pointed out. In these scenarios, we recommend that GS should be initially treated as an alternative branch in the conventional breeding programs, but it is necessary to include GS in breeding scenarios. This means considering small projects in the beginning, in order to understand the scenario that the crop is inserted.

In conclusion, we are highlighting the GS as a promising and innovative tool to be applied in plant breeding programs. However, to achieve this objective, a critical reflection about the problem of resource allocation is needed. The ideas regarded in this chapter addressed some practical and theoretical issues useful in this process of reshaping breeding.

References

Adams JU (2015) Genetics: big hopes for big data. Nature 527:S108–S109

Akanno EC, Schenkel FS, Sargolzaei M et al (2014) Persistency of accuracy of genomic breeding values for different simulated pig breeding programs in developing countries. J Anim Breed Genet 131:367–378

Arief VN, DeLacy IH, Crossa J et al (2015) Evaluating testing strategies for plant breeding field trials: redesigning a CIMMYT International wheat nursery. Crop Sci 55:164. doi:10.2135/cropsci2014.06.0415

Bassi FM, Bentley AR, Charmet G et al (2015) Breeding schemes for the implementation of genomic selection in wheat (Triticum spp.) Plant Sci 242:23–36. doi:10.1016/j.plantsci.2015.08.021

Bernardo R (2008) Molecular markers and selection for complex traits in plants: learning from the last 20 years. Crop Sci 48:1649. doi:10.2135/cropsci2008.03.0131

Bernardo R (2014) Genomewide selection when major genes are known. Crop Sci. doi:10.2135/cropsci2013.05.0315

Bernardo R, Yu J (2007) Prospects for Genome wide selection for quantitative traits in maize. Crop Sci 47:1082. doi:10.2135/cropsci2006.11.0690

Beyene Y, Semagn K, Mugo S et al (2016) Performance and grain yield stability of maize populations developed using marker-assisted recurrent selection and pedigree selection procedures. Euphytica 208:285–297

Boichard D, Ducrocq V, Croiseau P, Fritz S (2016) Genomic selection in domestic animals: principles, applications and perspectives. C R Biol 339:274–277

Burgueño J, de Los Campos G, Weigel K, Crossa J (2012) Genomic prediction of breeding values when modeling genotype × environment interaction using pedigree and dense molecular markers. Crop Sci 52:707. doi:10.2135/cropsci2011.06.0299

Butler DG, Cullis BR, Gilmour AR, Gogel BJ (2009) ASReml-R reference manual. Release 3.0. Technical report, Queensland Department of Primary Industries, Australia. Available: http://www.vsni.co.uk/downloads/asreml/release2/doc/asreml-R.pdf

Cabrera-Bosquet L, Crossa J, von Zitzewitz J et al (2012) High-throughput phenotyping and genomic selection: the Frontiers of crop breeding ConvergeF. J Integr Plant Biol 54:312–320. doi:10.1111/j.1744-7909.2012.01116.x
Carbonetto P, Stephens M (2012) Scalable variational inference for Bayesian variable selection in regression, and its accuracy in genetic association studies. Bayesian Anal 7:73–108
Coster A, Bastiaansen JWM, Calus MPL et al (2010) Sensitivity of methods for estimating breeding values using genetic markers to the number of QTL and distribution of QTL variance. Genet Sel Evol 42:1–11. doi:10.1186/1297-9686-42-9
Crossa J (2012) From genotype × environment interaction to gene × environment interaction. Curr Genomics 13:225–244. doi:10.2174/138920212800543066
Crossa J, de Los Campos G, Pérez P et al (2010) Prediction of genetic values of quantitative traits in plant breeding using pedigree and molecular markers. Genetics 186:713–724
Crossa J, Beyene Y, Kassa S et al (2013) Genomic prediction in maize breeding populations with genotyping-by-sequencing. G3 (Bethesda) 3:1903–1926. doi:10.1534/g3.113.008227
Crossa J, Pérez P, Hickey J et al (2014) Genomic prediction in CIMMYT maize and wheat breeding programs. Heredity (Edinb) 112:48–60. doi:10.1038/hdy.2013.16
Crossa J, de Los Campos G, Maccaferri M et al (2015) Extending the marker × environment interaction model for genomic-Enabled prediction and genome-wide association analysis in durum wheat. Crop Sci. doi:10.2135/cropsci2015.04.0260
Daetwyler HD, Calus MPL, Pong-Wong R et al (2013) Genomic prediction in animals and plants: simulation of data, validation, reporting, and benchmarking. Genetics 193:347–365. doi:10.1534/genetics.112.147983
de Los Campos G, Sorensen D (2014) On the genomic analysis of data from structured populations. J Anim Breed Genet 131:163–164. doi:10.1111/jbg.12091
de Los Campos G, Hickey JM, Pong-Wong R et al (2013) Whole-genome regression and prediction methods applied to plant and animal breeding. Genetics 193:327–345. doi:10.1534/genetics.112.143313
de Los Campos G, Sorensen D, Gianola D (2015a) Genomic heritability: what is it? PLoS Genet 11:e1005048
de Los Campos G, Veturi Y, Vazquez AI et al (2015b) Incorporating genetic heterogeneity in whole-genome regressions using interactions. J Agric Biol Environ Stat 20:467–490. doi:10.1007/s13253-015-0222-5
Dekkers JCM (2004) Commercial application of marker-and gene-assisted selection in livestock: strategies and lessons. J Anim Sci 82:E313–E328
Denis M, Bouvet J-M (2011) Genomic selection in tree breeding: testing accuracy of prediction models including dominance effect. BMC Proc 5:O13–O13. doi:10.1186/1753-6561-5-S7-O13
Desta ZA, Ortiz R (2014) Genomic selection: genome-wide prediction in plant improvement. Trends Plant Sci 19:592–601. doi:10.1016/j.tplants.2014.05.006
Doerge RW (2002) Mapping and analysis of quantitative trait loci in experimental populations. Nat Rev Genet 3:43–52. doi:10.1038/nrg703
Dudley JW, Johnson GR (2009) Epistatic models improve prediction of performance in corn. Crop Sci. doi:10.2135/cropsci2008.08.0491
Dufresne F, Stift M, Vergilino R, Mable BK (2014) Recent progress and challenges in population genetics of polyploid organisms: an overview of current state-of-the-art molecular and statistical tools. Mol Ecol 23:40–69. doi:10.1111/mec.12581
Elshire RJ, Glaubitz JC, Sun Q et al (2011) A robust, simple genotyping-by-sequencing (GBS) approach for high diversity species. PLoS One 6:e19379. doi:10.1371/journal.pone.0019379
Endelman JB (2011) Ridge regression and other kernels for genomic selection with R package rrBLUP. Plant Genome J 4(3):250–255. doi:10.3835/plantgenome2011.08.0024
Falconer DS, Mackay TFC (1996) Quantitative genetics. Pearson Education Limited, England
Fernando R, Garrick DJ (2009) GenSel – user manual for a portfolio of genomic selection related analyses animal breeding and genetics Iowa state university ames. http://www.biomedcentral.com/content/supplementary/1471-2105-12-186-S1.PDF

Ferrão LF V., Ferrão RG, Ferrão MAG, et al (2016a) Genomic prediction in Coffea canephora using Bayesian polygenic modeling. In: 5th International conference on quantitative genetics. Madison. p 203
Ferrão LF V., Ferrão RG, Ferrão MAG, et al (2016b) Mixed model to multiple harvestlocation trial applied to genomic prediction in Coffea canephora. Plant & Animal Genome Conference, San Diego, EUA
Finkel E (2009) With "Phenomics," plant scientists hope to shift breeding into overdrive. Science 80(325):380 LP–380381
Fisher RA (1919) XV.—The correlation between relatives on the supposition of Mendelian inheritance. Trans R Soc Edinburgh 52:399–433
Flint-Garcia SA, Thornsberry JM, Buckler ES IV (2003) Structure of linkage disequilibrium in plants. Annu Rev Plant Biol 54:357–374. doi:10.1146/annurev.arplant.54.031902.134907
Garcia AAF, Mollinari M, Marconi TG et al (2013) SNP genotyping allows an in-depth characterisation of the genome of sugarcane and other complex autopolyploids. Sci Rep 3:3399. doi:10.1038/srep03399
Garrick D, Dekkers J, Fernando R (2014) The evolution of methodologies for genomic prediction. Livest Sci:1–9. doi:10.1016/j.livsci.2014.05.031
Gelman A, Hill J (2007) Data analysis using regression and multilevel/hierarchical models. Cambridge University Press, Cambridge
Gelman A, Carlin JB, Stern HS, Rubin DB (2014) Bayesian data analysis. Taylor & Francis
Gianola D (2013) Priors in whole-genome regression: the Bayesian alphabet returns. Genetics 194:573–596. doi:10.1534/genetics.113.151753
Gianola D, Hill WG, Manfredi E, Fernando R (2009) Additive genetic variability and the Bayesian alphabet 363:347–363. doi: 10.1534/genetics.109.103952
Goddard ME, Hayes BJ (2007) Genomic selection. J Anim Breed Genet 124:323–330. doi:10.1111/j.1439-0388.2007.00702.x
Guan Y, Stephens M (2011) Bayesian variable selection regression for genome-wide association studies and other large-scale problems. Ann Appl Stat:1780–1815
Habier D (2010) More than a third of the WCGALP presentations on genomic selection. J Anim Breed Genet 127:336–337. doi:10.1111/j.1439-0388.2010.00897.x
Habier D, Fernando RL, Dekkers JCM (2007) The impact of genetic relationship information on genome-assisted breeding values. Genetics 177:2389–2397. doi:10.1534/genetics.107.081190
Habier D, Fernando RL, Dekkers JCM (2009) Genomic selection using low-density marker panels. Genetics 182:343–353. doi:10.1534/genetics.108.100289
Habier D, Fernando R, Kizilkaya K, Garrick D (2011) Extension of the bayesian alphabet for genomic selection. BMC Bioinform 12:186
He J, Zhao X, Laroche A, et al (2014) Genotyping by sequencing (GBS), an ultimate marker-assisted selection (MAS) tool to accelerate plant breeding. Front Plant Sci 5:484
Heffner EL, Sorrells ME, Jannink J-L (2009) Genomic selection for crop improvement. Crop Sci 49:1. doi:10.2135/cropsci2008.08.0512
Heffner EL, Lorenz AJ, Jannink JL, Sorrells ME (2010) Plant breeding with genomic selection: gain per unit time and cost. Crop Sci 50:1681–1690. doi:10.2135/cropsci2009.11.0662
Heffner EL, Jannink J-L, Iwata H et al (2011) Genomic selection accuracy for grain quality traits in Biparental wheat populations. Crop Sci 51:2597. doi:10.2135/cropsci2011.05.0253
Henderson CR (1949) Estimation of changes in herd environment. J Dairy Sci 32:706
Henderson CR (1950) Estimation of genetic parameters. Ann Math Stat 21:309–310
Heslot N, Yang H-P, Sorrells ME, Jannink J-L (2012) Genomic selection in plant breeding: a comparison of models. Crop Sci 52:146–160. doi:10.2135/cropsci2011.06.0297
Heslot N, Akdemir D, Sorrells ME, Jannink J-L (2014) Integrating environmental covariates and crop modeling into the genomic selection framework to predict genotype by environment interactions. Theor Appl Genet 127:463–480
Heslot N, Jannink J-L, Sorrells ME (2015) Perspectives for genomic selection applications and research in plants. Crop Sci 55:1–12. doi:10.2135/cropsci2014.03.0249

Hickey JM (2013) Sequencing millions of animals for genomic selection 2.0. J Anim Breed Genet 130:331–332

Holland JB (2001) Epistasis and plant breeding. In: Plant breeding reviews. Wiley, Oxford, pp 27–92

Hu Z, Li Y, Song X et al (2011) Genomic value prediction for quantitative traits under the epistatic model. BMC Genet 12:1–11. doi:10.1186/1471-2156-12-15

James G, Witten D, Hastie T, Tibshirani R (2013) An introduction to statistical learning. Springer

Jannink J-L, Lorenz AJ, Iwata H (2010) Genomic selection in plant breeding: from theory to practice. Brief Funct Genomics 9:166–177. doi:10.1093/bfgp/elq001

Jarquín D, Crossa J, Lacaze X et al (2013) A reaction norm model for genomic selection using high-dimensional genomic and environmental data. Theor Appl Genet. doi:10.1007/s00122-013-2243-1

Jarquín D, Kocak K, Posadas L et al (2014) Genotyping by sequencing for genomic prediction in a soybean breeding population. BMC Genomics 15:740. doi:10.1186/1471-2164-15-740

Jarquín D, Pérez-Elizalde S, Burgueño J, Crossa J (2016) A hierarchical Bayesian estimation model for Multienvironment plant breeding trials in successive years. Crop Sci. doi:10.2135/cropsci2015.08.0475

Jonas E, de Koning D-J (2013) Does genomic selection have a future in plant breeding? Trends Biotechnol 31:497–504

Kärkkäinen HP, Sillanpää MJ (2012) Back to basics for Bayesian model building in genomic selection. Genetics 191:969–987. doi:10.1534/genetics.112.139014

Kruschke JK (2011) Doing Bayesian data analysis. Elsevier, Langford

Kruschke JK, Aguinis H, Joo H (2012) The time has come: Bayesian methods for data analysis in the organizational sciences. Organ Res Methods 15:722–752. doi:10.1177/1094428112457829

Lado B, Barrios PG, Quincke M et al (2016) Modeling genotype × environment interaction for genomic selection with unbalanced data from a wheat breeding program. Crop Sci. doi:10.2135/cropsci2015.04.0207

Lande R, Thompson R (1990) Efficiency of marker-assisted selection in the improvement of quantitative traits. Genetics 124:743–756

Lopes MS, Bastiaansen JWM, Janss L et al (2015) Estimation of additive, dominance, and imprinting genetic variance using genomic data. G3 Genes|Genomes|Genetics 5:2629–2637. doi:10.1534/g3.115.019513

Lopez-Cruz M, Crossa J, Bonnett D et al (2015) Increased prediction accuracy in wheat breeding trials using a marker x environment interaction genomic selection model. G3{&}{#}58. Genes|Genomes|Genetics 5:569–582. doi:10.1534/g3.114.016097

Lorenz AJ, Chao S, Asoro FG et al (2011) Chapter 2: genomic selection in plant breeding: knowledge and prospects. Adv Agron. doi:10.1016/B978-0-12-385531-2.00002-5

Lorenzana RE, Bernardo R (2009) Accuracy of genotypic value predictions for marker-based selection in biparental plant populations. Theor Appl Genet 120:151–161. doi:10.1007/s00122-009-1166-3

Lynch M, Walsh B (1998) Genetics and analysis of quantitative traits, 1st edn. Sinauer Associates, Sunderland

Mackay I, Powell W (2007) Methods for linkage disequilibrium mapping in crops. Trends Plant Sci 12:57–63. doi:10.1016/j.tplants.2006.12.001

Mackay TFC, Stone EA, Ayroles JF (2009) The genetics of quantitative traits: challenges and prospects. Nat Rev Genet 10:565–577. doi:10.1038/nrg2612

MacLeod IM, Bowman PJ, Vander Jagt CJ et al (2016) Exploiting biological priors and sequence variants enhances QTL discovery and genomic prediction of complex traits. BMC Genomics 17:144. doi:10.1186/s12864-016-2443-6

Malosetti M, Ribaut J-M, van Eeuwijk FA (2013) The statistical analysis of multi-environment data: modeling genotype-by-environment interaction and its genetic basis. Front Physiol 4:44

Malosetti M, Bustos-Korts D, Boer MP, van Eeuwijk FA (2016) Predicting responses in multiple environments: issues in relation to genotype x environment interactions. Crop Sci 13:(accepted). doi: 10.2135/cropsci2015.05.0311

Meuwissen THE (2009) Accuracy of breeding values of 'unrelated' individuals predicted by dense SNP genotyping. Genet Sel Evol 41:1
Meuwissen T, Goddard M (2010) Accurate prediction of genetic values for complex traits by whole-genome resequencing. Genetics 185:623 LP–623631
Meuwissen TH, Hayes BJ, Goddard ME (2001) Prediction of total genetic value using genome-wide dense marker maps. Genetics 157:1819–1829
Meyer K (2007) WOMBAT—A tool for mixed model analyses in quantitative genetics by restricted maximum likelihood (REML). J Zhejiang Univ Sci B 8:815–821. doi:10.1631/jzus.2007.B0815
Mollinari M, Serang O (2015) Quantitative SNP genotyping of polyploids with MassARRAY and other platforms. Plant Genotyping Methods Protoc 1245:215–241
Moser G, Tier B, Crump R et al (2009) A comparison of five methods to predict genomic breeding values of dairy bulls from genome-wide SNP markers. Genet Sel Evol. doi:10.1186/1297-9686-41-56
Mrode RA, Thompson R (2005) Linear models for the prediction of animal breeding values. CABI, Wallingford
Nakaya A, Isobe SN (2012) Will genomic selection be a practical method for plant breeding? Ann Bot 110:1303–1316. doi:10.1093/aob/mcs109
Nielsen R, Paul JS, Albrechtsen A, Song YS (2011) Genotype and SNP calling from next-generation sequencing data. Nat Rev Genet 12:443–451
Nishio M, Satoh M (2014) Including dominance effects in the genomic BLUP method for genomic evaluation. PLoS One 9:e85792
O'Hara RB, Sillanpää MJ (2009) A review of bayesian variable selection methods: what, how and which. Bayesian Anal 4:85–118. doi:10.1214/09-BA403
Ott J, Kamatani Y, Lathrop M (2011) Family-based designs for genome-wide association studies. Nat Rev Genet 12:465–474
Ould Estaghvirou SB, Ogutu JO, Schulz-Streeck T et al (2013) Evaluation of approaches for estimating the accuracy of genomic prediction in plant breeding. BMC Genomics 14:1–21. doi:10.1186/1471-2164-14-860
Park T, Casella G (2008) The Bayesian lasso. J Am Stat Assoc 103:681–686. doi:10.1198/016214508000000337
Payne RW, Murray DA, Harding SA (2011) An introduction to the genstat command language (14th edn)
Pérez PR, de los Campos G (2014) Genome-wide regression & prediction with the BGLR statistical package. Genetics, genetics-114
Pérez P, de Los Campos G, Crossa J, Gianola D (2010) Genomic-Enabled prediction based on molecular markers and pedigree using the Bayesian linear regression package in R. Plant Genome 3:106–116. doi:10.3835/plantgenome2010.04.0005
Perez-Rodriguez P, Gianola D, Gonzalez-Camacho JM et al (2013) Comparison between linear and non-parametric regression models for genome-Enabled prediction in wheat. G3&#58. Genes|Genomes|Genetics 2:1595–1605. doi:10.1534/g3.112.003665
Poland JA, Rife TW (2012) Genotyping-by-sequencing for plant breeding and genetics. Plant Genome J 5:92. doi:10.3835/plantgenome2012.05.0005
Poland J, Endelman J, Dawson J et al (2012) Genomic selection in wheat breeding using genotyping-by-sequencing. Plant Genome J 5:103. doi:10.3835/plantgenome2012.06.0006
Rajsic P, Weersink A, Navabi A, Pauls KP (2016) Economics of genomic selection: the role of prediction accuracy and relative genotyping costs. Euphytica 210(2):259–276
Rencher AC, Schaalje GB (2008) Linear models in statistics. Wiley, Hoboken
Resende MFR, Muñoz P, Acosta JJ et al (2012a) Accelerating the domestication of trees using genomic selection: accuracy of prediction models across ages and environments. New Phytol 193:617–624. doi:10.1111/j.1469-8137.2011.03895.x

Resende MFR, Muñoz P, Resende MDV et al (2012b) Accuracy of genomic selection methods in a standard data set of loblolly pine (Pinus taeda L.) Genet 190:1503–1510. doi:10.1534/genetics.111.137026

Schulz-Streeck T, Ogutu JO, Gordillo A et al (2013) Genomic selection allowing for marker-by-environment interaction. Plant Breed 132:532–538. doi:10.1111/pbr.12105

Serang O, Mollinari M, Garcia AAF (2012) Efficient exact maximum a posteriori computation for bayesian SNP genotyping in polyploids. PLoS One 7:e30906. doi:10.1371/journal.pone.0030906

Simon N, Friedman J, Hastie T, Tibshirani R (2011) Regularization paths for Cox's proportional hazards model via coordinate descent. J Stat Softw 39:1–13

Smith AB, Cullis BR, Thompson R (2005) The analysis of crop cultivar breeding and evaluation trials: an overview of current mixed model approaches. J Agric Sci 143:449. doi:10.1017/S0021859605005587

Spindel J, Begum H, Akdemir D et al (2015) Genomic selection and association mapping in Rice (Oryza sativa): effect of trait genetic architecture, training population composition, marker number and statistical model on accuracy of Rice genomic selection in elite, tropical Rice breeding lines. PLoS Genet 11:1–25. doi:10.1371/journal.pgen.1004982

Stephens M, Balding DJ (2009) Bayesian statistical methods for genetic association studies. Nat Rev Genet 10:681–690. doi:10.1038/nrg2615

Tempelman RJ (2015) Statistical and computational challenges in whole genome prediction and genome-wide association analyses for plant and animal breeding. J Agric Biol Environ Stat 20:442–466. doi:10.1007/s13253-015-0225-2

Tibshirani R (1996) Regression shrinkage and selection via the lasso. J R Stat Soc Ser B:267–288

VanRaden PM (2008) Efficient methods to compute genomic predictions. J Dairy Sci 91: 4414–4423. doi: http://dx.doi.org/10.3168/jds.2007-0980

Vitezica ZG, Varona L, Legarra A (2013) On the additive and dominant variance and covariance of individuals within the genomic selection scope. Genetics 195:1223–1230. doi:10.1534/genetics.113.155176

Voorrips RE, Gort G, Vosman B (2011) Genotype calling in tetraploid species from bi-allelic marker data using mixture models. BMC Bioinform 12:1–11. doi:10.1186/1471-2105-12-172

Wang X, Yang Z, Xu C (2015) A comparison of genomic selection methods for breeding value prediction. Sci Bull 60:925–935. doi:10.1007/s11434-015-0791-2

Whittaker JC, Thompson R, Denham MC (2000) Marker-assisted selection using ridge regression. Genet Res 75:249–252

Xu S (2008) Quantitative trait locus mapping can benefit from segregation distortion. Genetics 180:2201–2208. doi:10.1534/genetics.108.090688

Xu S, Hu Z (2010) Methods of plant breeding in the genome era. Genet Res (Camb) 92:423–441. doi:10.1017/S0016672310000583

Zhou X, Carbonetto P, Stephens M (2013) Polygenic modeling with bayesian sparse linear mixed models. PLoS Genet 9:e1003264. doi:10.1371/journal.pgen.1003264

Part II
Crops

Chapter 3
Tropical Maize (*Zea mays* L.)

Gregory O. Edmeades, Walter Trevisan, B.M. Prasanna, and Hugo Campos

3.1 Introduction

3.1.1 History and Origin

Maize is the main tropical crop that has been domesticated in the American continent. Its origin has been the subject of much debate. Theories linking teosinte, *Tripsacum*, an unspecified common wild relative, and teosinte × *Tripsacum* crosses (tripartite hypothesis), have been proposed for the route taken by maize from wild plants to the domesticated crop that now depends on human intervention for survival (Wilkes 2004). In the last decade, molecular markers and ^{14}C analysis of maize cobs recovered from excavations in Mexico and Guatemala clarified the origin of maize (Matsuoka et al. 2002; van Heerwaarden et al. 2011). These indicated that maize was already cultivated as early as 8700 BP and perhaps originated from a single domestication event of the annual Balsas teosinte (*Zea mays* subspecies *parviglumis*) in mid-altitude areas of South-Central Mexico. From here maize diversified into the Mexican highlands by crossing with a highland-adapted teosinte, *Zea*

G.O. Edmeades (✉)
Independent Consultant, Cambridge, New Zealand
e-mail: greg_edmeades@msn.com

W. Trevisan
Independent Consultant, Dekalb, IL, USA
e-mail: Wtrevisan@comcast.net

B.M. Prasanna
International Maize and Wheat Improvement Center, ICRAF House, Gigiri, Kenya
e-mail: b.m.prasanna@cgiar.org

H. Campos
International Potato Center, Lima 12, Peru
e-mail: h.campos@cgiar.org

H. Campos, P.D.S. Caligari, *Genetic Improvement of Tropical Crops*,
DOI 10.1007/978-3-319-59819-2_3

mays subspecies *mexicana* (Warburton et al. 2011; Mir et al. 2013). Maize was therefore entirely tropical in origin. Its spread was via two major paths. The first was through northern Mexico to the southern USA and northward into the long photoperiods and short seasons of northern USA and Canada. The second path was through the lowlands of Mexico to Central America, the Caribbean, and thence to the Andes (Matsuoka et al. 2002). Peru has very similar geography/environments and diversity of maize races as highland Mexico and is often considered a subcenter of origin (Wilkes 2004).

The amazing diversity of maize in the Americas made its adaptation to other continents relatively rapid. The first introduction of Caribbean maize into Europe was in 1493, where it was used initially as a garden curiosity. It was joined there by northern US flints (Mir et al. 2013). McCann (2005) cites evidence that maize was present in Egypt in 1517, just 25 years after Columbus. The Portuguese were active in introducing maize to Africa through their colonies and trading posts in the Azores, Angola, Mozambique, Mombasa, and Zanzibar. Slavers and missionaries introduced maize to West Africa in the seventeenth century (McCann 2005). From these points of introduction, maize spread all over Africa and became a main staple.

The Turks and the Portuguese disseminated maize in the Asian continent. The crop was probably introduced to southern China in the late seventeenth century and spread to northern China in the late eighteenth century. Now in China, maize has overtaken rice in terms of cropped area, making it the second largest maize producer in the world after the USA.

The outcome of this global migration has been an extraordinary diversity of landraces with different ear shapes, ear sizes, grain colors, and textures and with a diversity of food, feed, and industrial uses. Maize is now found adapted to diverse environments from sea level to 4000 masl, from latitudes 0 to 57°, on soils with pH from 4 to 8, and in areas with annual rainfalls of 400–2500 mm.

3.1.2 Importance of Maize in the Tropics

Major maize production zones where tropically adapted germplasm is used (countries producing >100,000 tons of maize annually at latitudes <35°) are shown in Table 3.1. Around 30% of global maize production is from tropical areas and from tropically adapted germplasm, but occupying 49% of the global area planted to maize. Yields in temperate environments, led mainly by the USA, Europe, and China, averaged 7.2 t/ha vs. 3.3 t/ha for tropical regions. Nonetheless, yields have been increasing at about the same rate (74–75 kg/ha/year) in both ecologies, which translates to an annual increase of 1% in temperate and 2.3% in tropical regions (Fig. 3.1). Averages hide large variations in yield gain by regions, with the South American Cone and SE Asia leading the way (128–142 kg/ha/year) and the smallest increases in yield (27–40 kg/ha/year) being reported from the three African subregions and the Central American and Caribbean regions. Low yields in South Asia and sub-Saharan Africa (SSA) reflect lack of use of improved varieties, inadequate

Table 3.1 Maize production, area and yield, and time trends (2000–2014) in area and yield, for major production areas in tropical maize-producing regions, compared with the temperate world

	Average 2012–2014			Annual rate of increase	
Region	Production (Mt)	Area (Mha)	Yield (t/ha)	Area (%)	Yield (kg/ha/year) [%]
World	973.2	182.0	5.34	1.97	80 [1.49]
Temperate[a]	675.6	93.6	7.21	2.09	74 [1.02]
Tropical	294.5	89.1	3.30	1.89	75 [2.26]
Andean	8.1	2.2	3.72	0.70	76 [2.05]
Cent. America and Caribbean	27.1	9.4	2.87	−0.13	40 [1.40]
South. Cone	111.4	20.7	5.37	2.08	142 [2.65]
South Asia	33.5	11.7	2.86	2.10	86 [3.01]
Southeast Asia	38.6	9.2	4.18	1.37	128 [3.05]
N. Africa to W. Asia	10.2	1.7	5.88	1.15	38 [0.65]
W. and C. Africa	22.3	13.5	1.66	3.23	27 [1.64]
E. and S. Africa	43.3	20.2	2.14	2.05	42 [1.94]

Source: FAOSTAT (2016)
[a]Includes Mainland China
The definition of regions follows that of Pingali (2001)

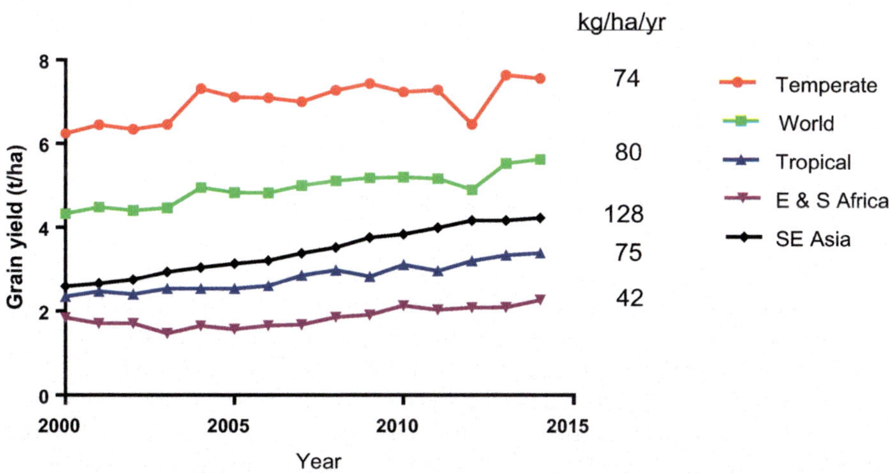

Fig. 3.1 Maize yields vs. time from 2000 to 2013 for the world, for temperate countries (including China), for tropical countries, for eastern and southern Africa, and for Southeast Asia. Corresponding linear gains with time are shown as numbers beside each graph (FAOSTAT 2016)

nutrients (poor soil fertility), and frequent occurrence of drought (Fischer et al. 2014), since more than 80% of maize in these areas is rainfed. Nonetheless, the contribution of yield increases in tropical areas to the 1 billion tons of maize grain produced annually has been significant. This is despite lower average yields, and the

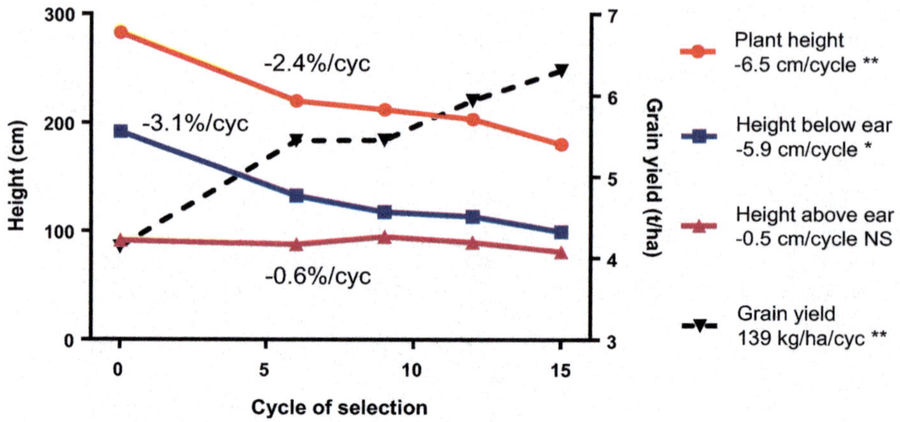

Fig. 3.2 Response of plant and ear height, stem height above the ear, and grain yield at or near the optimum plant density to 15 cycles of visual selection for reduced plant height in Tuxpeño Crema I (Johnston et al. 1986)

smaller research investment in tropical maize vs. temperate maize, especially by the private sector.

Time trends in grain yield (Fig. 3.1) also highlight fluctuations in yield with time, the most variable being in temperate regions, mainly due to the severe 2012 drought in the USA (Boyer et al. 2013) (Fig. 3.2).

3.1.3 Broad Patterns of Adaptation: Megaenvironments

In order to provide a systematic structure to maize breeding programs, the target environments have been broadly characterized into megaenvironments (MEs). An ME can be defined as an area growing at least 1 m ha of maize within which cultivar × environment interactions are relatively minor. They are often defined by temperature, altitude (e.g., highland vs. lowlands), rainfall (dry vs. wet and humid), and daylength, since tropical maize is usually very photoperiod sensitive and poorly adapted to latitudes >30°. CIMMYT has defined six major maize MEs for tropical environments for sub-Saharan Africa, but the descriptions apply also to other regions. Two temperate MEs at latitudes >30° can be added to these (Table 3.2). From a breeding perspective, each of these classes has been subdivided further by crop duration, incidence of specific diseases and pests, grain color, kernel texture, and protein quality.

The tropical and subtropical MEs differ mainly in the nature of disease pressure. Highland maize is unique – it grows exceptionally well at low temperatures, has a lower optimum temperature for development than maize of other adaptation classes (Ellis et al. 1992), and has a distinct morphology (Eagles and Lothrop 1994).

Table 3.2 Major maize megaenvironments showing approximate proportional areas, stratified by altitude, rainfall[a], temperature[b], and daylength[c]

Megaenvironment	Altitude (masl)	Proportion total area %	Potential yield t/ha	Example areas
MME1: Highland tropical	>2000	3	11	Ethiopia, Mexico, Andes
MME2: Wet upper mid-altitude subtropical	1600–2000	3	13	Ethiopia, Kenya, South Africa, Central America
MME3: Wet lower mid-altitude subtropical	1200–1600	5	13	Uganda, Kenya, Indian subcontinent (winter)
MME4: Dry mid-altitude subtropical	1200–2000	8	9	Tanzania, Western Kenya, central Mexico, Nepal
MME5: Wet lowland tropical	0–1200	15	9	Thailand, Nigeria, coastal Central America
MME6: Dry lowland tropical	0–1200	14	6	Coastal East Africa, Central America, India, NE Brazil
MME7: Wet temperate	0–1500	35	14	Corn Belt of the USA, western Europe, Argentina
MME8: Dry temperate	0–1500	17	9	Western USA, eastern Europe, northwest China

Adapted from Fischer et al. 2014
Potential yield is an estimate based on temperature and radiation receipt, as outlined by Muchow et al. (1990)
[a]Rainfall in growing season (the five consecutive months with the greatest P/PET ratio. For MME1, >350 mm; for MME4, 350–600 mm; for MME6, 350–800 mm; for MME8, 300–600 mm
[b]Average daily maximum temperature for the middle 70% of the growing season: MME1, 18–24 °C; MME2, 24–28 °C; MME3, 28–30 °C; MME4, 24–30 °C; MME5, 30–34 °C; MME6, 30–36 °C; MME7, 26–34 °C; and MME8, 26–36 °C.
[c]Daylength associated with the longest day during a summer growing season. MME1–MME6 are mainly in latitudes of <30°, while MME7 and MME8 are in latitudes 30–57°.

3.2 Tropical Maize Germplasm: Races, Landraces, and Germplasm Exchanges

3.2.1 Races of Maize: Packaging Diversity by Adaptation and Grain Type

A total of 285 maize races have been described (e.g., Wellhausen et al. 1952), though Hallauer and Miranda (1988) considered that perhaps only 130 were distinct entities. Of these, 71% originated in South America, 24% in Central America and Mexico, 2% in the Caribbean, and 3% in the USA (Serratos 2009). Adaptation of these can be considered 50% lowland tropical (<1200 masl), 10% mid-altitude (1200–1800 masl), and 40% highland (1800–2900 masl). In terms of grain type, around 40% are floury, 30% flints, 20% dent, 10% popcorns, and 3% sweet corn, and most are white grained (Pandey and Gardner 1992).

Molecular markers have largely confirmed racial groupings initially based on numerical taxonomy (e.g., Wellhausen et al. 1952). Reif and coworkers (2006) genotyped 497 plants drawn from 24 Mexican races with 25 SSR markers and grouped the accessions based on multivariate analysis. The accessions averaged almost eight alleles (range 4–14) per locus and so were highly variable. They reported that an average of 1.3 alleles per locus were unique to each accession. This and other studies on Mexican races have confirmed that there is a high level of variation within a race, and within individual landraces, and less among races and landraces (Warburton et al. 2002; Reif et al. 2004).

The large majority of tropical maize races are very sensitive to daylength and, when grown in photoperiods of greater than 14 h, become tall, unwieldy, late to flower, slow to dry down, and with much reduced grain yield (Stevenson and Goodman 1972; Edmeades et al. 2000a; Edmeades et al. 2000b). Since landraces are difficult to phenotype or use in crossing nurseries, temperate maize breeders have struggled to introgress tropical germplasm directly into Corn Belt maize, and a more staged approach has been proposed (Gerrish 1983; Holland et al. 1996). Private and public institutions in temperate regions have devoted considerable resources in adapting tropical and subtropical germplasm to temperate regions.

In summary, racial diversity is greatest where diverse natural selection environments are also found. Unique microclimates and pockets of disease in geographically isolated valleys, such as those found in Mexico and the Andean region, have resulted in a rich diversity of genotypes. The majority of this variation now resides in germplasm banks. In the interim, CIMMYT and several national breeding programs have captured a significant proportion of this variation in the most productive backgrounds in the form of pools, populations, and inbred lines that are now widely grown (Warburton et al. 2008).

3.2.2 The Search for Superior Alleles and Their Concentration: The Mexican Experience

Because of proximity to the center of origin of maize, there has been considerable focus on Mexican races. Climatic adaptation of Mexican races ranges from 0 to 2900 masl, an average annual temperature of 11–27 °C, from 400 to 3500 mm of seasonal rainfall, and average daylengths during the crop season of 12.5–13.5 h (Corral et al. 2008). Tuxpeño is reported by Corral et al. (2008) to have the greatest adaptability among Mexican races. Reif et al. (2006) studied 24 representative Mexican races using multivariate analysis of molecular marker data. They identified three racial complexes from (a) the high elevation, (b) medium to low elevation, and (c) narrow-eared races from NW Mexico. Highland tropical races (e.g., Chalqueño, Cacahuacintle, Palomero Toluqueño, Arrocillo Amarillo) typically have few tassel branches and are adapted to cool conditions. The medium to low elevation group includes Celaya, the popular, diverse, and more modern race Tuxpeño, and the long-eared Jala, Zapalote Chico and Grande, Bolita, Nal-Tel, and Pepitilla. The NW

Mexico group possesses long, thin eight-rowed ears from races such as Harinoso de Ocho and include Chapalote, Reventador, and Maíz Dulce adapted to slightly longer days and lower elevations. These have generally been shown to be the highest-yielding races (Crossa et al. 1990) and are the foundations on which CIMMYT's breeding program was originally based. Their analysis, based on molecular marker data and an earlier study using morphometric similarities (Goodman and Brown 1988), supports the proposal that parental races in Mexico are Chalqueño, Cónico Norteño, Bolita, and Celaya.

CIMMYT has tested and incorporated leading landraces from the Americas in pre-breeding gene "pools" and elite populations. Pools were later systematically arranged to address general requirements for adaptation (highland, lowland tropical, subtropical, and temperate), maturity (early, intermediate, late), grain color (yellow, white) and grain texture, and protein quality (flint, dent, and QPM[1]). Elite populations addressed similar niches, though some were direct introductions from national programs (e.g., Population 32, ETO Blanco from Colombia – Chavarriaga 1966) or composites of collections or varietal crosses from regions (e.g., Population 35, Antigua × Republica Dominicana) (Pandey and Gardner 1992). Pools were open ended, and pre-tested components were sown ear to row as females in a half-sib recurrent breeding scheme where the male pollinator was a balanced bulk of female entries. Though this scheme broke genetic linkages and mixed germplasm components under mild selection pressure for yield and agronomic traits, the flow of new materials into pools was relatively small, and limited attention was paid to heterotic groupings. Grain yields were rarely assessed directly but rather determined from an assumed shelling percentage of 80%, a practice that eliminated selection pressure for increased shelling percentages. These can be 5–8% greater in temperate hybrids vs. tropical hybrids when compared in a disease- and stress-free environment.

CIMMYT populations, on the other hand, were largely closed, with occasional introductions from matched genetic pools. From 1972 through the mid-1990s, they were improved through recurrent full-sib selection in which 250 families from each cycle were evaluated in a lattice design in two replications evaluated at six international locations. Selection was mild with the superior 40% families recombined, and cycles took 2 years to complete. During the 1990s within-family S_1 improvement occurred while progeny testing was underway, thereby reducing the load of undesirable recessives. Again, little attention was paid to heterotic group formation. Varieties were formed by recombining the ten best full-sib families at each test location or across locations and were supplied as open-pollinated varieties (OPVs) to National Agricultural Research System (NARS) scientists. During the 1990s, the populations were increasingly used to generate inbreds and resulted in a number of released CIMMYT maize lines (CMLs) that now number almost 600.

CIMMYT products, now mainly hybrids, inbred lines, and a small proportion of OPVs are available on request. Molecular data suggest that there is more genetic

---

[1] QPM is quality protein maize, a grain type with elevated levels of lysine and tryptophan caused by the presence of the *opaque-2* recessive gene, and improved for kernel hardness through the accumulation of modifier genes.

variation within landraces and populations than among them (Warburton et al. 2002; Reif et al. 2004), and the genetic variation among the CMLs is considered to encompass most of the variation present in the entire tropical gene pool (Warburton et al. 2008).

A recent initiative, Seeds of Discovery (SeeD) (http://seedsofdiscovery.org/), undertaken by CIMMYT and funded by the Government of Mexico, aims to characterize the genetic diversity of maize landraces with high-density SNP markers linked to phenotypes of landraces obtained in environments to which they are reasonably well adapted. In order to make this variability more useful, bridge populations are being developed where rare but useful alleles from landraces are being crossed into elite maize germplasm using markers to reduce linkage drag from remaining alleles.

3.2.3 Global Movements of Germplasm: Developing Other Sources

Tropical maize has moved successfully to latitudes, altitudes, and rainfall regimes similar to those where it originated. Examples would be lowland/mid-altitude Tuxpeños from Mexico introduced to Brazil and West Africa; Caribbean germplasm introduced into Thailand, the Philippines, Indonesia, and Brazil; and the Andean landrace Montana introduced in the highlands of Kenya. It was also transferred to higher latitudes, e.g., US Southern Dents introduced in southern Africa (Mir et al. 2013). However, successful movement across wide latitudes has only occurred when germplasm is essentially photoperiod insensitive as with temperate germplasm.

Breeders are always interested in adding useful genetic diversity to their breeding populations. That interest is especially acute among breeders of temperate germplasm in the USA where a large proportion of the hybrids sold today trace back to seven inbred lines developed ~50 years ago (Butruille et al. 2015). There have been systematic attempts to widen the genetic base of temperate maize using tropical germplasm, though photoperiod sensitivity has slowed this process. Two USDA-supported programs, the Latin American Maize Program (LAMP) and its successors GEM and G2F, have played a key role in linking tropical landraces with the mainstream temperate breeding programs. Under LAMP the USA and collaborating national institutions in 11 Latin American countries evaluated more than 12,000 accessions from 328 races from the Americas, systematically reducing these at successive stages of testing to 3000 and then to a core subset of 270 (Salhuana and Pollak 2006). Under GEM these elite landraces were crossed with elite US commercial germplasm to provide commercially useful germplasm. Other programs such as that at North Carolina State have used LAMP germplasm and commercial hybrids from the Caribbean and South America to generate adapted tropical populations. This was done by systematically crossing to temperate germplasm to introduce daylength insensitivity followed by backcrossing to the tropical source and line extraction (Holland et al. 1996; Goodman 1999). The result has been hybrids that have performed competitively with adapted commercial Corn Belt hybrids and production of inbreds that are essentially 100% tropical (Uhr and Goodman 1995). Others such as the Hallauer group at Iowa State University selected recurrently for earliness to flower for ten or more generations in tropical populations such as ETO, Tusón, Tuxpeño, and Suwan. By doing so, they steadily increased the frequency of alleles for daylength insensitivity in these broadly based populations (Teixeira et al. 2015). The breeding program based at the University of Hawaii in the subtropics, led by Dr. J Brewbaker for the past 50 years, produced a number of inbreds with excellent resistance to a number of tropical diseases (Brewbaker 2009), with some lines serving as sources of disease resistance in temperate zones as well. One such inbred (KS23-6) has been identified as resistant to maize chlorotic mottle virus (MCMV), one of the two important viruses in the devastating maize lethal necrosis (MLN) virus complex in eastern Africa. Clearly, tropical populations can be tamed for use in temperate breeding programs, but it is a long-term effort.

Today, the movements of germplasm around the world continue, though this has been significantly affected because of regulations on seed movement arising from intellectual property protection that are considered unduly restrictive by some private sector breeders (Butruille et al. 2015). CIMMYT and its sister CGIAR center in Nigeria, IITA, have played key roles in germplasm exchanges worldwide through their international testing programs. CGIAR germplasm transfers are managed under a Standard Material Transfer Agreement (SMTA) between the supplier and the recipient. Both centers have embarked on a deliberate policy of devolving international breeding programs to sub-Saharan Africa, the Andean zone, and Asia. Another important public source of elite temperate germplasm has been through inbred lines released from patent protection 20 years after their initial release in the USA and made available on request through the US National Plant Germplasm System (NPGS) (Kurtz et al. 2016). Many of these inbreds, although several decades old, are of considerable value to tropical maize breeders and carry a well-defined heterotic response.

3.3 Biology: Do Tropical and Temperate Maize Germplasm Groups Differ?

Tropical maize landraces have evolved under natural (and more recently human) selection to outcompete weeds and withstand leaf area reduction from insects and diseases. Accordingly, they are often tall, have an ear height/plant height ratio of 0.65, may tiller freely and be prolific, are excessively leafy with heavy husks to protect against insects and birds, have large tassels that ensure surplus pollen production, and have a low-moderate number of kernels (300–400) per plant that germinate and establish rapidly from a range of sowing depths. Harvest index in landraces is typically 0.25–0.40, and plants become barren under any form of stress at flowering. The process of improvement of tropical maize using modern improvement methods has brought about major changes in the morphology and partitioning of dry matter as yields and ability to withstand higher plant densities, moving the tropical phenotype in the direction of elite temperate hybrids.

Typically modern temperate maize hybrids have a HI of 0.5–0.55, exhibit vigorous ear growth, have strong stalks and small tassels and smaller but more erect leaves, are rarely barren, and have large ears with around 500–600 kernels/ear. Because of a long history of selection under high plant density and multiple test locations for stable grain yield and resistance to lodging, temperate maize hybrids seldom have more than one ear per plant at normal plant densities or become barren under stress.

3.3.1 Source/Sink Ratios, Ear Growth, and Yield Components

Early research on tropical maize populations within a decade of their formation from landraces concluded that the ratio of photosynthetic source to reproductive sink was significantly higher than that of temperate maize (Goldsworthy et al. 1974; Fischer and Palmer 1984). Ear growth in tropical varieties was not vigorous, and when grown at commercial densities, silks often emerged after pollen shed began. The trait ASI is an indicator of ear growth rate – a reflection of biomass partitioning to the developing ear. A large ASI is symptomatic of slow ear and silk growth. In fact, delayed silking is normally associated with any stress that reduces photosynthesis per plant since this reduces ear growth more than tassel growth and development (Edmeades et al. 1993, 2000a). ASI is strongly correlated with grain number per ear (Bolaños and Edmeades 1996). Delayed silking may help ensure cross-pollination, but when ASI is large (e.g., ASI >8 day), the plant aborts kernels and may become barren. In the height reduction study, ASI at optimum plant density for yield declined from 3.9 to 1 day after 15 cycles, and barrenness declined from 30% to 2% (Johnson et al. 1986). These correlated changes suggest that stem growth was competing for assimilates that promote ear growth at flowering.

Improved tropical maize hybrids are still characterized by heavy husks that protect the ear from birds and insects, large tassels, large leaves, and a harvest index in the range of 0.4–0.45 (Zaidi et al. 2003a). In Johnson's classic experiment, the source/sink ratio changed dramatically: the ratio of grain weight to leaf area increased from 85 g/cm^2 in C_0 to 225 g/cm^2 in C_{15}. In a separate study, Fischer et al. (1987) conducted six cycles of recurrent full-sib selection in three elite lowland tropical populations for leaf area density above the ear and/or reduced tassel primary branch number. Tassel size was highly heritable, and leaf area less so, but significant reductions in both were reported. Selection resulted in increases in both yield and the optimum density for grain yield and the proportion of biomass partitioned to the ear at flowering and maturity and in a reduction in ASI. There is little doubt that these changes in morphology have changed the tropical maize phenotype in the direction of temperate maize and increased the ability of tropical maize to withstand higher plant densities. At the same time, there is considerable room for further reduction in the grain yield/leaf area ratio of tropical maize through improvements in staygreen and in HI. The HI of tropical germplasm is still low and grain yields may be only 80–85% those of comparable temperate hybrids (Zaidi et al. 2003a; Zaidi et al. 2003b). Observations suggest that staygreen in tropical hybrids is lower under stress, tassels in tropical hybrids may be 50% larger, and standability and resistance to barrenness under high plant densities are poorer than in temperate counterparts. It is in stalk strength, density tolerance, kernels/m^2, and improved staygreen and stress tolerance that modern temperate cultivars outperform tropical landraces after a century of improvement. This increase in stress tolerance through the use of steadily increasing plant densities and extensive multilocation testing was described in a number of careful conducted studies (Duvick 1997, 2005; Tollenaar and Wu 1999; Tollenaar and Lee 2002, 2011; Campos et al. 2004, 2006; Barker et al. 2005; Cooper et al. 2014). While tropical maize has undergone extensive improvement in the last 50 years, tropical test sites are fewer and more variable, resulting in lower heritabilities than those in temperate environments, thus hampering genetic gains. Unfortunately, there are few documented comparisons of modern tropical vs. temperate hybrids in locations where both are adapted. There is still a lot of room for improvement of tropical germplasm from sustained testing under high plant densities and/or drought or by introgression of yield-efficient plant traits from temperate sources.

3.3.2 Photoperiod Response

Maize is a quantitative short-day plant, and tropical maize responds strongly when photoperiods are extended from around 13 to 16 h by delaying tassel initiation and adding additional nodes and leaves below the ear leaf. This in turn delays flowering but has little or no effect on length of grain filling. While photoperiods greater than 14.5 h do not occur naturally at latitudes <30°, they are features of temperate locations (maximum daylength at 40°N is about 15.5 h and at 50°N is 17 h). Growth distortions that result from exposure to long photoperiods make phenotyping of tropical germplasm difficult. If sensitive lowland tropical lines are sown in May in latitudes of ~40 °N, delays in flowering can be more than 30 days. Furthermore, the ASI also lengthens significantly, and kernels per ear decline drastically (Edmeades et al. 2000a). There is considerable genetic variation for photoperiod sensitivity (Table 3.3), but in general the order of sensitivity is lowland tropical > subtropical > highland > temperate. In comparison the well-known hybrid B73 × Mo17 averages a sensitivity of only 0.4 leaves/h (Edmeades et al. 1992).

3.3.3 Genetics

Within the *Zea* genus, there are five recognized species, *Z. diploperennis*, *Z. perennis*, *Z. luxurians*, *Z. nicaraguensis*, and *Z. mays*. Within the species *Zea mays*, there are four subspecies – *Z. m. huehuetenangensis*, *Z. m. mexicana*, *Z. m. parviglumis*, and

Table 3.3 Linear slopes of measures of sensitivity to photoperiod extensions from 13 to 15.5 h during the tropical summer season in Tlaltizapán, Mexico, with an average T_{max} = 31 °C and T_{min} = 18 °C

Germplasm type	*N*	Sensitivity TT to AD °Cdh^{-1}	Sensitivity time to AD dh^{-1}	Range dh^{-1}	Sensitivity final leaf no. leaves h^{-1}	Range leaves h^{-1}
Lowland tropical	33	124	7.3	3.0–12.3	2.3	1.1–3.5
Subtropical	24	108	6.4	3.1–11.2	2.5	0.3–3.5
Highland	13	89	4.9	2.3–7.8	1.3	0.9–2.0
Temperate	10	54	2.9	1.0–4.6	1.1	0.3–2.2

TT to AD = thermal time (TT) in degree days to 50% anthesis (AD), and sensitivity is measured as delays in AD in TT or days and in increased leaf number over hour of additional photoperiod (Edmeades et al. 1994)

Z. m. mays. The first four species and the first three subspecies of *Zea mays* are considered to be teosintes – wild grassy relatives of maize that often cross freely with maize in the highlands of Mexico. *Z. diploperennis* and *Z. perennis* are perennials, while the rest are annuals. All but *Z. perennis* are diploids (2n = 20). Although maize is monoecious (female and male floral organs on the same plant), it is a naturally outcrossing species. It can readily be self-pollinated, and there are no major crossing incompatibilities within the species.

There is considerable evidence that that modern maize underwent tetraploidy 5–12 M years ago, since there are remnants of two complete genomes within its current genome. Over time parts of the second genome have been preferentially expelled, leaving a fraction behind (Woodhouse et al. 2010). Schnable et al. (2011) suggests that the progressive loss of duplicate genes and overexpression associated with a duplicate gene pair are responsible for the remarkable array of genetic variation in this species.

Most quantitative traits such as grain yield and drought tolerance are controlled by additive gene action, while others such as tolerance to low *N* have a larger dominance component (e.g., Betrán et al. 2003a). Resistances to a few diseases (e.g., MSV – Semagn et al. 2014) or grain texture traits (e.g., *opaque-2* – Atlin et al. 2011) are controlled by single genes and are qualitative in nature.

Maize researchers worldwide have generated numerous reports of molecular markers tagging genes/QTLs for diverse traits of agronomic and scientific interest. QTLs for several important traits affecting maize have been mapped, including resistance to several diseases (e.g., downy mildews, northern corn leaf blight/turcicum leaf blight, common smut, *Fusarium* ear rot, banded leaf and sheath blight (BLSB), aflatoxins, etc.), abiotic stresses (e.g., drought, waterlogging, low nitrogen stress, etc.), and specialty traits (e.g., high oil content).

3.4 Heterosis

Here we define heterosis (or hybrid vigor) as the increase in growth, yield, fertility, or function of a progeny over the levels found in either parent. It is usually expressed as a percentage of either the mean of the parents (midparent heterosis) or occasionally the best performing parent (high-parent heterosis). Heterosis is the foundation of the successful maize seed industry, since it results in extra yield and is lost when offspring of the hybrid are planted in the next generation. The purchase of hybrid seed each crop season provides benefits for farmers and seed producers. Midparent heterosis between open-pollinated populations or landraces for yield can reach 15–20% and that between inbred lines often exceeds 100% (Tollenaar et al. 2004). Heterosis varies significantly among parental lines, and considerable research effort has been directed toward its prediction. Initial tests of parents focus first on general combining ability using tester lines before evaluating a series of specific crosses in the search for specific combining ability and unique crosses.

Underlying causes of heterosis are not fully understood, but they directly affect the fitness of individuals. Heterosis for a trait is a function of the square of the difference in allele frequency in the parents and the degree of dominance at those loci carrying alleles that differ (Lamkey and Edwards 1999). It is therefore specific to a particular cross. Lamkey and Edwards (1999) noted that randomly mating the F1 hybrid reduces heterosis by 50% and that inbreeding depression can be considered the converse of heterosis. Differences in the collinearity of genes between the two parents (Fu and Dooner 2002) may also be a significant source of heterosis.

There is general agreement that heterosis results in greater stress tolerance, especially to drought (Betrán et al. 2003a; Makumbi et al. 2011). Modern tropical hybrids will generally tolerate stress better than the OPVs from which their lines were originally derived. Heterosis, if its causes can be better understood, provides a unique model for selection for improved yields and resource use efficiency. In a comparison of hybrids and their inbred parents, Tollenaar et al. (2004) reported that heterosis was greatest (60–167%) for grain yield, biomass, kernel number/m^2, leaf area, plant height, and percent staygreen, 53% for harvest index, 12% for weight per kernel, and <10% or slightly negative for final leaf number, for ears/m^2, and for measures of crop duration. Zaidi et al. (2003a) observed similar results in tropical maize when comparing unrelated sets of hybrids, OPVs, and inbred lines. Hybrids showed better performance under drought, a shorter ASI, and 22–25% increase in grain yield when compared with OPVs. Leaf chlorophyll level showed 16–66% heterosis in hybrids compared with OPVs and inbreds. The largest effects of heterosis were seen on capacity to capture radiation (through leaf area and staygreen) and in the use of this assimilate to establish the size of the sink (kernels per unit area). Effects were less on HI, kernel weight, and crop duration.

Table 3.4 Heterotic combinations identified among CIMMYT populations

Ecology	Population A	Population B	Grain color
Lowland tropical	Tuxpeño (P21)	ETO (P32)	White
	Mezcla Tropical Blanca (P22)	La Posta (P43)	White
	Amarillo Dentado (P28)	Cogollero (P36)	Yellow
	Amarillo Cristalino-1 (P23)	Blanco Dentado-2 (P49)	Yellow
Subtropical	Amarillo Subtropical (P33)	Amarillo Bajio (P45)	Yellow
	A.E. Dent-Tuxpeño (P44)	ETO Illinois (P42)	White
	SIW-HG88A (P501)	SIW-HT88B (P502)	White
Highland	(P902)	(P903)	White

CIMMYT population numbers are shown in parenthesis (Vasal et al. 1999)

3.4.1 *Heterotic Patterns*

Differences in heterosis among germplasm groups have led to the identification of general heterotic patterns and partners. The common heterotic pattern in temperate maize is stiff stalk (females) × non-stiff stalk (males), developed from the Reid × Lancaster pattern identified in the 1940s. The well-known hybrid B73 × Mo17 is an example of this pattern. In tropical maize the patterns are more diverse, though often not well defined. Studies of heterotic responses among CIMMYT populations over environments showed that the level of heterosis between populations rarely exceeded 15% (Vasal et al. 1992), but reinforced the importance of the Tuxpeño × ETO (coastal tropical flint) combination. Under subtropical conditions subtropical × tropical population crosses showed consistent heterosis. Table 3.4 shows population pairs improved by reciprocal recurrent selection aimed at increasing heterosis as well as yield per se.

In Kenya, Kitale Composite was a broad-based population developed mostly from a number of US white dent varieties adapted to the East African mid-altitudes. Hybrids became important with the introduction of Ecuador 573, which, when crossed with Kitale II, gave excellent heterosis and a pleasing plant type. These types of hybrids currently occupy about 70% of the Kenya highlands and are well disseminated in the highlands of Tanzania, Ethiopia, and Uganda.

In South and Southeast Asia, most of the maize area is grown during the monsoon season and requires a high level of foliar and ear disease resistance. Suwan 1 by harder dent Tuxpeños and Suwan 1 by other tropical flints are the most used heterotic partners.

Today, CIMMYT recognizes two main heterotic groups, A and B, where A is considered Tuxpeño types and B as non-Tuxpeños. Suwan 1 provides a third major heterotic group that combines with both A and B groups. However, because these groups are not well defined and were derived from broad-based populations, there is often as much heterosis among lines within A or B groups as there is between A and B lines (Reif et al. 2003) IITA initially developed populations with their own heterotic groupings but is gradually aligning these with CIMMYT's A × B heterotic pattern.

Duvick (2005) maintained that heterotic patterns are consolidated or even generated by selection, and Tracy and Chandler (2006) agreed with this assertion. Reciprocal recurrent selection has increased heterosis in some studies (Eberhart et al. 1995; Hallauer and Carena 2012). Molecular data shows genetic distance between stiff stalk and non-stiff stalk lines that has increased steadily since the 1960s in pioneer germplasm as each group was evaluated and improved by testers from the opposing group (Cooper et al. 2004). Remarkably, evaluations of temperate inbreds and their crosses over breeding eras have shown that heterosis per se has increased at only 22% of the rate of hybrid yield, while the rate of yield improvement of inbred parents was 80% that of the hybrid (Fig. 3.3). A similar response would be expected in tropical germplasm.

3.5 Breeding

3.5.1 The Genetic Gain Equation

This is most usefully written as $\Delta G = i\sigma_a h/t$, where ΔG is the yearly genetic gain; i, the standardized selection differential; h, the square root of narrow sense heritability (the ratio of additive genetic variance to phenotypic variance); σ_a, the square root of additive genetic variance; and t, the number of years per breeding cycle (Butruille et al. 2015). The variable i is related to the selection intensity expressed as percent selected individuals but is not the same. A decrease in selected individuals from 20% to 5% changes i from 1.40 to 2.06 and increases genetic gains by only 47%. In recurrent selection schemes, the selected fraction should not be reduced below five to ten individuals because of the risks of inbreeding during recombination and, in the early stages of population formation, the risk of recovering parental genotypes because of inadequate recombination. Typically, t in commercial pedigree breeding is 5–10 years, but can be as little as 6 months for half-sib recurrent selection and 1 year for full-sib recurrent selection. Annual gains therefore are greatest when heritability and additive genetic variance are large, the proportion of selected individuals is small, and time per cycle is short. Additive genetic effects are also termed the breeding value and are increased by allelic dosage or allelic substitution such as occur with marker-assisted selection (Moose and Mumm 2008). Where field trials are used to identify superior progenies, the presence of random variability in the trial caused by soil variability or pests results in a reduction of h (sometimes referred to as repeatability) and hence in ΔG.

Genetic gain equations can only be used as a guide. In most breeding situations, selection is for several key traits or for an index of traits, and genetic gain must take these traits into account. Finally, the proof of effectiveness lies not in the predictions but in realized gains under realistic field conditions.

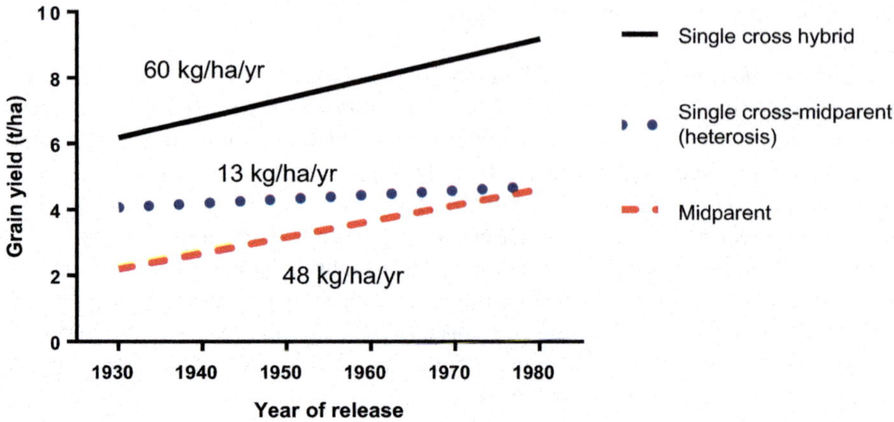

Fig. 3.3 Yields of single crosses, average yields of their parents, and the difference estimated as heterosis, as affected by the year of release of commercial hybrids. Data are from six hybrids and 12 lines from each 10-year period (Adapted from Duvick (2005))

3.5.2 Development of Useful Genetic Sources

Sources that include a high frequency of alleles affecting the expression of a specific trait have an increasing role in providing key genetic variation that can be introgressed into elite inbreds used in marker-assisted selection (MAS). CIMMYT has used two approaches in developing sources. The drought-tolerant population (DTP) is an example of the first approach. DTP was formed from 13 of the best putative sources of drought tolerance. Three hundred landraces and a number of diverse but improved sources of drought tolerance were compared with this base population, and 40 components were introgressed via a half-sib mixing phase followed by S1 recurrent selection using a progeny test under drought and heat. In C5, the population was split into yellow- and white-grained fractions and at C9 inbred lines were extracted (Monneveux et al. 2006) and tested for heterotic response and adaptation. A second approach used full-sib recurrent selection under managed drought stress within elite populations to increase the frequency of drought tolerance alleles in germplasm already adapted to the lowland tropics (e.g., Edmeades et al. 1999). Both approaches have generated lines that have become important sources of drought and heat tolerance, such as DTPYC9-F46-1-2-1-2 and La Posta Sequía C7F64-2-6-2-2 (Cairns et al. 2012). Thus, population formation and improvement have resulted in an increase in the frequency of drought-adaptive alleles and identification of superior sources of drought tolerance. The first approach is slow and should have been structured more strongly around heterotic groups and grain color. The second approach generates a useful product more rapidly and is the preferred route. It reinforces the assertion by Blum (1988) that stress tolerance alleles exist in low frequencies in most elite breeding populations, resulting in directly usable sources of tolerance to the key stresses of drought. Low *N*, heat, and acid soils, alone or in combination, are being identified among elite inbred lines. Trachsel et al. (2016b) identified a number of such lines, concluding that "it will be possible to develop hybrids tolerant to multiple abiotic stresses without incurring yield penalty under unstressed conditions using these lines."

3.5.3 Population Improvement Methods

Although population improvement has become much less used over the past 20 years, it has had a useful role in building sources with good agronomic performance with relatively broad adaptation prior to extracting inbred lines. Population improvement methods have been extensively reviewed by Hallauer and Miranda (1988) and specifically for tropical germplasm by Pandey and Gardner (1992). The most effective methods involve ear-to-row family structures, where a progeny test, usually in the form of a yield trial, is followed by a recombination step of a superior fraction that generates the progenies for the next round of testing. This systematically increases the frequencies of favorable alleles. Population formation and improvement normally do not involve tracking and use of pedigrees and in CIMMYT's case did not consider heterotic responses in its initial stages. The emphasis on compatibility of components used to form populations, however, provided some selection for general combining ability.

Population improvement methods are described by their family structures, and the most commonly use are half-sib, full-sib, and S_1 recurrent schemes (Paterniani 1990; Pandey and Gardner 1992). Most populations are structured around 200–1500 families, depending on objectives and resources:

- Half-sib recurrent schemes usually involve a series of females planted ear to row and detasseled, with a male pollinator comprised of a balanced bulk of all females or a subset of females. The male serves as a visual check against which female rows are selected, and three to five ears per selected female are chosen to use as the progenies for the next round of selection. The block must be isolated from other pollen sources by time or distance. It is usually unreplicated, though it can be repeated in other locations or planting dates. In tropical environments where irrigation is available, two cycles of half-sib selection can be completed per year. This same field layout can be used during pedigree breeding as an isolated crossing block where the male is the topcross parent and the females are inbreds selected as candidates for topcross testing.
- Full-sib recurrent schemes involve plant-to-plant crosses among 30–50 selected progenies, so both parents are known. Typically 200–300 of these crosses are grown ear to row in a replicated yield trial, though numbers of replications and plot size are limited by seed supply from single ears (or from two ears where reciprocal crosses are made). Selections are based on family performance at several sites. Remnant seed of the family is used during a single step of

recombination and progeny formation among the 30–50 selected families. In tropical locations one complete cycle of full-sib recurrent selection is possible per year, thus allowing progeny testing in the normal crop season and recombination during the dry season.

- S_1 recurrent selection schemes require three crop seasons to complete a selection cycle. Selfs are made in the first season and are phenotypically evaluated as S_1 families in the second season. In the third season, the superior families (20–40) are recombined by plant-to-plant crosses in all possible combinations, and a balanced bulk is prepared for selfing the following season. Superior S_1 families can immediately be advanced by selfing to form inbred lines. Since three cycles are not often possible in 1 year, one modification used when selecting for drought tolerance was to generate 1500 S_1 families and prescreen them in an unreplicated trial under heat and drought in NW Mexico in the summer. Numbers were then reduced to 200–250 for winter testing in replicated tests in the rain-free winter season in central Mexico, followed by recombination of the best 40 families from remnant seed (Edmeades et al. 1999). Seed per ear quickly becomes a limitation if more test locations are used.

Predicted gains under these three schemes indicate that ΔG is surprisingly similar from each, especially where genotype × environment interaction (GEI) is small. Reported gains for yield are around 3–10%/cycle for half-sib, 2–8%/cycle for full-sib, and 1–8% for S_1 selection (Johnston et al. 1986; Paterniani 1990; Pandey et al. 1991; Pandey and Gardner 1992; Bolaños and Edmeades 1993; Edmeades et al. 1999; Monneveux et al. 2006; Hallauer and Carena 2012). When progeny tests are conducted in several diverse environments, GEI decreases h, and gains can be less than expected, as in full-sib selection in CIMMYT's elite populations subject to international progeny testing (Pandey and Gardner 1992). Yield gains will decline when other traits such as disease resistance become priorities. Percent gains can also be high when the populations are relatively unimproved for the target trait or when mean yields are low. A better measure in such cases is gain in yield per unit area. As a rule of thumb, gains for drought tolerance from recurrent selection in germplasm previously unimproved for the trait have averaged 100 kg/ha/year (or 5% per year) when selection was conducted largely in a single dry environment (Bänziger et al. 2006). Because yield levels were low, these gains fall in the upper range of 2–5% gain/cycle expected from recurrent selection (Hallauer and Carena 2012). Gains in OPVs generated over the past 30 years in eastern and southern Africa in more advanced germplasm show sustained average annual gains under MSV infection (151 kg/ha), optimal conditions (95 kg/ha), and low N (69 kg/ha) but reduced gains under random drought stress (36 kg/ha) and no change under managed drought stress (Masuka et al. 2017b). These results suggest that gains under managed drought stress may decline as variability for ASI and barrenness become exhausted.

In summary, intrapopulation improvement has allowed rapid improvement in performance, provided a tool for testing effects of selection for a specific trait, and developed useful source germplasm. As the population has been improved, it has served as the source of an increasing proportion of elite inbred lines. Randomly extracted S_2 lines from three populations selected for drought tolerance vs. their conventionally selected equivalents, when topcrossed to a common tester, showed on average an improvement in drought tolerance that reflected the differences between their parent populations (Edmeades et al. 1997). The choice of a population in which to begin selfing is therefore a critically important decision that directly determines the probability of extracting high-performing inbreds (Hallauer and Miranda 1988).

A widely used interpopulation improvement scheme is reciprocal recurrent selection (RRS) (Paterniani 1990; Pandey and Gardner 1992; Eberhart et al. 1995). Here one population of a heterotic pair is used as the tester of the other (and vice versa) in a half- or full-sib mating system, with an emphasis on improving the performance of the cross between populations. Hallauer (1999) reported an average increase in midparent heterosis in four temperate population pairs from 9 in C_0 to 43% after an average of nine cycles of selection, essentially by increasing the divergence of allele frequencies. Improvement in the population yields per se, however, averaged only about 1%/cycle. Interpopulation schemes are efficient especially where dominance effects are large and increase the probability of extracting high-performing lines with an established heterotic response.

3.5.4 Pedigree Breeding

Pedigree breeding is by far the most common form of maize improvement and resembles RRS in that pedigrees that define ancestry and heterotic response are maintained and used to predict performance. Testers, usually inbred lines or single-cross hybrids, are selected from the opposing heterotic group to screen for general combining ability. The remaining lines are subject to specific combining ability tests with a small group of elite inbreds from the opposing group. Cooper et al. (2014) describe these steps and note that superior new inbreds are rapidly recycled to create new combinations within each heterotic group. Testing of topcross progenies is within the target population of environments (TPEs), using relatively few sites for GCA tests and many more locations for SCA testing. Field testing is followed by rapid data analysis and information extraction so lines can be advanced in off-season nurseries. Based on selection data, promising lines are advanced to the next level of multilocation testing.

The rate of gain from pedigree breeding has been evaluated mainly in temperate breeding programs and is summarized by Fischer et al. (2014). Gains in researcher-managed trials have averaged 100 kg/ha/year or around 0.8% annually. Given that the cycle length of most pedigree breeding programs is 5–10 years, this equates to a per cycle gain of 4–8% or 400–800 kg/ha. Correlated changes resulting from selection over time in temperate germplasm have been largely in traits that are associated with maintenance of leaf area through improved staygreen plus a very significant increase in ability to tolerate stresses, especially high plant density and drought. Resistance to lodging and to barrenness has increased significantly, and ASI and grain protein concentration have declined. Leaves have become more upright and tassel size has reduced, but there have been little or no changes in yield potential per plant (Egli 2015), plant height, leaf area, or time to flower (Tollenaar and Wu 1999; Tollenaar and Lee 2002; Duvick 2005; Barker et al. 2005; Campos et al. 2004, 2006). Regrettably, few assessments have been made of changes in biomass production and HI. Recent assessments of gains in conventional selection programs in eastern and southern Africa show gains in grain yield in CIMMYT-bred germplasm of 109, 33, 23, 21, and 141 kg/ha/year under optimal, managed drought stress, random drought stress, low *N* stress, and MSV infection, respectively (Masuka et al. 2017a). Relative gains in grain yield ranged from 0.6 (low *N*) to 2%/year (MSV), and ASI and barrenness decreased over time (Masuka et al. 2017a; Semagn et al. 2014). Correlations between inbred line and test cross performance, while normally low (c. 0.3) under optimal conditions, generally increase as abiotic stress levels increase to values around 0.5. This suggests a possible role for evaluations of inbred lines per se under severe drought stress conditions (Kebede et al. 2013).

3.5.5 Doubled Haploids

The process of developing inbred lines is time-consuming and, in many tropical maize breeding programs, is subject to error and loss through seed and pollen mixtures and inadequate field and seed storage facilities. The production of doubled haploid (DH) lines sharply reduces the time taken to develop homozygous lines with proven performance and reduces losses from seed identification errors. Doubled haploids have been in routine use commercially for the past decade to generate >0.5 million lines annually for each leading multinational seed company. Doubled haploids increase the rate of genetic gain by reducing t, the time to complete a breeding cycle, by 1–2 years, and produce lines whose uniformity makes them easier to phenotype and which are well suited to molecular marker applications.

The technology involves the use of a haploid inducer line (used as pollen parent) in crosses with desired source populations (as female parent). The inducer line carries phenotypic markers that enable differentiation of haploids from diploids (at the seed stage) in the induced progeny. Inducers show induction rates of 6.7–11.3% (Prigge et al. 2011), and the haploid induction rate is under polygenic control (Geiger and Gordillo 2009). The most widely used phenotypic marker is an *R1-nj* aleurone coloration visible in seed or a liguleless gene *lg2* that can be detected in seedlings (Melchinger et al. 2016). *R1-nj* anthocyanin marker inhibition is quite common in tropical maize germplasm and significantly reduces efficiency of haploid identification. Molecular markers that reliably differentiate germplasm carrying the anthocyanin color inhibitor have been identified by Chaikam et al. (2014). The *R1-nj* marker is also ineffective in germplasm with natural anthocyanin expression in pericarp tissue. Given these limitations, the CIMMYT team developed haploid inducer lines with triple anthocyanin color markers, including the expression of anthocyanin coloration in the seedling roots and leaf sheaths, in addition to the Navajo marker on the seed (Chaikam et al. 2016).

Tropically adapted inducers with high haploid induction rate are being developed by CIMMYT, in collaboration with the University of Hohenheim, Germany. CIMMYT, in partnership with Kenya Agricultural and Livestock Research Organization (KALRO), established a maize DH facility at Kiboko (Kenya) in 2013; the facility, developed through the financial support of Bill and Melinda Gates Foundation and the Syngenta Foundation for Sustainable Agriculture, offers DH development services to national programs and to small- and medium-enterprise seed companies in sub-Saharan Africa. The use of DH lines, along with marker-assisted selection (MAS), will increase rates of genetic gain from many tropical pedigree breeding programs.

3.5.6 Secondary Traits and Their Use in Selection

Secondary traits are often used in selection and frequently form part of a selection index along with grain yield. A secondary trait could give greater gains for the primary trait (grain yield) than selection for yield alone when $h_{\mathrm{GY}} < |r_G h_{\mathrm{ST}}|$ (Falconer and McKay 1996), where h_{GY} and h_{ST} are the square roots of heritability of grain yield and the secondary trait, and r_G is the genetic correlation between grain yield and the secondary trait. This condition is rarely met except when yield is low and the secondary trait is expressed best under stress. However, in most cases, secondary traits are added to a selection index along with the primary trait in the belief that the heritability of the index will exceed that of the primary trait and yield.

A useful secondary trait should be (1) genetically associated with grain yield under the target stress and be genetically variable and more heritable than yield; (2) cheap and fast to measure; (3) observed at or before flowering, so that undesirable parents are not crossed; and (4) not associated with yield loss under unstressed conditions. The value of a secondary trait can be assessed by analyses of correlation and heritability, by divergent selection for that trait, by modeling, or by statistical procedures based on selection index theory. Using this last approach, Bänziger and Lafitte (1997) determined that the use of secondary traits plus yield during selection for tolerance of maize to low soil *N* was about 20% more efficient than selection for yield alone, and this benefit increased as yield levels declined. Needless to say, very few secondary traits proposed mainly by nonbreeders have passed these tests!

3.5.7 Participatory Plant Breeding

Many large centralized breeding programs lack meaningful farmer contact during selection, despite advantages of scale, wide area testing, and access to germplasm (Morris and Bellon 2004). There is often inadequate farm-level testing, and frequently consumer response to disease reaction, grain texture and color, and ease of harvest and shelling are ignored. Differences in maturity, cooking quality, taste, and especially in stover quantity and quality can also be overlooked when breeding for grain yield (Witcombe 2006). Sometimes ear characteristics are more important than yield to farmer families (Louette and Smale 2000). Farmers have a unique comparative advantage at some stages in the selection process, and plant breeding without adoption of the product by the many small-scale farmers operating in target areas is one measure of failure.

The key breeding issue is the stage in product development when farmer input is most valuable. Farmer participation in goal setting and determining selection criteria may be adequate during the product development stage, but assessing the suitability of the finished variety as a preliminary to its release should also be a priority. During the selection process at researcher-managed locations, it is important to bear in mind the effect of the selection environment vs. farmers' fields. Gains in the selection environment must be related to gains in the target on-farm environments and rank similarly. Theory developed by Falconer and McKay (1996) shows that selection response in farmers' fields (R_{FF}) is greatest when genetic variance σ_{G}^2, selection intensity i, heritability in farmers' fields h^2_{FF}, and the genetic correlation r_G between selection environment and farmer's fields are all high, since $R_{\mathrm{FF}} = \sigma_G^2 i\, h^2_{\mathrm{FF}} r_G$ (Bänziger and Cooper 2001). Heritability and genetic variance for yield generally fall as yield levels decline relative to a high-yielding selection environment, since error variance relative to genetic variance rises, i.e., GEI between selection and target environments becomes important, and gains in farmers' fields will decline. If r_G is negative, gains in selection environments will lead to losses in yield in farmers' fields – but fortunately this is seldom the case. There is no substitute for on-farm and consumer testing of tropical maize varieties prior to their release, and increasingly variety release committees are demanding such data. Harvest field days, and ratings of varieties by farmers, both male and female, are an important part of product development and delivery (Bänziger et al. 2000). CIMMYT's maize product advancement process typically includes not only regional on-station trials of promising pre-commercial hybrids coming out of the breeding pipeline vis-à-vis internal genetic gain checks and commercial checks but also extensive regional on-farm varietal trials to ascertain the performance of the promising pre-commercial hybrids under farmer-managed conditions. This also provides opportunity for the socioeconomic team to assess farmers' own product as well as their trait preferences. The best entries coming out of this rigorous process are then allocated to public/private sector partners for varietal registration, scale-up, and delivery in the target geographies.

3.6 Traits

3.6.1 Determinants of Yield

The primary trait during selection is almost always grain yield, and much of the discussion in previous sections pertain directly to selecting for yield. The major drivers of grain yield are assimilates generated from radiation captured by a healthy and effective canopy and their efficient partitioning to grain. In the absence of any stress, grain yield in the tropics varies in proportion to radiation received (Jong et al. 1982) and is modulated by genetic potential. Increasing temperature speeds development and shortens crop duration, but has much less effect on net photosynthesis, so high temperatures will generally reduce yield by reducing the time available for assimilation (Muchow et al. 1990).

Grain yield (GY) can be defined by the following identity (Edmeades et al. 2000b):

$$GY = RAD^{*}RI^{*}GLD^{*}RUE^{*}HI \tag{3.1}$$

where RAD is the incident radiation received per day; RI, fraction of radiation intercepted over the crop's life; GLD, green leaf duration; RUE, radiation use efficiency; and HI, harvest index. Using typical values, GY = 23MJ/m^{2}*0.55*103d*1.7g/MJ*0.45 = 10.0t/ha.

Where water is the limiting production, Passioura (1977) proposed a parallel expression: GY = W*WUE*HI, where W, water transpired by the crop; WUE, water use efficiency; or biomass/unit water transpired (e.g., GY = 445 mm*50 kg biomass/mm × 0.45 = 10.0 t/ha). A similar identity can be used when N is the limiting growth: GY = N uptake × N use efficiency × HI (e.g., GY = 200 kg N uptake*111 kg biomass/kg/N*0.45 = 10 t/ha).

From Eq. 3.1, breeders can increase grain yield through the last four variables – by ensuring that canopy closure is rapid, by boosting RUE and delaying leaf senescence through adequate nutrition and resistance to foliar disease and insects, and by selecting for traits like ASI that maintain a high HI. Grain yield under drought can be reduced because of direct effects on stand, leaf area, and RI, from accelerated senescence (reduced GLD) and from drought-induced barrenness which in turn reduce HI. Drought, and nitrogen deficiency, may also reduce RUE and possibly increase WUE, though effects on these are often less important than those on intercepted radiation and harvest index. If it occurs early in the crop life, then the leaf area is reduced and ear size declines. If later in the crop's life cycle, it will induce premature leaf senescence.

Grain yield can also be considered the product of its yield components:

GY = plants/m^2*EPP*GPE*WPG

where plants/m^2, established stand density; EPP, ears per plant; GPE, grains per ear; and WPG, weight per grain (g). Using typical values, GY = [5.5*1.1*500*0.33] = 10 t/ha. Although selection affects each component, they are normally negatively correlated. Increases in prolificacy are associated with a reduced ear size. Similarly, an increase in kernels per ear can be associated with reduced kernel weight (e.g., Uribelarrea et al. 2008) such that yield is little affected.

3.6.2 *Increased Yield Potential*

The steady increase in grain yield with selection in all classes of maize has been mainly associated with increases in kernels per plant and kernels/m^2 and not with changes in weight per kernel (Bolanos and Edmeades 1996; Chapman and Edmeades 1999; Duvick 2005). Changes that have occurred in grain yield with selection in temperate maize are summarized in Table 3.5.

Most of the changes in temperate maize were correlated with selection for high and stable yields through multilocation testing rather than direct responses to selection. Thus, when selecting for increased grain yield potential and stability, correlated increases occurred in leaf angle, staygreen (including foliar pathogen and insect pest resistance), vigorous silking with minimal delay under stress, and standability, all at high plant density. It is likely that tropical maize will ultimately mimic temperate maize in trait changes under selection, leading to increased partitioning to the ear, accelerated by selection for reduced leaf and tassel size, increased kernel number/m^2, tolerance of high plant density, and staygreen. These traits lead to general stress tolerance, i.e., an ability to withstand any stress that acts primarily by reducing photosynthesis per plant.

Evidence suggests that prolificacy is not a trait that imparts stress tolerance per se, and Echarte and Andrade (2003) concluded that HI was higher in non-prolific vs. prolific hybrids. Prolificacy is an opportunistic trait and may have a role in exploiting good growing conditions when maize at low plant density is intercropped with lower-growing crops. It is also a useful trait when maize is planted at low densities in the expectation of severe drought stress, as in semiarid parts of South Africa. Husk cover remains a high priority as ear size increases to ensure that infestation by diseases and insects and damage from birds are minimized, and rapid dry down minimizes risks of pest infections or mold building up prior to harvest or in storage.

3.6.3 *Selecting for Abiotic Stress Tolerance*

Drought and low *N* are the most common of these stresses and fortunately the responses of the plant to moderate levels of each are similar. In most production environments, drought, unlike *N* stress, occurs unpredictably throughout the season, and its variability will likely worsen with global climate change. Spatial variability caused by changes in soil texture means that a typical variety will be exposed to varying levels of drought or *N* deficit within the same field. Variation in grain yield under both is strongly associated with kernel number per plant, and this is directly

Table 3.5 Rates of change in specific morphological and stress-related traits under selection in temperate maize, their approximate heritabilities in tropical maize, and relative selection priorities (L, low; M, medium; **H**, high)

Trait	Temperate			Tropical			
	Δ/year	Comment	Ref[a]	h^2	Priority	Comment	Ref[a]
Morphology and phenology							
Leaf angle score[b]	0.1	Leaves more erect	2	H	M	Erect leaves for intercropping, high density tolerance	1
Tassel weight (g)	−0.05	Smaller tassels	2	H	**H**	Smaller tassels give stress tolerance, reduce shading	7, 25
Plant height (cm)	ns	Maintained	2	H	**H**	Maintain height below 2.5 m to reduce lodging	8
Time to 50% anthesis	ns	Maturity maintained	2,5	H	L	Fitted to suit cropping system	11
Grain fill duration	na	Increased	1	?	M	Increase; raises yield potential and HI	22
Husk cover	na			H	**H**	Increase tip cover to prevent pathogen/pest invasion	23
Grain dry down rate	na			?	M	Increase; reduces drying time in field and pest risks	22
Leaf area	na	Less gives density tolerance	6	M	M	Reduce leaf size to increase optimum plant density	7
Productivity of grain and biomass							
GY at opt density (kg/ha)	90	Yield increased	1	M	H	Yield potential and density tolerance increased	7,8
Staygreen score[b]	0.12	Senescence is delayed	4	L	H	Increase; need smaller leaves that live longer	11
Kernel weight (mg)	0.7	Increased	2	M-H	L	Maintained or increased	11,21
Prolificacy	0.002	Barrenness decreased	2	M	L	Increase for intercropping systems and yield stability	15, 16

(continued)

Table 3.5 (continued)

Trait	Temperate			Tropical			
	Δ/year	Comment	Ref[a]	h^2	Priority	Comment	Ref[a]
HI, unstressed (%)	0.1	Slight increase	2	M-H	H	Increase except where value of stover is high	8,17,19
Photosynthesis, no stress	ns	Unchanged	3	?	L	Increase and stabilize across stresses	20
Stress tolerance – abiotic							
GY, irrigated control	196	High-yield environment	5	M	M	Maintain or increase yield potential	9,10
GY, flowering drought (kg/ha)	120	60% of irrigated gain	5	M-L	H	Increase; decrease aborted ears and kernels	9,10
GY, grain filling drought, kg/ha	50	25% of irrigated gain	5	L	M-H	Increase; concurrent with improved staygreen	9,10
ASI, drought (°Cd)	−2.6	Synchrony improved	5	M	H	Decrease; improve floral synchrony, kernel set	11, 25
Ears per plant	0.002	Less barrenness	2	M-L	H	Increase; reduce barrenness under stress	11, 25
Kernels per ear	1.6	More under drought	5	M	M	Increase; reduce kernel abortion under stress	11, 25
Lodging (%)	−0.9	Stands better	2	L	H	Increase stalk and root strength	8, 24
Leaf rolling score[b]	0.035	Rolls more readily	4	H	L	Limited importance under drought; sheds radiation	11
Heat tolerance		Improved	1,3	?	H	Increase at flowering	12,14
Grain protein (%)	−0.03	Less protein, more starch	2	?	M	Maintain or increase under low soil *N*	18

See footnotes for sources. Modified from Fischer et al. (2014)
[a]Sources: 1, Duvick (2005); 2, Duvick (1997); 3, Tollenaar and Lee (2011); 4, Barker et al. (2005); 5, Campos et al. (2006); 6, Lambert et al. (2014); 7, Fischer et al. (1987); 8, Johnson et al. (1986); 9, Edmeades et al. (1999); 10, Bolaños and Edmeades (1993); 11, Bolaños and Edmeades (1996); 12, Cairns et al. (2012); 13, Zaidi and Singh (2005); 14, Cairns et al. (2013); 15, Motto and Moll (1983); 16, de Leon and Coors (2002); 17, Worku and Zelleke (2007); 18, Lafitte et al. (1997); 19, Echarte and Andrade (2003); 20, Echarte et al. (2008); 21, Campos et al. (2004); 22, Cross (1975); 23, Demissie et al. (2008); 24, Pandey and Gardner (1992); and 25, Chapman and Edmeades (1999)
[b]Scores are from 1 (least desirable) to 9 (most desirable)

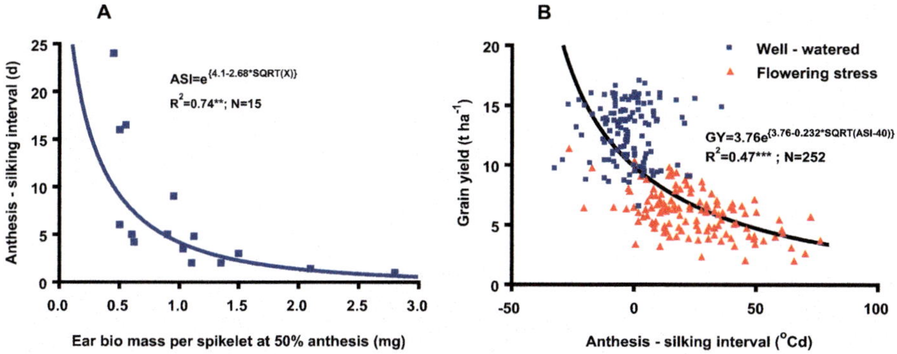

Fig. 3.4 (**a**) ASI vs. ear biomass per spikelet at 50% anthesis in a tropical population grown at high plant density in Mexico; (**b**) grain yield vs. ASI in 126 elite temperate hybrids under either severe drought at flowering or no stress, Woodland, CA (Edmeades et al. 1993; Edmeades 2008)

affected by stress that occurs 10–14 days either side of anthesis. ASI is a morpho-physiological trait that can be used to predict kernel set, barrenness, and grain yield under stresses (Chapman and Edmeades 1999), including those associated with high plant density (Dow et al. 1984). The correlation between ASI and grain yield (GY) under stress at flowering is often −0.4 to −0.7 (Bolaños and Edmeades 1996). This association of grain yield with ASI has been observed in landraces, populations, lines, and even elite Corn Belt hybrids and reflects rate of growth of ovules and hence of silks (Fig. 3.4).

Having ears and tassels develop synchronously has stabilized kernel set under stress and lifted HI (Chapman and Edmeades 1999). A reduction in the size of tropical tassels would reduce shading and competition during ear growth, but how far can it be taken? In three tropical populations, recurrent selection reduced tassel branch number by 7–9% per cycle, and over six cycles tassel biomass declined from 7.1% to 5.5% of shoot biomass at 50% silking. Tassel weight fell from 9.3% to 5.4 g/plant for the population Tuxpeño, and concomitantly optimum density increased by 16% and yield by 13% (Fischer et al. 1987). Although comparisons of tassel weight are complicated by pollen shed and senescence, tassel weight in temperate maize declined from 5 g/plant in 1933 to 2 g/plant in 1993 as a result of selection for higher yields (Duvick and Cassman 1999). Reduced tassel size, if accompanied by decreased ASI, should not jeopardize kernel set because of reduced pollen supply in modern stress-tolerant tropical hybrids. Selection for reduced tassel size continues to be a useful route to higher yields and improved density tolerance in tropical maize.

Unlike grain yield, secondary traits such as ASI, barrenness, and staygreen scores have stable or even increasing heritabilities as drought stress at flowering intensifies, and when combined with yield in a selection index, they improve the heritability of that index. Indices such as these were applied to several maize populations as proof of concept and to a single population selected under low *N* (Lafitte and Edmeades 1994). Selection outcomes for drought in tropical populations (Table 3.6)

Table 3.6 Selection gains in six tropical maize populations

		Yield			ASI SS	Ears/plant
		SS	WW	Low *N*	SS	SS
Population	Cycles selected	kg/ha/cycle			d/cycle	no./cycle
La Posta Sequía	3	229**	53 ns	233**	−1.2**	0.07**
Pool 26 Sequía	3	288**	177**	207**	−1.5**	0.08**
Tuxpeño Sequía	8	80**	38**	86**	−0.4**	0.02**
Pool 18 Sequía	2	146**	126**	190**	−2.1**	0.05**
DTP1	6	160*	80 ns	210*	−0.6**	0.03**
DTP2	9	80*	120 ns	60 ns	−0.3**	0.01*
Mean gain		164	99	164	−1.0	0.04

Sources: Edmeades (2012), Monneveux et al. (2006)
Four were evaluated at three to eight drought sites and two low *N* sites, and two (DTP1, DTP2) were evaluated at one low *N*, one severe stress, or one well-watered location. Yields relative to unstressed levels were 30% under drought stress (SS) and 59% under low *N*
Symbols *, **, and ns signify significant rate of change per selection cycle at $P < 0.01$, $P < 0.05$, or $P > 0.05$, respectively

indicate significant gains also under low *N*, first reported by Bänziger et al. (1999). Tolerances to these two stresses are related – through the common mechanisms of partitioning to the ear under stress and through staygreen. It is not until *N* deficiency becomes very severe with yields reduced by >65% that *N*-specific traits become important, and the genetic correlation between yield under drought and low *N* decreases to non-significance. As previously noted, gene action for drought tolerance is generally additive, while tolerance to low *N* has a greater dominance component (Betrán et al. 2003a; Makumbi et al. 2011). Inbred line performance is therefore a better guide to hybrid performance under low *N* than under drought (Makumbi et al. 2011) with the line/hybrid correlation under low *N* sometimes reaching 0.65 (Zaidi et al. 2003b).

A useful manual (Bänziger et al. 2000) outlined practical steps to manage stress levels and improve precision of trials conducted under drought or low *N*. These methodologies have been refined and updated recently for several abiotic stresses (e.g., for drought tolerance, see Zaman-Allah et al. 2016). Improved phenotyping methods have been augmented with modern molecular selection tools (Semagn et al. 2014; Beyene et al. 2015, 2016) and with more precise phenotyping methods that rely on uniform field conditions and remote sensing (Lu et al. 2011; Masuka et al. 2012; Araus and Cairns 2013; Trachsel et al. 2016).

Temperatures during the main cropping season are predicted to increase in most tropical maize-growing areas (Jones and Thornton 2003). The change is predicted to be greatest in night temperatures and accompanied by increased vapor pressure deficits. The development of fertile pollen and silks, and therefore kernel set, is threatened by temperatures >38–40 °C (Schoper et al. 1987; Westgate and Bassetti 1990; Cicchino et al. 2010). Heat tolerance is often associated with drought tolerance, but recent reports indicate they are largely independent traits (Cairns et al. 2012). Edreira et al. (2011) observed that temperate hybrids were more susceptible than tropical hybrids to heat stress at flowering and noted that this stress delayed anthesis and in so doing shortened ASI. If selection environments include hot dry summers typical of Mediterranean climates, then the two traits can be improved simultaneously. Global warming predictions indicate both traits will be increasingly important in the future (Lobell et al. 2011a). Conversely, cold tolerance may be required as maize continues its progress as a winter crop in the Indian subcontinent (Zaidi et al. 2010a), as well as heat tolerance to cope with very high pre-monsoon temperatures after flowering. Although considerable genetic variation for cold tolerance exists in highland germplasm (Eagles and Lothrop 1994), it is not a global priority trait for tropical maize.

Acid soils are relatively common in the tropics, usually where rainfall is high and soils are weathered and leached, or they are associated with specific parent material. Recent assessments of progress are very promising (Pandey et al. 2007). Under pH 4.7, and Al^{3+} saturation of 60%, yields of OPVs in 1993 were around 3 t/ha but had increased to 9 t/ha in hybrids by 2008.

Waterlogging tolerance is of greatest value in low-lying and poorly drained areas such as the Indo-Gangetic Plain of India and where maize is cultivated in rice paddies in Southeast Asia and Brazil. Here heavy monsoon rains or typhoons frequently leave summer crops with their roots underwater – something that is often fatal for maize. Waterlogged plants wilt as if droughted, and symptoms are exacerbated under bright sunlight. The ability to maintain chlorophyll and a short ASI and the formation of aerenchyma and brace roots under anaerobic soil conditions are adaptive responses associated with grain yield in waterlogged conditions. Gene action is mainly additive and good progress can be expected from recurrent selection (Zaidi et al. 2010b).

3.6.4 Biotic Stresses

An assessment of the major diseases and insect pests of tropical maize (Table 3.7) shows that priority diseases are maize lethal necrosis (MLN), maize streak virus (MSV), turcicum leaf blight (TLB), gray leaf spot (GLS), and ear rots in Africa; post-flowering stalk rots, downy mildew, banded leaf and sheath blight (BLSB), and *Fusarium* and *Diplodia* ear rots in Asia; and tar spot complex, TLB, and GLS in Latin America. The customary approach to breeding for resistance is to expose segregating progenies or inbred lines to each disease in "hotspots" where they occur naturally and repeatedly at a high level. Here we focus on MLN because it is relatively new to sub-Saharan Africa and seriously threatens maize production and productivity in the continent.

Maize lethal necrosis (MLN) occurs when maize is coinfected with two viruses, maize cholorotic mottle virus (MCMV) and a potyvirus, most frequently the sugarcane mosaic virus (SCMV). SCMV has been present in Africa for perhaps 50 years, but MCMV is much more recent and is more dangerous of the two (Mahuku et al. 2015). The MLN disease is spread by seed contamination and was first reported in

Table 3.7 Major tropical maize diseases and insect pests, their areas of importance, and their breeding priority (L, low; M, medium; *H*, high)

Scientific name	Common name	Area of importance in tropics[a]	Priority	Ref[b]
Diseases				
Exserohilum turcicum	Turcicum leaf blight	ST and HL globally	**H**	1,3,4
Bipolaris maydis	Maydis leaf blight	LT and ST globally	M	1,8
Puccinia sorghi	Common rust	ST and HL globally	M	1,3
Cercospora zeae-maydis	Gray leaf spot	ST of E, S. Africa	**H**	1,3,4
Sphacelotheca reiliana	Head smut	ST in hot dry areas	M	3
Phaeosphaeria maydis	*Phaeosphaeria*	ST of the Americas	M	1,3
Maize streak virus	Streak	LT, ST of SS Africa	**H**	2,3
Maize lethal necrosis	MLN	ST of East Africa	**H**	2,9
Peronosclerospora spp.	Downy mildew spp.	LT Asia and Africa	**H**	1
Phyllachora maydis and Monographella maydis	Tar spot complex	LT Central America	M	1,4
Rhizoctonia solani f. sp. *sasakii*	Banded leaf and sheath blight (BLSB)	LT, ST of South Asia	**H**	1
Fusarium moniliforme	Ear and stalk rots	LT and ST globally	**H**	1,3,4
Stenocarpella maydis	*Diplodia* ear and stalk rot	ST and cool areas	M	1,3
Aspergillus flavus	Ear rot, aflatoxins	ST and dry areas	**H**	1,4
Insects				
Spodoptera frugiperda	Fall armyworm	LT, ST in Americas	M	7,12
Diatraea saccharalis	Sugarcane borer	LT, ST in Americas	M	7,12
Ostrinia furnacalis	Asian maize borer	LT, ST of Asia	M	7,12
Chilo partellus	Spotted stem borer	LT, ST Africa, S. Asia	**H**	1
Busseola fusca	African maize stem borer	ST of S. and E. Africa	**H**	7,12
Sesamia calamistis	African pink borer	LT of Africa	M	1,12
Heliothis zea	Corn earworm	ST, HL globally	M	7,12
Sitophilus zeamais	Grain weevil	Global	M	6,10
Prostephanus truncatus	Larger grain borer (LGB)	C. America, Africa	**H**	5
Parasites				
Striga hermonthica	*Striga*	Sub-Saharan Africa	**H**	11, 13

See footnotes for sources

[a]*LT* lowland tropics, *ST* subtropical areas, *HL* highland tropics

[b]1, Zaidi and Singh (2005); 2, Semagn et al. (2014); 3, Vivek et al. (2010); 4, Cairns et al. (2012); 5, Kumar (2002); 6, Abebe et al. (2009); 7, Ortega (1987); 8, CIMMYT (2004); 9, Mahuku et al. (2015); 10, Demissie et al. (2008); 11, Makumbi et al. (2015); 12, Mihm (1997); and 13, Ejeta and Gressel (2007)

the Rift Valley of Kenya in 2011 (Wangai et al. 2012) and since then has caused considerable losses to maize production in several countries in eastern Africa, including Kenya, Ethiopia, Rwanda, Tanzania, and Uganda. Though thrips and leaf beetles (MCMV) and aphids (SCMV and other potyviruses) are the principal vectors, infection via soil contamination, especially with MCMV in debris in the soil, also appears possible.

An early infection of MLN at the seedling stage results in necrosis and death and zero grain yield. Infection later in crop development also results in leaf death but some grain may form. Based on survey data, de Groote (2016) estimated annual yield losses to MLN in Kenya of up to 90% in infected areas and grain losses countrywide at 500,000 tons worth US $180 million. Incidence varies markedly by season and location with more than 50% of farmers affected in Western Kenya. With external donor support, CIMMYT together with KALRO established an MLN screening facility at Naivasha, Kenya, in 2013. The facility enables screening of the elite lines and hybrids from African public and private institutions under artificial infection. All older commercial cultivars are susceptible to varying degrees. Through intensive screening of >75,000 germplasm entries, CIMMYT identified sources of tolerance/resistance to MLN as well as to MCMV. Five MLN-tolerant hybrids have been released, and seed is being scaled-up or commercialized by seed companies in Kenya, Tanzania, and Uganda. A further 22 MLN-tolerant/MLN-resistant hybrids were tested in National Performance Trials (NPTs) in East Africa in 2016 (Prasanna 2016). CIMMYT has also made significant progress in identification, validation, and deployment of molecular markers for resistance to MLN, though genetic studies have shown that resistance to MLN is controlled by multiple loci with relatively small effects (Gowda et al. 2015). Pyramiding sources of MLN resistance using molecular markers to minimize linkage drag is a high priority, and around 50 widely used CIMMYT lines are presently being converted to MLN-tolerant versions using marker-assisted backcrossing. The current outbreak of MLN may cost 5 years of genetic progress for yield and other traits, while sources of MLN tolerance/resistance are identified and introgressed – a price worth paying to avoid a pandemic.

CIMMYT is presently focusing on developing breeder-ready markers for improving specific disease resistance traits. A recent example is MSV, a major disease affecting maize productivity in several African countries. Nair et al. (2015) fine-mapped, identified, and validated a set of SNP markers for a major QTL for MSV resistance (*msv1*). The validated MSV resistance haplotype is now used in forward breeding imparting MSV resistance to elite inbred lines with tolerance to drought, heat, or MLN and for deriving improved biotic and abiotic stress-tolerant lines. Production of SNP markers is also currently being developed by CIMMYT for resistance to other key diseases, especially TSC, TLB, and GLS and for improved nutritional characteristics.

Insects Maize insect pests are ubiquitous and cause considerable losses. The most serious (Table 3.7) are the stem borers of Asia and Africa that often consume foliage and damage ears as well as cause extensive stalk breakage. The larger grain borer of Central America and more recently of Africa continues to cause extensive loss to inadequately stored grain. Since natural infection is spatially variable, the infestation with artificially reared larvae of the target insect is an important step to developing host plant resistance (see Mihm 1997). While there is genetic variation in resistance to all of these insects, there is no naturally occurring immunity. Commercial maize transgenic events have been successfully deployed to control lepidopteran pests in maize grown in South Africa, Argentina, Paraguay, Brazil, and the Philippines. There is, however, no equivalent transgene that can effectively control storage insects.

Striga The parasitic weed *Striga hermonthica* represents a special type of biotic stress. It produces about 40,000 tiny seeds per plant, and each year a fraction germinate and infect roots of hosts such as maize and sorghum. Symptoms of infection include stunting, wilting, loss of chlorophyll, and in some cases complete loss of developing ears, especially where soil *N* levels are low. It affects much of the savanna zone of West Africa and the middle altitudes of East Africa (Ejita and Gressel 2007). No immune sources of maize have been identified. Lane et al. (1997) reported that some plants in a wild progenitor of maize, *Zea diploperennis*, restricted parasite penetration of its roots and impaired the development and survival of *Striga*. IITA scientists developed *Striga*-tolerant inbred lines and hybrids using this natural genetic resistance from teosinte (Amusan et al. 2008) and have deployed it through commercial varieties that combine drought and *Striga* tolerance in West Africa (Badu-Apraku 2010; Badu-Apraku and Fakorede 2013). The deployment of imidazolinone-resistant maize hybrids whose seed is coated with imazapyr herbicide prior to planting provides good protection from *Striga* (Kanampiu et al. 2007). This naturally occurring gene has been incorporated into commercial varieties/hybrids in East Africa (Makumbi et al. 2015).

3.6.5 Grain Nutritional Quality

3.6.5.1 Provitamin A-Enriched Maize

Several countries in sub-Saharan Africa and Central America grow white maize, a consumption of which could result in a relatively high level of vitamin A deficiency, especially among those who treat it as a staple food. Also, most yellow maize grown and consumed throughout the world has only 2 μg/g or less of provitamin A carotenoids. CIMMYT has been successful in developing an array of provitamin A-enriched maize germplasm. Under the HarvestPlus-Maize Program, where the primary target is improving provitamin A concentration in the endosperm, considerable progress has been achieved to date at CIMMYT and IITA on developing provitamin A-enriched maize germplasm, in active collaboration with several institutions/universities worldwide (Prasanna et al. 2014). The first-generation provitamin A-enriched hybrids developed by CIMMYT have about 6–9 μg/g of provitamin A; three of such hybrids have been released in 2012 in Zambia. Eight second-generation provitamin A-enriched maize hybrids (with >10 μg/g of provitamin A) have been released in 2015 in Zambia, Zimbabwe, and Nigeria. A clinical trial conducted on 140 children in Zambia demonstrated that high provitamin A (17–24 μg/g) orange maize grain was as effective as vitamin A supplementation in correcting deficiency (Gannon et al. 2014).

3.6.6 Trait × Management Interactions

The history of temperate maize improvement is characterized by positive trait × crop management interactions that have led to steady increases in productivity at the farm level (Fischer et al. 2014). Increasing tolerance to high plant densities was accompanied by a doubling of planting density over the past 70 years in the USA and a concomitant rise in the optimum density for grain yield. Improved drought tolerance in temperate maize hybrids allowed them to withstand drought far better in the 2012 US drought than during a drought of comparable severity in 1988 (Boyer et al. 2013). Similar interactions have been exploited in tropical maize, but to a lesser degree. Excellent progress has been made in improving drought tolerance in tropical germplasm (Edmeades et al. 1999; Bänziger et al. 2006; Semagn et al. 2014). More than 230 drought-tolerant varieties and hybrids have been released in sub-Saharan Africa over the last 10 years and in 2016 are being grown by more than five million farm families on more than two million hectares. These new cultivars have stabilized and increased yields in some countries, such as Ethiopia (Abate et al. 2015). However, increased tolerance to high plant densities has not been fully exploited in tropical germplasm to date since most cultivars are being developed and evaluated under densities of 40–70,000 plants/ha – a density that has not changed in the last 40 years. A notable exception is in the northwest of Mexico where maize is planted up to 120,000 plants/ha in irrigated areas. While lower densities can be justified because intercropping is practiced in some maize fields in the tropical world, plant density should be considered a selection tool for increasing abiotic stress tolerance and improving resistance to barrenness and lodging.

Harvest index remains stubbornly low in tropical hybrids at around 0.42 (Zaidi et al. 2003a). In some areas such as the Indian subcontinent and East Africa, the value of stover may approach that of grain, and for these areas, a lower HI may result in better retention of nutrients in stover, thereby increasing its market value. There is considerable genetic variation for stover quality and production in modern tropical genotypes, and it is possible to combine high grain yield with excellent in vitro digestibility of stover (Zaidi et al. 2013). However, low HI may also reflect reduced sink strength by the ear, which may be associated with reduced biomass production. There is still considerable room for improvement in HI in tropical maize germplasm.

3.7 Field Trials and Phenotyping

3.7.1 *Multilocation Testing vs. Managed Stress Environments*

Multilocation testing (MLT) of progenies and especially of advanced selections has a proven track record in maize, as attested by the steady improvements in yield and in stress tolerance in temperate germplasm developed using this testing method (Duvick 2005; Cooper et al. 2014). Because successful varieties and hybrids rely on an integrated array of traits giving rise to stable and high yield, MLTs will continue to play a critically important role in their identification. Evaluation in a randomly selected subsample of the target population of environments (TPEs) has gradually increased allele frequencies for stress tolerance and identified a number of genotypes with tolerance to drought (Campos et al. 2006), high density, and unidentified stresses that have led to increased plant-to-plant uniformity (Edmeades 2012). TPEs within megaenvironments defined by G × E interaction (GEI) patterns are often identified by geography (e.g., Löffler et al. 2005; Cairns et al. 2012), but can also be identified by yield level within a geography. Weber et al. (2012) and Windhausen et al. (2012) noted that GEI for yield could be reduced by subdividing test sites in eastern and southern Africa into low- (<3 t/ha) vs. high-yielding environments, rather than by dividing them geographically into eastern vs. southern Africa. Data from test sites do not need to be treated equally – the incidence of known stresses at specific sites within the MLT system can be used to weight results from those locations more heavily than those from other test sites (Löffler et al. 2005).

The success of the MLT approach is unchallenged, but the use of randomly selected test locations is a costly approach, in part because common stresses such as drought are spatially variable and stochastic in nature. The use of well-characterized managed stress environments (MSEs) for traits such as drought allows the stress intensity and timing to be repeatable from year to year and easily measured. In addition, the development of MSEs for acid soils, low *N*, and *Striga* increases the spatial uniformity within screening trials. For example, most tropical maize breeders now opt for a proportion of MSEs that represent the intensity and timing of an important type of drought stress in that TPE. These are usually rain-free, irrigated locations that allow stringent control of the nature, timing, and intensity of water stress. Low *N* MSEs where the soil *N* level is reduced by cutting and removal of crop residues now play an important part in regional testing networks in sub-Saharan Africa and South Asia. It is important that MSEs maintain a significant positive genetic correlation for yield with the TPE (Cooper et al. 2014), something that low *N* sites used during the normal crop season in Africa that appear to do better than drought screens conducted in the dry season (Weber et al. 2012).

Theory for assessing gains in the target environment from MSEs conducted in a winter dry environment or in a managed "hotspot" indicates that gains are greatest when genetic variance σ_G^2, heritability in the MSE, and the genetic correlation between the MSE and the TPE are all high. Managed stress normally ensures that heritability and genetic variance for the trait are maximized, and data from MSEs can be weighted more heavily during selection than data from randomly stressed sites. Comparisons of a drought MSE vs. the TPE for a tropical population (Byrne et al. 1995) suggest that about 80% of gains in yield under drought observed in the MSE were also observed in the TPE. The combination of screening in MSEs with testing via MLTs in the target environment provides an important insurance against bias arising from excessive dependence on MSEs that are consistently conducted out of season. The judicious use of this combination of test sites has reduced testing costs and resulted in improved performance under drought (Bänziger et al. 2006).

3.7.2 Designs and Plot Management

As stress levels rise in field experiments, underlying spatial variability in soil texture and depth that affects plant-available water or *N* becomes apparent. This produces visible spatial variation in stress symptoms and a pattern of correlations among plot residuals. During screening, numbers of genotypes under test often range from 150 to 1000, so block sizes (replications) are large. Obvious trends in soil texture, historical fertility, and weed populations should be anticipated when trials are laid out. Replications properly located across such gradients can partially remove the effects of trends during analysis. While there is no complete substitute for soil of uniform depth and texture or uniform incidence of the pest of interest, row-column designs and suitable incomplete block designs (e.g., alpha (0,1) lattice designs) help to block and account for variation that occurs within large replications (Barker et al. 2005; Brown et al. 2014; Cooper et al. 2014). Bordering requirements when testing genotypes of different height also imply that single-row plots should be avoided, though seed and land supplies often dictate that the larger numbers of entries that can be screened in single-row plots can result in an overall increase in gain. Smaller plots may also result in less exposure of replicates to large-scale soil variation and therefore increase the precision of experiments conducted under severe stress (Bänziger et al. 1995).

A good plot practice is essential so that heritabilities are maximized (Bänziger et al. 2000). Great care should be exercised when planting trials under abiotic stress to avoid missing hills and plots, since the absence of complete competition increases the availability of radiation, water, and *N* to adjacent plots as well. For the same reasons, bordering increases in importance as stress levels rise, and end plants near wide alleys should be discarded from each plot because increased access to inputs may render them virtually unstressed. Mechanization of planting, side-dressing, and harvesting generally reduces error and contributes to increased heritabilities, while allowing a significant expansion in the numbers of genotypes than can be screened or tested. Mechanical shelling in the field also allows indirect pressure for increased shelling percentage as a component of increased HI. When screening for drought, it is important to stratify entries by flowering date where possible, since water stress increases with time at this sensitive growth stage, and differentially penalizes later flowering entries. Heritabilities are higher in trials of homozygous vs. segregating genotypes, but inbred lines are more difficult to manage (Bolaños and Edmeades 1996). Finally, the severity of stress imposed is important. If grain yields fall to <15% of unstressed levels, the heritability for yield falls and secondary traits become erratic. On the other hand, if stress levels are insufficient, then the genetic correlations between stress and unstressed are high, and little new information is obtained. In general, the target yield reduction should be around 50% of potential, and two stress levels are often used to bracket this goal and ensure that the appropriate level of stress is obtained in at least one of them. It is always wise to include an unstressed repetition of the trial so that changes in yield potential can be monitored during selection. The use of drip irrigation techniques and the judicious use of *N* additions or lime can be used to modify the severity and uniformity of specific stresses more accurately (Bänziger et al. 2000). A very effective tool to enhance the quality of data gathered under drought or any other source of stress is the use of mixed models to analyze data, since they represent a cost-effective way to reduce the impact of soil and experimental heterogeneity on field data and to increase the genetic gain to environmental noise ratio, in other words heritability, and therefore expected genetic gains.

3.7.3 High-Throughput Phenotyping

Advances in genotyping have led to dramatic reductions in cost per data point, but this has not been matched by a corresponding decline in the cost of phenotypic data until recently when remote sensing of traits has been deployed. The traditional emphasis on collection of quality data from field trials remains as important as ever for traits such as grain yield and yield components, dates of 50% silking and anthesis, biomass, plant height, staygreen and leaf rolling scores, and disease and insect scores. To reduce error rate from transcription errors, plots should be identified by weather proof bar-coded tags, and data should be collected directly on a handheld tablet that reads the bar code to identify the plot. This electronic data record should allow for notes and comments against plot numbers as well as numerical data and scores, but if not then these should be recorded in a field book that includes maps, instructions, randomizations, etc.

Remote-sensed data is increasingly important when phenotyping, is nondestructive, and can be repeated as often as necessary. In its simplest form, it involves the use of handheld sensors such as infrared thermometers, digital cameras, and GreenSeeker® sensors (Cairns et al. 2012). In recent years, there has been a sharp increase in the use of unmanned aerial vehicles such as tethered balloons, regular aircraft, and recently miniature helicopter and fixed-wing drones fitted with multispectral cameras (Araus and Cairns 2013). These new methods of hyperspectral analyses are both fast and cheap to use and take readings over a short interval under stable atmospheric and crop conditions (Zaman-Allah et al. 2015), and repeated passes 5–15 days apart provide an understanding of how traits change over time.

Remote sensing methods can be divided into those depending on spectral reflectance (e.g., SPAD, NDVI), digital imagery (e.g., leaf color and leaf area), and thermal imagery (e.g., canopy temperature) (Masuka et al. 2012; Zaman-Allah et al. 2015; Vergara-Diaz et al. 2016). Good relationships between grain yield, biomass, and NDVI have been reported for maize (Lu et al. 2011; Cairns et al. 2012; Trachsel et al. 2016). Other traits include ear imagery to estimate yield components and silk number per ear and measurements of plant height and flowering date. The rapid rise in the number, utility, and cost-effectiveness of drones equipped with standard digital cameras will certainly multiply options for using red/green/blue (RGB) wavelength ratios in the near future (e.g., Vergara-Diaz et al. 2016). Although the rate of data collection is very high using remote sensing, images are typically data dense. The processing of these in real time is computationally demanding but is needed to generate a representative value that differentiates among genotypes and can be used for selection. There is an extensive effort underway to determine what additional vegetation indices can be obtained from remote sensing of segregating progenies in small plots and what these indices can contribute to selection. Near-infrared (NIR) analysis can also be used to detect concentrations of specific metabolites in tissues, in some cases nondestructively (Araus and Cairns 2013).

3.7.4 Data Management

Data processing can become a bottleneck in breeding. For many maize breeders from smaller institutions, the easiest software to access is the Breeding Management System (BMS), formerly known as the Integrated Breeding Platform (https://www.integratedbreeding.net/breeding-management-system). The IBP was developed by the Generation Challenge Program of the CGIAR for a diverse array of tropical crops. It has now evolved into the BMS supported by the Bill and Melinda Gates Foundation as a stand-alone software suite that currently can be downloaded free of charge. BMS is automatically linked to the GENSTAT® suite of analysis programs and has dedicated cloud computing or stand-alone options for database management. BMS is being adopted by a number of leading public sector breeding programs. The suite of programs comprising this package is well suited to MLT data analysis and MAS as well. The BMS offers to many national breeders the real prospect that MAS (including marker-assisted backcrossing) can be implemented in real time in NARS' breeding programs. It also includes information on how to outsource genotyping, something that makes real-time MAS possible for many programs.

It is important to emphasize the importance of database management and archival of annotated results in a machine-readable and easily accessible form in maize breeding programs that are committed to long-term crop improvement. Such databases facilitate any study of genetic gain vs. time, allow head-to-head comparisons of hybrids over time and space, and allow meta-analysis of large, diverse, and often unbalanced datasets.

3.8 Applied Biotechnology

The broad goals of biotechnology in crop improvement relate to discovery and location of new and useful genetic variation and to an accelerated rate of genetic gain. For a general overview of DNA technologies and molecular marker types, see Brown et al. (2014).

3.8.1 Association Mapping

A common means of gene discovery is through genome-wide association study (GWAS). GWAS or association mapping through analysis of linkage disequilibrium is a powerful tool for dissecting complex traits and identification of potential favorable alleles that can contribute to the enhancement of target traits. Association mapping can establish marker-trait associations in panels of inbreds. Typically, it is applied to association mapping panels that consist of several hundred diverse homozygous lines. Random association between alleles is reduced by genetic linkage, creating disequilibrium. Heavily selected temperate inbreds show a high level of LD (many alleles linked in blocks of ~100 kb and moving as a unit during selection), whereas tropical maize shows a rapid decay of LD (2–5 kb) because of its long history of recombination as OPVs and relatively recent history of selection. Lines are generally genotyped with several thousand SNP markers, though increasingly >500,000 polymorphic SNPs are being generated through genotyping-by-sequencing methods and utilized in association and selection studies. GWAS generally has low statistical power for associating rare alleles with phenotypic differences (Yan et al. 2011). The outcome from GWAS is a series of precisely defined genomic regions associated with the trait of interest that can be linked to candidate genes identified through resequencing those regions. This can be used to develop simple PCR gene-based markers for marker-assisted selection (MAS). While GWAS studies have been useful at identifying regions of interest across diverse genetic backgrounds, outcomes of these studies have produced relatively few useful candidate genes or regions for subsequent use in MAS. Bernardo (2008) observed that GWAS involving poorly adapted genotypes has provided relatively little useful information for breeders. This is in part because of false positives arising from existing relationships among lines in the study and because of the very considerable challenges of accurately phenotyping a diverse set of lines in any single environment. Yan et al. (2011) noted that increasing the numbers of genotypes has a much greater effect on the efficiency of GWAS than increases in marker density. GWAS leads are usually validated by analysis of biparental mapping populations.

3.8.2 Accelerating Genetic Gain Through Marker-Assisted Selection

Gene-phenotype associations form the basis of MAS. These are normally established through careful phenotyping and genotyping with molecular markers of a segregating $F_{2:3}$ population or a set of recombinant inbred lines (RILs) from a biparental cross. Relatively few QTLs have been used in MAS, in part because their phenotypic effects are dependent on the genetic background of the lines in the study and may interact with the environment (e.g., Jiang et al. 1999). Furthermore, some are associated with minor genetic effects, and the cost of the technology exceeds the benefits from added genetic gain. A problem occasionally still arises from false positives because too few genotypes were involved in the study (Beavis 1994). A further logistical issue has been a failure in some programs to have genotypic data available at the time selection decisions are made. If selection decisions are delayed by one crop cycle, most of the benefits in MAS are lost (Bernardo 2008).

Despite these challenges, a number of tropical maize programs such as CIMMYT's are using production SNP markers to ensure that specific regions associated with disease resistance or grain quality are present in selected lines, as previously noted (Prasanna et al. 2014). Marker-assisted backcrossing (MABC) is also widely used in the transfer of these traits and is routine in commercial companies in the transfer of key QTL and transgenes to elite inbred lines. QTLs are increasingly being identified that function in a range of genetic backgrounds. Recently a meta-QTL analysis across three tropical maize biparental populations (RILs) showed six constitutive genomic regions associated with drought tolerance (Almeida et al. 2014) and identified an 8 Mb region delimited in 3.06 that harbored most of the morphophysiological traits associated with improved performance under drought.

Marker-assisted recurrent selection (MARS) has been used on a wider scale to accelerate breeding (see Sect. 7.5) in large commercial breeding programs in the USA (Crosbie et al. 2006; Edgerton 2009). In brief, MARS functions by establishing gene-phenotype associations among F_3 progeny topcrosses of a biparental cross and uses these to guide selections for three additional selection cycles. It is a scheme that is heavy on phenotyping and functions by identifying those QTLs with significant effects on the trait of interest. Although gains from MARS can be double those of conventional selection (Eathington et al. 2007; Beyene et al. 2015, 2016), in practice it has been a resource-intensive process involving progenies of biparental, but in some cases multiparental, crosses, which in retrospect, were not always the best choices. For these reasons, and the development of genotyping-by-sequencing methods producing up to a million SNP-based markers, MARS has been largely subsumed into genomic selection (GS) (Chap. 2).

The comparative effectiveness of MARS and GS was tested by CIMMYT in a study of gains under drought in East Africa. Gains were evaluated from MARS in ten biparental tropical crosses, using 148–184 F2:3 progenies crossed to a single tester. Lines were genotyped with 190–225 SNP markers, and a selection index based on phenotypic and marker data was applied in order to select families for recombination in C0. In each successive cycle, the selection index was applied to plants that had been genotyped with 55–87 SNPs (Beyene et al. 2015, 2016). Gains under optimal conditions were 93 kg/ha/year and under water stress were 46 kg/ha/year, vs. 50 and 15 kg/ha/year, respectively, from conventional selection under similar conditions (Semagn et al. 2014). A comparison between conventional selection and GS was undertaken in 13 biparental populations using 191–326 SNPs (i.e., relatively few for GS), and the best 10% based on GEBVs were intermated. Overall genetic gain from GS under drought was 52 kg/ha/year vs. 16 kg/ha/year using conventional selection. MARS and GS therefore appear able to increase genetic gain by a factor of two to three times that of conventional selection under drought stress.

Building on its effectiveness in temperate maize breeding, GS would also become a key component of the toolbox tropical maize breeders have at their disposal. Its potential would be fully exploited when, GS is also used to predict parental combinations and therefore to increase heterosis.

3.8.3 Transgenics

Transgenic technologies have received tremendous attention by the commercial seed sector since the first transgenic hybrid became commercial in 1996. The majority are herbicide-resistant (mainly glyphosate (RoundUp Ready®, RR)) or insect-resistant (*Bt*) cultivars. Recently, a transgenic event encoding a cold shock protein from the bacteria *Bacillus subtilis* and providing drought tolerance in maize, MON87460, has been approved for release in South Africa, and breeders are seeking approval for its release in Kenya and Uganda. By 2015 transgenic cultivars of all crops were planted on 180 M ha annually, of which 53% were in developing countries and 29% were maize – mainly Brazil, Argentina, Paraguay, South Africa, Uruguay, Bolivia, and the Philippines (James 2015). The adoption of transgenic varieties has been a sound investment for the vast majority of farmers, with farmers in developing countries receiving $4.22 for every dollar they invested in transgenic seed in 2013 (Brookes and Barfoot 2015). The release of transgenic crops is subject to extensive regulation, and they cannot be tested or grown without a legal and functional regulatory system in place. Transgenic maize imported from producing countries to the developed world is also subject to stringent safety testing. These are indications of public wariness of this technology and concerns that it is owned by a few large multinational seed companies. Yet 20 years after the launch of the first commercial Bt hybrids, there have been no validated cases of health-related problems among animals and humans.

A major challenge for breeders is stewardship of the transgene to ensure that it remains in its designated genetic background. This means that transgenic and conventional germplasm must be separated during seed processing and preparation and especially in the field where stray pollen can lead to adventitious presence of the transgene in conventional grain. A further issue with *Bt* genes is preventing the buildup of resistance to the *Bt* toxin in local lepidopteran insect populations, and this requires the planting of 20% of the cropped area to non-*Bt* hybrids. Enforcement of this refugia requirement is essential but challenging.

One example of public breeder access to transgenes is provided through the Gates Foundation-funded project Water Efficient Maize for Africa (WEMA). Here Monsanto's drought tolerance transgenic event MON87460 (Castiglioni et al. 2008) and their widely used *Bt* gene MON810 are available for royalty-free use under license by local maize seed companies in Kenya, Uganda, Tanzania, and Mozambique, once the transgenes have been deregulated in those countries. In South Africa where the transgenic event conferring tolerance to some lepidopteran insects MON810 was released 18 years ago, target insects have developed a significant level of tolerance to this *Bt* toxin (van Rensberg 2007) necessitating its replacement by the MON89034 transgenic event. CIMMYT is also involved in the extensive field testing of conventionally improved drought-tolerant hybrids, since transgenic and conventional improvements are thought to be additive in effectiveness. WEMA is being executed by the African Agricultural Technology Foundation (AATF) based in Nairobi, Kenya. It can be expected that MON87460 and MON810 will be detected in local maize varieties quite rapidly after their open release, simply because of gene flow via pollen. This may have implications for transboundary movement of produce and seed to countries unwilling to import genetically modified maize. Nonetheless, the prospect of future use of these transgenes in sub-Saharan Africa is a significant and exciting development.

3.9 Seed Production and Marketing

Plant breeding is never an end by itself: its return on investment occurs in farmers' fields. Too often maize breeders have declared "mission accomplished" at product release and have failed to engage in ensuring that seed of their improved varieties reaches intended users. This process starts with insisting that hybrids and varieties are tested in farmers' fields during development so that possible reasons for non-adoption are recognized before a variety is released. The second step, well beyond breeders' control, is the development of the seed industry needed to deliver to end users the efforts of plant breeders.

A national seed industry typically passes through various stages of maturation. Government-sponsored seed production usually gives way to many small start-up seed companies that use publicly available germplasm from CIMMYT, IITA, or NARS and a few multinationals who often do not breed in-country but import seed for testing. Competition narrows the field to a few successful national companies and multinationals that may purchase smaller companies to increase sales volume or enter into research agreements with them to access germplasm and technologies. Sub-Saharan Africa is currently characterized by many small start-up seed companies that struggle to maintain sales volumes and seed quality and cannot afford an agronomist/breeder on staff or a large demonstration program for new products (MacRobert 2009; Langyintuo et al. 2010).

Although the hybrid seed business has been important in some countries like Brazil and Argentina since the 1940s, only in the early 1970s did it begin to spread among other Latin American countries. Pioneer Hi-Bred Int. established its first non-USA breeding station in Jamaica in 1964. In tropical Southeast Asia, the seed business became more competitive in the early/mid-1980s when several large seed companies established hybrid breeding research. Counting on Suwan 1 as a good source of adaptation, three-way hybrids and then single crosses started to compete with this excellent OPV and occupy significant area, particularly in Thailand, Indonesia, and the Philippines. The single-cross hybrid DK888 was released by DeKalb in Thailand in 1993 and became a widely planted cultivar for several decades. The spread of hybrid seed businesses in Thailand, the Philippines, and Indonesia sparked a similar response in India, Vietnam, Myanmar, and other Asian countries. Today, the hybrid maize seed business in these countries constitutes an area of rapid growth.

3.9.1 Product Types

In tropical areas, particularly in marginal production environments or where few inputs are used and agronomy is poor, the choice of OPVs vs. hybrids is still strongly debated. The analysis hinges around relative seed prices vs. yield differences, since OPV seed price may be that of commercial grain if farmers retain their own seed. Pixley (2006) concluded that the best hybrids in southern Africa outyielded the best OPVs by around 18% across a range of yield levels and that hybrids deliver greater value to farmers than OPVs when farmer yield is consistently over 2 t/ha. This figure is a guide only and depends on prices of grain and seed, the yield advantage of hybrids over OPVs under stress, tolerance of farmers to risk, and the availability of quality hybrid seed. Nonetheless improved OPVs are still the most reliable option for farmers that cannot reach the 2 t/ha yield level, while for hybrids their advantage became clear at yields >3 t/ha. Pixley's analysis showed that recycling hybrid seed (sowing F_2 seed) was least profitable at all yield levels because it yielded an average of 32% less than F_1 seed. Unfortunately the practice of recycling hybrid seed can reach 50% in drought-prone areas of East Africa (de Groote, 2013, personal communication). However, a viable seed industry depends ultimately on the annual sales of hybrid seed, and the use of hybrid seed has been shown to contribute to increased farmer welfare (Mathenge et al. 2014). Furthermore, the rate of turnover of varieties is likely to be greater when hybrid seed is grown, and this brings benefits to seed companies and farmers alike (Gaffney et al. 2016). Most successful companies in the tropics market a small amount of improved OPV seed as a service and make their main income from hybrids. There is room for both products, and a mature seed industry operating in a risky production environment like sub-Saharan Africa should be encouraged to focus on both.

What sort of hybrid is most appropriate? The hybrid seed industry began with double crosses (i.e., a cross between two single crosses) but noticed that three-way crosses were around 6% and single crosses around 19% higher yielding (Pixley 2006). There are benefits in three-way crosses. The single-cross seed parent is high yielding and stress tolerant, so costs per unit of seed are low, and the male inbred pollinator can be destroyed after flowering, thus making theft of inbred lines difficult. However, the increasing yields possible on inbreds designated as females, improved security based on DNA fingerprinting, and increasing skills in the seed sector make single crosses increasingly attractive. There is little doubt that single crosses will ultimately replace three-way crosses in all but the toughest seed production environments, provided the cost of seed of single crosses can remain within reach of resource-poor farmers.

3.10 Future Prospects and Outlook

The future of tropical maize breeding looks very promising, though the challenges of the changing physical environment will be a major headwind against future yield increases (Thornton et al. 2011). Important for increasing and stabilizing maize yields in the tropics will be traits related to tolerance of high temperature and drought, along with horizontal resistance to changes in virulence of pests as temperatures rise. Many of the developments that will lead to increased and stable yields may occur in the emerging private sector, and in temperate areas private sector investments have driven a steady increase in maize yields over the past 25 years that significantly exceeds that of rice and wheat (Fischer et al. 2014). There continues to be a need for balance in terms of the roles of public sector international research-for-development efforts on maize for smallholders in low-yielding environments with that of private sector technological edge for increasing and stabilizing yields of tropical maize.

The following appear to be major opportunities in tropical maize:

- Density tolerance is needed. The key developments of increased tolerance to high plant density and improved stalk strength in temperate maize (Fischer et al. 2014) have yet to occur in the majority of tropical germplasm. A systematic planting of all experimental plots at 30–50% higher density than farmer's fields would be a good starting point, even if lodging and barrenness are occasionally severe. Tollenaar and Lee (2011) also conclude that increased stress tolerance is the key to further gains in yield and yield stability in temperate maize.
- Adoption and rate of turnover of tropical varieties need to increase. The rapid replacement of existing varieties with new stress-tolerant hybrids is perhaps the best way of ensuring that genetic changes in varieties can keep up with climate change (G. Atlin, 2014, personal communication).
- Yield potential needs to increase. This can be from obvious changes in partitioning such as reductions in tassel size and leafiness and increases in HI and shelling percentage. The size of individual ears and yield per plant at low densities have changed little with selection in temperate maize (Duvick 2005; Egli 2015), so

increased yields have come from a steady rise in ears per unit area. However, increases in kernel number per ear in tropical maize growing in optimal environments could come from increased emphasis on synchrony of pollination and kernel development (Cárcova and Otegui 2007). Yield potential increases might also arise from longer-term changes in key enzymes such as rubisco (Parry et al. 2003) and its activator, rubisco activase (Salvucci et al. 2008) whose temperature responses and kinetics may well be altered through gene editing procedures in the future. Such changes would enhance fitness in hot dry environments that are predicted to become the norm.

- Phenotyping is growing in importance and is an essential component of future breeding activities. Tropical soils are generally more variable than temperate soils, so physical and statistical techniques that can be used to minimize these effects should always be used (Barker et al. 2005). The basics of uniform stands and adequate bordering are still being neglected in some tropical breeding programs, and continued investments in high-quality field research facilities, mechanization to plant and shell, and managed stress environments are needed in increasing numbers. Remote sensing for estimating traits that will increase the heritability of a selection index should be energetically explored and embraced (Lu et al. 2011; Cairns et al. 2012; Araus et al. 2012; Araus and Cairns 2013; Zaman-Allah et al. 2015; Vergara-Diaz et al. 2016). The routine use of remote-sensed traits such as plant height is sharply reducing labor requirements in the field, and anthesis and perhaps silking date observations are currently under evaluation.
- Molecular breeding will play an increasingly important part in tropical maize improvement. The most promising emerging technologies today are genomic selection linked with doubled haploid production. The availability of doubled haploid facilities, high-density genotyping as well as high-throughput and low-cost genotyping capacities, marker resources, and analytical capability through the offices of CIMMYT and the Breeding Management System put these methods increasingly within reach of medium-sized tropical maize breeding programs. Transgenics will have an increased role as trust in the technology increases (Edgerton 2009), and there is ample scope in the tropics for more extensive use of herbicide and insect resistance technologies, as well as those relating to disease and abiotic stress tolerance. Gene editing will undoubtedly have significant impact, though this may be 5–10 years away in tropical maize.
- The selection cycle will continue to shorten. Doubled haploids are clearly here to stay. Prediction of performance will help narrow the numbers of genotypes evaluated in the field. Genomic selection and the use of production markers that identify specific haplotypes and provide a framework for prediction shorten the breeding cycle by reducing phenotyping needs for several generations. In fact, genotypes can be assessed from seed chips of doubled haploids and known susceptible segregants can be eliminated before a seed is sown.
- Identification of new and useful genetic variability within the species is a likely outcome of investments in screening maize landraces in initiatives such as the

Seeds of Discovery (SeeD) project, and these will likely find future uses as defensive traits (Tester and Langridge 2010; Prasanna 2013).

- Tools for data management and information extraction in real time will become increasingly necessary as both phenotyping and genotyping move into the millions of data points per genotype. There will be a growing need for efficient algorithms that reduce this *tsunami* of data to an index used to rank genotypes. The development and use of a mixed model framework for the analysis of multi-location and managed stress datasets will allow information to be mined from older as well as current datasets (Cooper et al. 2014).
- Training of a new generation of field-oriented breeders is essential. Firsthand knowledge of germplasm, environments, and their interactions remains at the heart of successful tropical maize breeding despite the sometimes cosmetic appeal of new technologies and tools. There is no substitute for trained staff observing genotype and consumer reactions in the field, especially in farmers' fields. The current shortage of young graduates in field plant breeding, particularly in tropical crops such as maize, represents a significant threat that must be addressed and overcome in order to maintain, and ideally increase, the current rates of genetic gain in this fascinating crop.

References

Abate T, Shiferaw B, Menkir A et al (2015) Factors that transformed maize productivity in Ethiopia. Food Secur 7:965–981

Abebe F, Tefera T, Mugo S et al (2009) Resistance of maize varieties to the maize weevil *Sitophilus zeamais* (Motsch.) (Coleoptera: Curculionidae). Afr J Biotechnol 8:5937–5943

Almeida GD, Nair S, Borem A et al (2014) Molecular mapping across three populations reveals a QTL hotspot region on chromosome 3 for secondary traits associated with drought tolerance in tropical maize. Mol Breed. doi:10.1007/s11032-014-0068-5

Amusan IO, Rich PJ, Menkir A et al (2008) Resistance to *Striga hermonthica* in a maize inbred line derived from *Zea diploperennis*. New Phytol 178:157–166

Araus JL, Cairns JE (2013) Field high-throughput phenotyping: the new crop breeding frontier. Trends Plant Sci. doi.org/10.1016/j.tplants.2013.09.008

Araus JL, Serret MD, Edmeades GO (2012) Phenotyping maize for adaptation to drought. Front Physiol. doi:10.3389/phys.2012.00305

Atlin GN, Palacios N, Babu R et al (2011) Quality protein maize: progress and prospects. Plant Breeding Rev 34:83–131

Badu-Apraku B (2010) Effects of recurrent selection for grain yield and Striga resistance in an extra-early maize population. Crop Sci 50:1735–1743

Badu-Apraku B, Fakorede MAB (2013) Breeding early and extra-early maize for resistance to biotic and abiotic stresses in sub-Saharan Africa. Plant Breed Rev 37:115–197

Bänziger M, Cooper M (2001) Breeding for low input conditions and the consequences for participatory plant breeding: examples from tropical maize and wheat. Euphytica 122:503–519

Bänziger M, Lafitte HR (1997) Efficiency of secondary traits for improving maize for low-nitrogen target environments. Crop Sci 37:1110–1117

Bänziger M, Lafitte HR, Edmeades GO (1995) Intergenotypic competition during evaluation of maize progenies under limited and adequate N supply. Field Crops Res 44:25–31

Bänziger M, Edmeades GO, Lafitte HR (1999) Selection for drought tolerance increases maize yields across a range of nitrogen levels. Crop Sci 39:1035–1040
Bänziger M, Edmeades GO, Beck DL, Bellon M (2000) Breeding for drought and nitrogen stress tolerance in maize: from theory to practice. CIMMYT, Mexico DF, 68 pp
Bänziger M, Setimela PS, Hodson D, Vivek B (2006) Breeding for improved abiotic stress tolerance in maize adapted to southern Africa. Agric Water Manag 80:212–224
Barker TC, Campos H, Cooper M et al (2005) Improving drought tolerance in maize. Plant Breed Rev 25:173–253
Beavis W (1994) The power and deceit of QTL experiments: lessons from comparative QTL studies. Proc 49th ASTA meetings. ASTA, Chicago, pp 250–266
Bernardo R (2008) Molecular markers and selection for complex traits in plants: learning from the last 20 years. Crop Sci 48:1649–1664
Betrán FJ, Beck D, Bänziger M, Edmeades GO (2003a) Genetic analysis of inbred and hybrid grain yield under stress and nonstress environments in tropical maize. Crop Sci 43:807–817
Beyene Y, Semagn K, Mugo S et al (2015) Genetic gains in grain yield through genomic selection in eight bi-parental maize populations under drought stress. Crop Sci 55:154–163
Beyene Y, Semagn K, Crossa J et al (2016) Improving maize grain yield under drought stress and non-stress environments in sub-Saharan Africa using marker-assisted recurrent selection. Crop Sci 56:1–10
Blum A (1988) Plant breeding for stress environments. CRC Press, Boca Raton
Bolaños J, Edmeades GO (1993) Eight cycles of selection for drought tolerance in lowland tropical maize. I. Responses in grain yield, biomass, and radiation utilization. Field Crops Res 31:233–252
Bolaños J, Edmeades GO (1996) The importance of the anthesis-silking interval in breeding for drought tolerance in tropical maize. Field Crops Res 48:65–80
Bolaños J, Edmeades GO, Martinez L (1993) Eight cycles of selection for drought tolerance in lowland tropical maize. III. Responses in drought- adaptive physiological and morphological traits. Field Crop Res 31:269–286
Boyer JS, Byrne P, Cassman KG et al (2013) The U.S. drought of 2012 in perspective: a call to action. Glob Food Secur 2:139–143
Brewbaker JL (2009) Registration of nine tropical maize populations resistant to tropical diseases. J Plant Registration 3:10–13
Brookes G, Barfoot P (2015) GM crops: global socioeconomic and environmental impacts 1996–2013. PG Economics Ltd, Dorchester. (2015globalimpactstudyfinalMay2015%20(2).pdf)
Brown J, Caligari P, Campos H (2014). Plant breeding, 2nd ed. Wiley Blackwell, Oxford
Butruille DV, Birru FH, Boerboem ML et al (2015) Maize breeding in the United States: views from within Monsanto. Plant Breed Rev 39:199–282
Byrne PF, Bolaños J, Edmeades GO, Eaton DL (1995) Gains from selection under drought versus multilocation testing in related tropical populations. Crop Sci 35:63–69
Cairns JE, Sonder K, Zaidi PH et al (2012) Maize production in a changing climate: impacts, adaptation, and mitigation strategies. Adv Agron 114:1–58
Cairns JE, Crossa J, Zaidi PH et al (2013) Identification of drought, heat, and combined drought and heat tolerant donors in maize. Crop Sci 53:1–12
Campos H, Cooper M, Habben JE et al (2004) Improving drought tolerance in maize: a view from industry. Field Crops Res 90:19–34
Campos H, Cooper M, Edmeades GO et al (2006) Changes in drought tolerance in maize associated with fifty years of breeding for yield in the US Corn Belt. Maydica 51:369–381
Cárcova J, Otegui ME (2007) Ovary growth and maize kernel set. Crop Sci 47:1104–1110
Castiglioni P, Warner D, Bensen RJ et al (2008) Bacterial RNA chaperones confer abiotic stress tolerance in plants and improved grain yield in maize under water-limited conditions. Plant Physiol 147:446–455
Chaikam V, Nair SK, Babu R et al (2014) Analysis of effectiveness of *R1-nj* anthocyanin marker for in vivo haploid identification in maize and molecular markers for predicting the inhibition of *R1-nj* expression. Theor Appl Genet 128:159–171

Chaikam V, Martinez L, Melchinger A et al (2016) Development and validation of red root marker-based haploid inducers in maize. Crop Sci 56:1678–1688
Chapman SC, Edmeades GO (1999) Selection improves drought tolerance in tropical maize populations: II. Direct and correlated responses among secondary traits. Crop Sci 39:1315–1324
Chavarriaga EM (1966) Maize ETO, una variedad producida en Colombia. Separata de la Revista ICA 1:5–30
Cicchino M, Edreira JIR, Uribelarrea M, Otegui ME (2010) Heat stress in field-grown maize: response of physiological determinants of grain yield. Crop Sci 50:1438–1448
CIMMYT (2004) Maize diseases: a guide for field identification. CIMMYT, Mexico DF, 119pp
Cooper M, Smith OS, Graham G et al (2004) Genomics, genetics and plant breeding: a private sector perspective. Crop Sci 44:1907–1913
Cooper M, Messina CD, Podlich D et al (2014) Predicting the future of plant breeding: complementing empirical evaluation with genetic prediction. Crop Pasture Sci 65:311–336
Corral JAR, Puga ND, Gonzalez JJS et al (2008) Climatic adaptation and ecological descriptors of 42 Mexican maize races. Crop Sci 48:1502–1512
Crosbie TM, Eathington SR, Johnson GR et al (2006) Plant breeding: past present and future. In: Lamkey KR, Lee M (eds) Plant breeding: The Arnel R Hallauer International Symposium. Blackwell, Iowa, pp 3–50
Cross HZ (1975) Diallel analysis of direction and rate of grain filling of seven inbred lines of corn. Crop Sci 15:532–535
Crossa J, Taba S, Wellhausen EJ (1990) Heterotic patterns among Mexican races of maize. Crop Sci 30:1182–1190
De Groote H, Tongruksawattana S, Oloo F et al (2016) Community-survey based assessment of the geographic distribution and impact of maize lethal necrosis (MLN) disease in Kenya. Crop Prot 82:30–35
DeLeon N, Coors JG (2002) Twenty-four cycles of mass selection for prolificacy in the Golden Glow maize population. Crop Sci 42:325–333
Demissie G, Tefera T, Tadesse A (2008) Importance of husk covering on field infestation of maize by *Sitophilus zeamais* Motsch (Coleoptera: Curculionidae) at Bako, western Ethiopia. Afr J Biotechnol 7:3777–3782
Dow EW, Daynard TB, Muldoon JF et al (1984) Resistance to drought and density stress in Canadian and European maize (*Zea mays* L.) hybrids. Can J Plant Sci 64:575–585
Duvick DN (1997) What is yield? In: Edmeades GO et al (eds) Developing drought- and low N-tolerant maize, Proceedings of a Symposium. CIMMYT, El Batan, pp 332–335
Duvick DN (2005) The contribution of breeding to yield advances in maize (*Zea mays* L.) Adv Agron 86:83–145
Duvick DN, Cassman KG (1999) Post-green revolution trends in yield potential of temperate maize in the north-Central United States. Crop Sci 39:1622–1630
Duvick DN, Smith JCS, Cooper M (2004) Long-term selection in a commercial hybrid maize breeding program. Plant Breed Rev 24:109–151
Eagles HA, Lothrop JE (1994) Highland maize from Central Mexico – its origin, characteristics, and uses in breeding programs. Crop Sci 34:11–19
Eathington SR, Crosbie TM, Edwards MD et al (2007) Molecular markers in a commercial breeding program. Crop Sci 47:S154–S163
Eberhart SA, Salhuana W, Sevilla R, Taba S (1995) Principles for tropical maize breeding. Maydica 40:339–355
Echarte L, Andrade FH (2003) Harvest index stability of Argentinean maize hybrids released between 1965 and 1993. Field Crops Res 82:1–12
Echarte L, Rothstein S, Tollenaar M (2008) The response of leaf photosynthesis and dry matter accumulation to nitrogen supply in an older and a newer maize hybrid. Crop Sci 48:656–665
Edgerton MD (2009) Increasing crop productivity to meet global needs for feed, food, and fuel. Plant Physiol 149:7–13
Edmeades GO (2008) Drought tolerance in maize: an emerging reality. In: James C (ed) Global status of commercialized biotech/GM crops: 2008, ISAAA Brief 39. ISAAA, Ithaca, pp 197–217

Edmeades GO (2012) Progress in achieving and delivering drought tolerance in maize – an update. In: James C (ed) Global status of commercialized biotech/GM crops: 2012, ISAAA Brief 44. ISAAA, Ithaca, pp 239–272
Edmeades GO, Ellis RH, Lafitte HR (1992) Photothermal responses of tropically-adapted maize. Agron Abstr 84:124
Edmeades GO, Bolaños J, Hernandez M, Bello S (1993) Causes for silk delay in a lowland tropical maize population. Crop Sci 33:1029–1035
Edmeades GO, Chapman SC, Lafitte HR (1994) Photoperiod sensitivity of tropical maize cultivars is reduced by cool night temperatures. Paper presented at 86th annual meeting of the American Society of Agronomy, Seattle, November 13–18, 1994
Edmeades GO, Bänziger M, Cortes M, Ortega A (1997) From stress-tolerant populations to hybrids: the role of source germplasm. In: Edmeades GO et al (eds) Developing drought and low N tolerant maize. CIMMYT, El Batan, pp 263–273
Edmeades GO, Bolaños J, Chapman SC et al (1999) Selection improves drought tolerance in tropical maize populations: I. Gains in biomass, grain yield, and harvest index. Crop Sci 39:1306–1315
Edmeades GO, Bolaños J, Elings A et al (2000a) The role and regulation of the anthesis-silking interval in maize. In: Westgate ME, Boote KJ (eds) Physiology and modeling kernel set in maize, CSSA Special Publication No. 29. CSSA, Madison, pp 43–73
Edmeades GO, Bänziger M, Ribaut JM (2000b) Maize improvement for drought-limited environments. In: Otegui ME, Slafer GA (eds) Physiological bases for maize improvement. Haworth, Binghampton
Edreira JIR, Carpicia EB, Sammarro D, Otegui ME (2011) Heat stress effects around flowering on kernel set of temperate and tropical maize hybrids. Field Crops Res 123:62–73
Egli DB (2015) Is there a role for sink size in understanding maize population-yield relationships? Crop Sci 55:2453–2462
Ejeta G, Gressel J (eds) (2007) Integrating new technologies for *Striga* control: towards ending the witch-hunt. World Scientific, New Jersey
Ellis RH, Summerfield RJ, Edmeades GO, Roberts EH (1992) Photoperiod, leaf number, and interval from tassel initiation to emergence in diverse cultivars of maize. Crop Sci 32:398–403
Falconer DS, MacKay TFC (1996) Introduction to quantitative genetics, 4th edn. Prentice Hall, London
FAOSTAT (2016) http://faostat3.fao.org/home/E. Accessed 1 Feb 2016
Fischer KS, Palmer AFE (1984) Tropical maize. In: Goldsworthy PR, Fisher NM (eds) The physiology of tropical field crops. Wiley, Oxford, pp 213–247
Fischer KS, Johnson EC, Edmeades GO (1987) Recurrent selection for reduced tassel branch number and reduced leaf area density above the ear in tropical maize populations. Crop Sci 27:1150–1156
Fischer RA, Byerlee D, Edmeades G (2014) Crop yields and global food security: will yield increases continue to feed the world? ACIAR Monograph No. 158. Australian Centre for International Agricultural Research, Canberra
Fu H, Dooner HK (2002) Intraspecific violation of genetic colinearity and its implications in maize. PNAS 99:9573–9578
Gaffney J, Anderson F, Franks C et al (2016) Robust seed systems, emerging technologies, and hybrid crops for Africa. Glob Food Sec 9:36–44
Gannon B, Kaliwile C, Arscott SA et al (2014) Biofortified orange maize is as efficacious as vitamin A supplement in Zambian children even in the presence of high liver reserves of vitamin A: a community-based, randomized placebo-controlled trial. Am J Clin Nutr. doi:10.3945/ajcn.114.087379
Gerrish EE (1983) Indications from a diallel study for interracial maize hybridization in the Corn Belt. Crop Sci 23:1082–1084
Goldsworthy PR, Palmer AFE, Sperling DW (1974) Growth and yield of lowland tropical maize in Mexico. J Agric Sci 83:223–230

Goodman MM (1999) Broadening the genetic diversity in maize breeding by use of exotic germplasm. In: Coors JG, Pandey S (eds) The genetics and exploitation of heterosis in crops. ASA, CSSA, Wisconsin, pp 139–148
Goodman MM, Brown WL (1988) Races of corn. In: Sprague GF, Dudley JW (eds) Corn and corn improvement. Crop Science Society of America, Wiaconsin, pp 33–79
Gowda M, Das B, Makumbi D et al (2015) Genome-wide association and genomic prediction of resistance to maize lethal necrosis disease in tropical maize germplasm. Theor Appl Genet. doi:10.1007/s00122-015-2559-0
Hallauer AR (1999) Temperate maize and heterosis. In: Coors JG, Pandey S (eds) The genetics and exploitation of heterosis in crops. ASA, CSSA, Wisconsin, pp 353–361
Hallauer AR, Carena MJ (2012) Recurrent selection methods to improve germplasm in maize. Maydica 57:266–283
Hallauer AR, Miranda Fo JB (1988) Quantitative genetics in maize breeding, 2nd edn. Iowa State University, Ames
van Heerwaarden J, Doebley J, Briggs WH et al (2011) Genetic signals of origin, spread, and introgression in a large sample of maize landraces. PNAS 108:1088–1092
Holland JB, Goodman MM, Castillo-Gonzalez F (1996) Identification of agronomically superior Latin American maize accessions via multistage evaluations. Crop Sci 36:778–784
J. C. Reif, M. L. Warburton, X. C. Xia, D. A. Hoisington, J. Crossa, S. Taba, J. Muminović, M. Bohn, M. Frisch, A. E. Melchinger, (2006) Grouping of accessions of Mexican races of maize revisited with SSR markers. Theoretical and Applied Genetics 113 (2):177–185
James C (2015) Global status of commercialized biotech crops: 2015. ISAAA Brief No 51. ISAAA, Ithaca
Jiang C, Edmeades GO, Armstead I et al (1999) Genetic analysis of adaptation differences between highland and lowland tropical maize using molecular markers. Theor Appl Genet 99:1106–1119
Johnson EC, Fischer KS, Edmeades GO, Palmer AFE (1986) Recurrent selection for reduced plant height in lowland tropical maize. Crop Sci 26:253–260
Jones PG, Thornton PK (2003) The potential impacts of climate change on maize production in Africa and Latin America in 2055. Glob Environ Chang 13:51–59
Jong SK, Brewbaker JL, Lee CH (1982) Effects of solar radiation on the performance of maize in 41 successive monthly plantings in Hawaii. Crop Sci 22:13–18
Kanampiu F, Diallo A, Burnet M et al (2007) Success with the low biotech of seed-coated imidazolinone-resistant maize. In: Ejeta G, Gressel J (eds) Integrating new technologies for *Striga* control. World Scientific, New Jersey, pp 145–158
Kebede AZ, Melchinger AE, Cairns JE et al (2013) Relationship of line per se and testcross performance for grain yield of tropical maize in drought and well-watered trials. Crop Sci 53:1228–1236
Kumar H (2002) Resistance in maize to the larger grain borer, *Prostephanus truncatus* (horn) (Coleoptera: Bostrichidae). J Stored Prod Res 38:267–280
Kurtz B, Gardner CAC, Millard MJ et al (2016) Global access to maize germplasm provided by the US national plant germplasm system and by US plant breeders. Crop Sci 56:931–941
Lafitte HR, Edmeades GO (1994) Improvement for tolerance to low soil nitrogen in tropical maize II. Grain yield, biomass production, and N accumulation. Field Crops Res 39:15–25
Lafitte HR, Edmeades GO, Taba S (1997) Adaptive strategies identified among tropical maize landraces for nitrogen-limited environments. Field Crops Res 49:187–204
Lambert RJ, Mansfield BD, Mumm RH (2014) Effect of leaf area on maize productivity. Maydica 59:58–64
Lamkey KR, Edwards JW (1999) Quantitative genetics of heterosis. In: Coors JG, Pandey S (eds) The genetics and exploitation of heterosis in crops. ASA, CSSA, Madison, pp 31–48
Lane JA, Child DV, Moore THM et al (1997) Phenotypic characterization of resistance in *Zea diploperennis* to Striga. Maydica 42:45–51
Langyintuo AS, Mwangi W, Diallo AO et al (2010) Challenges of the maize seed industry in eastern and southern Africa: a compelling case for private-public intervention to promote growth. Food Policy 35:323–331

Lobell DB, Bänziger M, Magorokosho C, Vivek B (2011a) Nonlinear heat effects on African maize as evidenced by historical maize yields. Nat Clim Chang 1:42–45. doi:10.1038/NCLIMATE1043

Löffler CM, Wei J, Fast T et al (2005) Classification of maize environments using crop simulation and geographic information systems. Crop Sci 45:1708–1716

Louette D, Smale M (2000) Farmers' seed selection practices and traditional maize varieties in Cuzalapa, Mexico. Euphytica 113:25–41

Lu Y, Hao Z, Xie C, Crossa J, Araus J-L, Gao S, Vivek BS, Magorokosho C, Mugo S, Makumbi D, Taba S, Pan G, Li X, Rong T, Zhang S, Xu Y (2011) Large-scale screening for maize drought resistance using multiple selection criteria evaluated under water-stressed and well-watered environments. Field Crop Res 124(1):37–45

MacRobert JF (2009) Seed business management in Africa. CIMMYT, Harare Zimbabwe

Mahuku G, Lockhart BE, Wanjala B et al (2015) Maize Lethal Necrosis (MLN), an emerging threat to maize-based food security in sub-Saharan Africa. Phytopathology 105:956–965

Makumbi D, Betran J, Bänziger M, Ribaut J-M (2011) Combining ability, heterosis and genetic diversity in tropical maize (*Zea mays* L.) under stress and non-stress conditions. Euphytica 180:143–162

Makumbi D, Diallo A, Kanampiu F et al (2015) Agronomic performance and genotype x environment interaction of herbicide-resistant maize varieties in East Africa. Crop Sci 55:540–555

Masuka B, Araus JL, Das B, Sonder K, Cairns JE (2012) Phenotyping for abiotic stress tolerance in maize. F J Integr Plant Biol 54(4):238–249

Masuka B, Atlin GN, Olsen M et al (2017a) Gains in maize genetic improvement in eastern and southern Africa: I. CIMMYT hybrid breeding pipeline. Crop Sci 57:1–12. doi: 10.2135/cropsci2016.05.0343

Masuka B, Magorokosho C, Olsen M et al (2017b) Gains in maize genetic improvement in eastern and southern Africa: II CIMMYT open-pollinated variety breeding pipeline. Crop Sci 57. doi: 10.2135/cropsci2016.05.0408

Mathenge MK, Smale M, Olwande J (2014) The impacts of hybrid maize seed on the welfare of farming households in Kenya. Food Policy 44:262–271

Matsuoka Y, Vigouroux Y, Goodman MM et al (2002) A single domestication for maize shown by multilocus microsatellite genotyping. PNAS 99:6080–6084

McCann J (2005) Maize and grace- Africa's encounter with a new world crop, 1500–2000. First Harvard University Press, Massachusetts, 289p

Melchinger AE, Brauner PC, Böhm J, Schipprack W (2016) In vivo haploid induction in maize: comparison of different testing regimes for measuring haploid induction rates. Crop Sci 56:1127–1135

Mihm JA (ed) (1997) Insect resistant maize: recent advances and utilization. CIMMYT, Mexico DF

Mir C, Zerjal T, Combes V et al (2013) Out of America: tracing the genetic footprints of the global diffusion of maize. Theor Appl Genet 126:2671–2682. doi:10.1007/s00122-013-2164-z

Monneveux P, Sánchez C, Beck D, Edmeades GO (2006) Drought tolerance improvement in tropical maize source populations: evidence of progress. Crop Sci 46:180–191

Moose SP, Mumm RH (2008) Molecular plant breeding as the foundation for 21st century crop improvement. Plant Physiol 147:969–977

Morris ML, Bellon MR (2004) Participatory plant breeding research: opportunities and challenges for the international crop improvement system. Euphytica 136:21–35

Motto M, Moll RH (1983) Prolificacy in maize: a review. Maydica 28:53–76

Muchow RC, Sinclair TR, Bennett JM (1990) Temperature and solar radiation effects on potential maize yields across locations. Agron J 82:338–343

Nair S, Babu R, Magorokosho C et al (2015) Fine mapping of *Msv1*, a major QTL for resistance to Maize Streak Virus leads to development of production markers for breeding pipelines. Theor Appl Genet 128:1839–1854

Ortega AC (1987) Insect pests of maize: a guide for field identification. CIMMYT, Mexico DF, 106pp

Pandey S, Vasal SK, Deutsch JA (1991) Performance of open-pollinated maize cultivars selected from 10 tropical maize populations. Crop Sci 31:285–290
Pandey S, Gardner CO (1992) Recurrent selection for population, variety, and hybrid improvement in tropical maize. Adv Agron 48:1–87
Pandey S, Narro LA, Friesen DK, Waddington SR (2007) Breeding maize for tolerance to soil acidity. Plant Breed Rev. 28:59–100
Parry MAJ, Andralojc PJ, Mitchell RAC et al (2003) Manipulation of rubisco: the amount, activity, function and regulation. J Exp Bot 54:1321–1333
Passioura JB (1977) Grain yield, harvest index, and water use of wheat. J Aust Inst Agric Sci 43:117–120
Paterniani E (1990) Maize breeding in the tropics. Crit Rev Plant Sci 9:125–144
Pingali P (ed) (2001) CIMMYT 1999/2000 world maize facts and trends, Meeting world maize needs: technological opportunities and priorities for the public sector. CIMMYT, Mexico DF
Pixley KV (2006) Hybrid and open-pollinated varieties in modern agriculture. In: Lamkey KR, Lee M (eds) Plant breeding: the Arnel R Hallauer International Symposium. Blackwell, Iowa, pp 234–250
Prasanna BM (2013) Diversity in global maize germplasm: characterization and utilization. J Biosci 37:1–13
Prasanna BM (2016) Maize lethal necrosis (MLN) in eastern Africa: an update on R4D efforts led by CIMMYT. The African Seed (issue #2, march 2016), pp. 18–21
Prasanna BM, Babu R, Nair S et al (2014) Molecular breeding for tropical maize improvement. In: Wusirika R, Bohn M, Lai J, Kole C (eds) Genetics, genomics and breeding of maize. Science Publishers/CRC Press, Boca Raton, pp 89–118
Prigge V, Sanchez C, Dhillon B et al (2011) Doubled haploids in tropical maize: I. Effects of inducers and source germplasm on in vivo haploid induction rates. Crop Sci 51:1498–1506
Reif JC, Melchinger AE, Xia XC et al (2003) Genetic distance based on simple sequence repeats and heterosis in tropical maize populations. Crop Sci 43:1275–1282
Reif JC, Xia XC, Melchinger AE et al (2004) Genetic diversity determined within and among CIMMYT maize populations of tropical, subtropical, and temperate germplasm by SSR markers. Crop Sci 44:326–334
Reif JC, Warburton ML, Xia XC, Hoisington DA, Crossa J, Taba S, Muminović J, Bohn M, Frisch M, Melchinger AE (2006) Grouping of accessions of Mexican races of maize revisited with SSR markers. Theor Appl Genet 113(2):177–185
van Rensberg JBJ (2007) First report of field resistance by the stem borer *Busseola fusca* (fuller) to Bt-transgenic maize. S Afr J Plant Soil 24:147–151
Salhuana W, Pollak L (2006) Latin American maize project (LAMP) and germplasm enhancement of maize (GEM) project: generating useful breeding germplasm. Maydica 51:339–355
Salvucci M (2008) Association of rubisco activase with chaperonin-60β: a possible mechanism for protecting photosynthesis during heat stress. J Exp Bot 59:1923–1933
Schnable JC, Springer NM, Freeling M (2011) Differentiation of the maize subgenomes by genome dominance and both ancient and on-going gene loss. PNAS 108:4069–4074
Schoper JB, Lambert RJ, Vasilas BL (1987) Pollen viability, pollen shedding, and combining ability for tassel heat tolerance in maize. Crop Sci 27:27–31
Semagn K, Beyene Y, Babu R et al (2014) Quantitative trait loci mapping and molecular breeding for developing stress resilient maize for sub-Saharan Africa. Crop Sci 55:1–11
Serratos JAH (2009) The origin and diversity of maize in the American continent. Greenpeace (eds) (http://www.greenpeace.org/mexico/PageFiles/44856/el-origen-y-la-diversidad-del-2.pdf)
Stevenson JC, Goodman MM (1972) Ecology of exotic races of maize. I. Leaf number and tillering of 16 races under four temperatures and two photoperiods. Crop Sci 12:864–868
Teixeira JEC, Weldikidan T, de Leon N et al (2015) Hallauer's Tusón: a decade of selection for tropical-to-temperate phenological adaptation in maize. Heredity 114:229–240
Tester M, Langridge P (2010) Breeding technologies to increase crop production in a changing world. Science 327:818–822

Thornton PK, Jones PG, Ericksen PJ, Challinor AJ (2011) Agriculture and food systems in sub-Saharan Africa in a 4°C+ world. Phil Trans R Soc A 369:117–136
Tollenaar M, Lee EA (2002) Yield potential, yield stability and stress tolerance. Field Crops Res 75:161–169
Tollenaar M, Lee EA (2011) Strategies for enhancing grain yield in maize. Plant Breed Rev 34:38–82
Tollenaar M, Wu J (1999) Yield improvement in temperate maize is attributable to greater stress tolerance. Crop Sci 39:1587–1604
Tollenaar M, Ahmadzadeh A, Lee EA (2004) Physiological basis of heterosis for grain yield in maize. Crop Sci 44:2086–2094
Trachsel S, Levya M, Lopez M (2016) Identification of tropical maize germplasm with tolerance to drought, nitrogen deficiency, and combined heat and drought stresses. Crop Sci 56:3031–3045. doi:10.2135/cropsci2016.03.0182
Tracy WF, Chandler MA (2006) The historical and biological basis of the concept of heterotic patterns in Corn Belt dent maize. In: Lamkey KR, Lee M (eds) Plant breeding: the Arnel R Hallauer International Symposium. Blackwell, Iowa, pp 219–233
Uhr DV, Goodman MM (1995) Temperate maize inbreds derived from tropical germplasm: II. Inbred yield trials. Crop Sci 35:785–790
Uribelarrea M, Cárcova J, Borras L, Otegui ME (2008) Enhanced kernel set promoted by synchronous pollination determines a tradeoff between kernel number and kernel weight in temperate maize hybrids. Field Crop Res 105:172–181
Vasal SK, Srinivasan G, Gonzalez F et al (1992) Heterosis and combining ability of tropical x subtropical maize germplasm. Crop Sci 32:1483–1489
Vasal SK, Cordova H, Pandey S, Srinivasan G (1999) Tropical maize and heterosis. In: Coors JG, Pandey S (eds) The genetics and exploitation of heterosis in crops. ASA, CSSA, Wisconsin, pp 363–373
Vergara-Díaz O, Zaman-Allah MA, Masuka B, Hornero A, Zarco-Tejada P, Prasanna BM, Cairns JE, Araus JL (2016) A novel remote sensing approach for prediction of maize yield under different conditions of nitrogen fertilization. Front Plant Sci 7:666
Vivek BS, Odongo O, Njuguna J et al (2010) Diallel analysis of grain yield and resistance to seven diseases of 12 African maize (*Zea mays* L) inbred lines. Euphytica 172:329–340
Wangai AM, Redinbaugh MG, Kinyua ZM et al (2012) First report of *Maize chlorotic mottle virus* and maize lethal necrosis in Kenya. Plant Dis 96:1582
Warburton ML, Xia X, Crossa J et al (2002) Genetic characterization of CIMMYT inbred maize lines and open pollinated populations using large scale fingerprinting methods. Crop Sci 42:1832–1840
Warburton ML, Reif JC, Frisch M et al (2008) Genetic diversity in CIMMYT nontemperate maize germplasm: landraces, open pollinated varieties, and inbred lines. Crop Sci 48:617–624
Warburton ML, Wilkes G, Taba S et al (2011) Gene flow among different teosinte taxa and into the domesticated maize gene pool. Genet Res Crop Evol 58:1243–1261
Weber VS, Melchinger AE, Magorokosho C et al (2012) Efficiency of managed-stress screening of elite maize hybrids under drought and low nitrogen for yield under rainfed conditions in southern Africa. Crop Sci 52:1011–1020
Wellhausen EJ, Roberts LM, Hernandez E, Mangelsdorf PC (1952) Races of maize in Mexico: their origin, characteristics and distribution. The Bussey Institute, Harvard University, Cambridge, MA, 223 p
Westgate ME, Bassetti P (1990) Heat and drought stress in corn: what really happens to the corn plant at pollination? In: Wilkinson D (ed) Proc 45th Annual Corn and Sorghum Res. Conf. ASTA, Washington, DC, pp 12–28
Wilkes G (2004) Corn, strange and marvelous: but is a definitive answer known? In: Smith CW, Betrán J, Runge ECA (eds) Corn – origin, history, technology and production. Willey, New Jersey, pp 3–63

Windhausen VS, Wagener S, Magorokosho C et al (2012) Strategies to subdivide a target population of environments: results from the CIMMYT-led maize hybrid testing programs in Africa. Crop Sci 52:2143–2152
Witcombe J, Virk DS, Goyal SN et al (2006) Participatory plant breeding: a market oriented cost-effective approach. In: Lamkey KR, Lee M (eds) Plant breeding: the Arnel R Hallauer international symposium. Blackwell, Oxford, pp 107–119
Woodhouse MR, Schnable JC, Pedersen BS et al (2010) Following tetraploidy in maize, a short deletion mechanism removed genes preferentially from one of the two homeologs. PLoS Biol. dx.doi.org/10.1371/journal.pbio.10000409
Worku M, Zelleke H (2007) Advances in improving harvest index and grain yield of maize in Ethiopia. East Afr J Sci 1:112–119
Yan J, Warburton M, Crouch J (2011) Association mapping for enhancing maize (*Zea mays* L) genetic improvement. Crop Sci 51:433–449
Zaidi PH, Srinivasan G, Sanchez C (2003a) Morpho-physiological traits associated with variable field performance of different types maize germplasm across multiple environments. Maydica 48:207–220
Zaidi PH, Srinivasan G, Sanchez C (2003b) Relationship between line *per se* and cross performance under low N fertility in tropical maize (*Zea mays* L). Maydica 48:221–231
Zaidi PH, Singh NN (2005) Stresses on maize in tropics. Directorate of Maize Research, New Delhi, p 500
Zaidi PH, Yadav M, Maniselvan P et al (2010a) Morpho-physiological traits associated with cold stress tolerance in tropical maize (*Zea mays* L.) Maydica 55:201–208
Zaidi PH, Maniselvan P, Srivastava A et al (2010b) Genetic analysis of water-logging tolerance in tropical maize (*Zea mays* L.) Maydica 55:17–26
Zaidi PH, Vinayan MT, Blummel M (2013) Genetic variability of tropical maize stover quality and the potential for genetic improvement of food-feed value in India. Field Crop Res 153:94–101
Zaman-Allah M, Vergara O, Araus JL et al (2015) Unmanned aerial platform-based multispectral imaging for field phenotyping of maize. Plant Methods 11. doi:10.1186/s13007-015-0078-2
Zaman-Allah M, Zaidi PH, Trachsel S et al (2016) Phenotyping for abiotic stress tolerance in maize – drought stress, A field manual. CIMMYT, Mexico

Chapter 4
Genetic Improvement of Rice (*Oryza sativa* L.)

Kshirod K. Jena and Eero A.J. Nissila

4.1 Introduction

Rice (*Oryza sativa* L.) is the most important cereal among the agricultural crops as it feeds more than three billion people worldwide and is the staple food of over 55% of the population in Asia and Africa (IRRI 2016: http://www.grisp.net/file_cabinet/files/813972/download/RICE%20and%20FPs(%201April%202016).pdf?m=1459997988). Rice is cultivated across a wide range of agroclimatic conditions, from river deltas to mountainous regions and from lower latitudes (35°S) to higher latitudes (53°N) in tropical and temperate countries. It is grown below, at, or near sea level and at an elevation of 2000 m. Of the four major ecosystems in which rice is cultivated and produced, the irrigated rice ecosystem dominates in Asia (about 75% production area), while the rainfed ecosystem dominates in Africa (Khush 2005). A higher demand for rice is predicted for the twenty-first century because of increasing global population (Godfray et al. 2010). A conservative estimate states that an additional 116 million tons of rice needs to be produced by 2035 from the current production of 740 million tons to feed the growing human population (Seck et al. 2012). Keeping in view the impacts of climate change and shrinking arable lands, an annual yield increase of about 1.2–1.5% equivalent to an average yield increase of 0.6 t ha^{-1} is needed worldwide (Seck et al. 2012). The genetic improvement of rice is one of the imperative strategies to achieve this goal.

K.K. Jena (✉)
Plant Breeding Division, International Rice Research Institute,
DAPO Box 7777, 1301 Metro Manila, Philippines
e-mail: k.jena@irri.org

E.A.J. Nissila
Plant Breeding Division, International Rice Research Institute,
DAPO Box 7777, 1301 Metro Manila, Philippines

Newfields Consulting Ltd., 1440 Murronharjuntie, 31950 Punkalaidun, Finland
e-mail: e.nissila@newfields-consulting.com

H. Campos, P.D.S. Caligari, *Genetic Improvement of Tropical Crops*,
DOI 10.1007/978-3-319-59819-2_4

4.2 History of Rice Cultivation and *Oryza* Classification and Domestication

Rice belongs to the genus *Oryza* under the grass family *Poaceae* and has two cultivated species, *O. sativa* and *O. glaberrima*, and 22 wild species (Marathi and Jena 2015). These species are either annual or perennial or diploid or tetraploid and are distributed across the tropical, subtropical, and temperate regions of Asia, Africa, Central and South America, and Australia. *O. sativa*, with its two subspecies (indica and japonica), is widely cultivated in Asia, Africa, Europe, Oceania, and North, Central, and South America, while *O. glaberrima* is mainly cultivated in West Africa (Yoshida 1981; Jena 2010). *O. sativa* was domesticated around 10,000 years ago in the foot hills of the Himalayas in India and southern China, while *O. glaberrima* was domesticated around 5000 years ago in the swampy basins of the upper river delta of Niger in West Africa (Chang 1976). Due to continuous selection, wild progenitors of rice have undergone a series of adaptive changes during domestication, leading to the development of fixed rice lines for cultivation (Ehsan and Eizenga 2014). Following Mendelian genetic principles, rice breeding has been improved using variations in yield-improving traits, leading to increased rice production. However, a narrow genetic variability exists in the improvement and development of high-yielding cultivars because of genetic erosion in the germplasm used.

Using the semi-dwarfism (*sd1*) and photoperiod insensitive traits of Dee-Geo-Woo Gen, the International Rice Research Institute (IRRI) at Los Baños, Laguna, Philippines, developed the first semi-dwarf high-yielding rice variety, IR8. This led to the green revolution in rice production in the 1960s because of the increased yield potential of the developed cultivars. However, rice yield has not increased significantly during the past three decades. The genetic improvement of rice genotypes by applying conventional and modern technologies is, therefore, needed to further increase rice yield.

4.3 Rice Breeding at IRRI and Value-Added Traits for Grain Quality Improvement

To keep up with food requirements of an ever-increasing human population, it is imperative to increase rice production but at the same time improve grain quality to give higher nutritional value. Conventional rice breeding programs apply two phases of breeding: one is evolutionary phase where variable populations are created by the breeders and the second is the evaluation phase where superior genotypes are selected. Genetic variability for agronomic traits is created using pedigree breeding, backcross breeding, and mutation breeding strategies followed by selection and fixation of traits toward development of new cultivars. Initial breeding efforts following green revolution rice development breeders were mainly exploiting

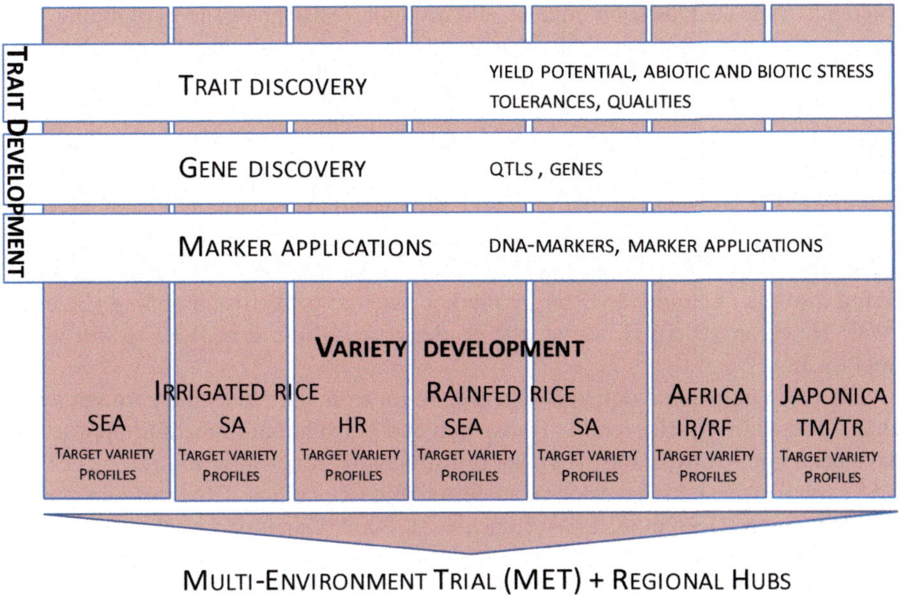

Fig. 4.1 The current plant breeding structure at IRRI. Varietal development groups include irrigated, rainfed, Africa, and japonica rice. Trait development pipeline includes trait discovery, gene discovery, and marker applications. SEA, South East Asia; SA, South Asia; HR, hybrid rice; *TM* temperate japonica; *TR* tropical japonica

cultivated rice germplasm. Breeding program at IRRI is structured based on rice-growing ecosystems to increase indica rice production such as irrigated rice, rainfed lowland rice, abiotic stress-tolerant breeding for drought, submergence, salinity and problem soils, hybrid rice improvement, temperate japonica rice, and multi-environment testing program through international network for genetic evaluation of rice (INGER). Since 2012, rice breeding program has restructured into two development pipelines such as trait development and varietal development. To strengthen variety development pipeline, the trait development pipeline identifies novel gene discoveries and DNA marker-applications. A cross-cutting team has been associated with breeding program to facilitate transfer of new traits into elite breeding lines as well as evaluation of breeding materials for biotic and abiotic stress resistance or tolerance to produce desirable products (Fig. 4.1). The genetic variability for many traits for higher rice production in different environments required exploitation of other gene pools including wild relatives as well as different emerging technologies such as wide hybridization for biotic stress resistance, cytoplasmic diversification for hybrid rice, double-haploid breeding for breeding cycle reduction through another culture and rapid generation advance (RGA), and transgenic rice for novel gene introgression for yield and nutritional quality improvement. Subsequently, rice breeding has been strengthened by increasing selection efficiency using DNA marker-assisted breeding technology, mapping value-added genes for disease and insect resistance, QTL mapping particularly for abiotic stress tolerance traits such as submergence and drought, resistance gene pyramiding, and the use of selected genes with their functions known through map-based gene cloning (Khush and Brar 1998a, b).

Rice grain does not have b-carotene (provitamin A) precursors in its endosperm. For example, millions of poor children in rice farming regions of Asia suffer from vitamin A deficiency as rice is the major staple food in most Asian countries. The so-called golden rice with provitamin A (b-carotene) gene has been developed through genetic transformation of three genes (phytoene synthase, phytoene desaturase, and lycopene cyclase) into rice variety Taipei 309 and later transferred into the rice cultivar IR64 by marker-assisted backcross breeding (Ye et al. 2000; Hefferon 2015). However, the golden rice produced at IRRI is still under field testing.

The other nutritional deficiency genes are for iron and zinc which are not available significantly in rice grain. Transgenic and conventional breeding approaches have been applied. Recently, soybean ferritin gene (*Sferit-1*) for high iron (Fe) and *NAS* gene for zinc (Zn) have been transferred into rice cultivar IR64 through genetic transformation (Goto et al. 1999; Lucca et al. 2002; Trijatmiko et al. 2016). Thus, Fe and Zn content in seeds of transgenic rice has increased to 15 ugg-1 and 45ugg-1, respectively, under flooded field condition compared to normal rice grains (Trijatmiko et al. 2016). Besides the nutritional deficiency gene manipulation in rice, major research activities were undertaken to improve amylose content in rice cultivars suitable for different countries. A great deal of genetic variation in amylose content has been achieved through creating allelic variation in waxy (wx) locus of rice (Khush 2001).

4.4 Genetic Improvement for Yield

Yield is a complex trait which is coordinated by several characteristics of rice. This review focuses on the traits directly associated with rice yield such as the size and number of reproductive organs (Peng et al. 2004). Modern rice cultivars with the semi-dwarf gene *sd1* are associated with an increase in yield as the gene encodes a genetic factor that interferes with the signal transduction pathway of the phytohormone gibberellin (Hedden 2003). Other traits related to enhancing the yield potential of rice are grain number per panicle, panicle size and branching, grain size, days to heading, thousand grain weight, and resistance to lodging. These traits are not controlled by a single genetic effect but are rather controlled by a large number of small effect genes arising from coordinated component traits and underlying genetic predispositions.

These yield-enhancing traits are controlled by quantitative trait loci (QTL). Through QTL fine mapping, several yield-enhancing genes in rice have been identified (Miura et al. 2011; Huang et al. 2013; Liu et al. 2015), six of which are isolated in breeding programs for increasing seed size. The grain width gene, *GW2*, which encodes a RING-type E3 ubiquitin ligase and loss of its function by a 1-bp deletion, increases grain width and yield (Song et al. 2007). The other genes, *GS3* and *qSW5/GW5* for grain size and grain width, respectively, have loss of function mutations to increase yield. However, the genes *GS5* and *OsSPL16/qGW8* control grain size and grain width, respectively (Wang et al. 2012). The thousand grain weight gene (*TGW6*) encodes indole-3 acetic acid (IAA)-glucose hydrolase and regulates grain weight. Loss of function through 1-bp deletion in the coding DNA sequence showed increased grain weight (Ishimaru et al. 2013).

Several key yield-enhancing genes regulating grain number and panicle architecture are of primary importance in modern breeding programs. The grain number gene, *Gn1a*, encodes cytokinin oxidase/dehydrogenase 2 (OSCKX2), and the reduced function of this enzyme enhances the production of the phytohormone cytokinin and increases the grain number (Ashikari et al. 2005). Other genes such as *Ghd7, DEP1,* and *SPIKE1* regulate an increase in grain number per panicle with distinct functional mechanisms. However, an ideal plant architecture (*IPA1*) gene, *OsSPL14/WFP* encoding squamosa promoter-binding protein-like 14 in young panicles, increases panicle branching, thereby increasing grain number per panicle and resulting in higher yield (Miura et al. 2010). The rice cultivars will be improved using modern genomics technologies with a clear understanding of the function of several yield component traits which will help make the cultivars more productive. The lodging resistance gene, *SCM2/APO1*, encoding the F-box-containing protein, controls grain number per panicle and culm diameter. An abundant transcript of the *SCM2* gene causes strong culms, and its expression in developing panicles increases grain number (Ookawa et al. 2010; Zhang et al. 2016).

QTL fine mapping and the cloning and functional characterization of several genes have provided opportunities for plant breeders to correct the weak points of rice cultivars and introduce genes to develop an "ideal plant" and panicle architecture. These plants will have erect leaves for harnessing sufficient solar radiation to increase grain-filling rate, better photosynthesis efficiency, increased grain number per panicle, increased panicle branching, and harvest index. At IRRI, the genetic potential of 12 elite rice cultivars was improved by introducing several QTL-derived functional genes such as *Gn1a, SPL14, SCM2, TGW6,* and *SPIKE1* (Jena K. K., Kim S.R., Ramos J. personal communication). The improved lines exhibited high yield potential, lodging resistance, and ideal plant and panicle architecture (Fig. 4.2). These lines expressed the yield-enhancing genes in different genetic backgrounds that were developed through marker-assisted selection (MAS) and background analysis using 6 K Infinium SNP chips (Kim et al. 2016). The new breeding lines developed through this strategy of a combination of genomics and conventional plant breeding have shown promising results in increasing the yield potential of rice. It is estimated that the yield potential can be increased by 9.09–31.80% due to the coordination of the major yield-enhancing genes that play a major role in the expression of traits in indica rice cultivars (Table 4.1). It is, therefore, necessary to further explore and increase yield potential as well as stabilize the yield of the improved lines through pyramiding several major QTLs/genes contributing similar or additional traits for yield enhancement (Fig. 4.3a and b).

Fig. 4.2 Breeding lines and varieties showing improved plant architecture, increased panicle branching, and grain number

Table 4.1 Genetically improved breeding lines showing superior traits and yield gain over NSIC Rc238 in 2015 WS and 2016 DS

Variety/breeding lines	Possible gene(s)	Primary branching number (No.)		Secondary branching number (No.)		Grain numbers per panicle (No.)		1000 grain wt (g)		Yield gain over NSIC Rc238[a] (%)	
		WS	DS	WS	DS	WS	DS	WS	DS	WS	DS
NSIC Rc238 (check)	*SCM2, Ghd7*	11	10	24	38	238	172	22	22	–	–
IR 118289-4-56-78	*Gn1a, SPL14, SCM2, Ghd7*	16	14	72	65	282	254	28	29	27.30	31.80
IR 118289-3-2-8	*SPL14, SCM2, Ghd7*	17	17	65	63	277	258	25	27	13.63	22.72
IR 111662-3-42-1	*Gn1a, SPL14, SCM2, Ghd7*	18	16	72	91	324	355	25	25	13.63	13.63
IR 118290-4-78-B	*Gn1a, SPL14, SCM2, Ghd7*	21	19	79	107	363	372	25	24	13.63	9.09

[a]*DS* dry season, *WS* wet season

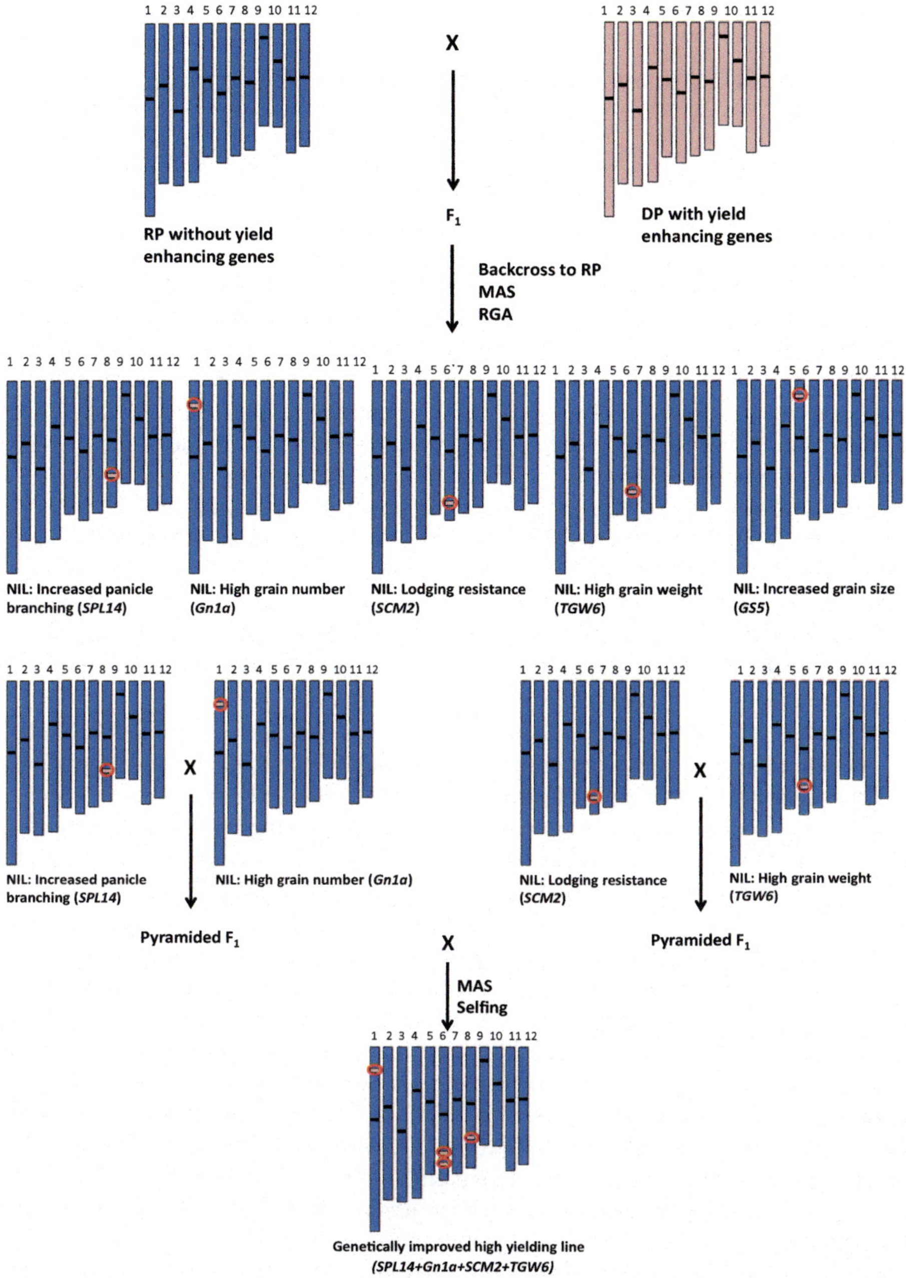

Fig. 4.3 Strategy for the genetic improvement of rice yield. (**a**) Scheme for the development of near-isogenic lines (NILs) for various yield-enhancing genes (*SPL14*, *Gn1a*, *SCM2*, *TGW6*, and *GS5*) through several backcrossing to recurrent parent (RP) by marker-assisted selection. (**b**) Marker-assisted pyramiding genes using NILs with yield-enhancing loci. Color coding is as follows: *black rectangles* for centromere and *pink* for yield-enhancing loci/genes. *RP* recurrent parent; *DP* donor parent; *MAS* marker-assisted selection; *RGA* rapid generation advance; *SPL14* squamosa promoter-binding protein-like 14; *Gn1a* grain number; *SCM2* strong culm mutant 2; *TGW6* thousand grain weight 6; *GS5* grain size 5

4.5 Genetic Improvement for Biotic Stress Resistance

The production and productivity of rice crops worldwide are affected by a diverse set of bacterial, fungal, and viral pathogens which cause diseases such as bacterial leaf blight (BB), bacterial leaf streak, bacterial leaf scald, brown spot, leaf blast, neck blast, sheath blight, and tungro virus. Of these diseases, the two major diseases are BB and leaf blast which affect the foliar tissues of the plant at the vegetative stage. These are vascular diseases that spread through the stomata and xylem vessels of the leaves. The bacterial pathogen *Xanthomonas oryzae* pv. *oryzae* (*Xoo*) causes BB at the maximum vegetative growth stage, while the fungal pathogen *Magnaporthe oryzae* causes leaf blast at the vegetative and reproductive stages. BB mainly affects rice crops in irrigated ecosystems. These two diseases are of high economic importance due to their severity in rice fields in the tropics as well as temperate regions worldwide. A conservative estimate of yield loss due to BB is pegged at 40–80% every year (Jeung et al. 2006; Srinivasan and Gnanamanickam 2005; Kim et al. 2015). On the other hand, the fungal blast disease kills enough rice that could feed 60 million people each year and, in severe cases, completely destroys the rice crop (Zeigler et al. 1994; Jeung et al. 2007). The need for more research on BB disease resistance and the development of new cultivars that are resistant to BB are, therefore, of high importance.

Breeding for biotic stress resistance is very important for rice production to be sustainable. Biotic stresses caused by the changing climate which bring new pests and diseases and the evolution of pathogens and insects which seriously damage the rice crop as a result of the breakdown of resistance in elite cultivars. Although a number of native resistance genes have been transferred into cultivars, most of the developed cultivars are still susceptible to new pathotypes and biotypes of diseases and insects. Several genes conferring broad-spectrum resistance to major diseases and insects which are fixed in the background of elite indica cultivars are available (Table 4.2). Of the 60 blast resistance genes identified to date, only the genes *Pi2*, *Pi9,* and *Pi40* confer broad-spectrum and durable resistance to virulent blast pathotypes (Jeung et al. 2007; Suh et al. 2009). At IRRI, breeding lines with pyramided blast resistance genes have been developed which confer resistance to diverse pathotypes of *Magnaporthe oryzae* (Jena K.K. and Hechanova S.L. personal communication). Breeding lines with strong blast resistance should be used in MAS to incorporate resistance genes into different elite cultivars.

Majority of the 40 BB resistance genes identified so far in rice cultivars are becoming susceptible to new pathotypes or races of BB, one of which is the *Xa21* gene which confers a broad-spectrum of resistance to different BB races (Ronald et al. 1992; Kim et al. 2015). The *Xa21* gene has also been pyramided with the genes *Xa4* and *xa5* in the background of IR72 that also confers a broad-spectrum of resistance to BB races. The new *Xa40* gene also confers resistance to a broad-spectrum of BB races (Kim et al. 2015). These genes and their pyramids can be transferred into different elite rice cultivars for genetic improvement and for the development of high-yielding rice varieties. Since only a few genes have a broad-

Table 4.2 Potential genes/QTLs conferring broad-spectrum resistance to biotic and abiotic stresses with their chromosome location and function

Biotic/abiotic stresses	Genes/QTL	Chr. No.[a]	Function/encoded protein	References
Bacterial blight	*xa5*	5	Gamma subunit of transcription factor IIA	Iyer-Pascuzzi and McCouch 2007
	Xa21	11	Receptor kinase-like protein carrying leucine-rich repeat motif and a serine-threonine kinase-like domain	Song et al. 1995
	Xa40	11	WAK 3	Kim et al. 2015
Blast	*Pi40*	6	Nucleotide-binding site leucine-rich repeat	Jeung et al. 2007
Brown planthopper	*Bph18*	12	CC-NBS-NBS-LRR	Ji et al. 2016
Drought	*DRO1*	9	Conserved hypothetical protein, *deeper rooting* 1	Uga et al. 2013
	$qDTY_{12.1}$	12	Transcription factor "no apical meristem" and its co-localized target genes	Dixit et al. 2015
Salinity	*SKC1*	1	High-affinity K^+ transporter	Ren et al. 2005
Submergence	*Sub1a*	9	Ethylene response factor-related protein	Xu et al. 2006

[a]*Chr* chromosome

spectrum of resistance to BB like these genes, additional efforts are needed to improve rice cultivars with durable resistance genes. The pyramiding effect of different BB resistance genes and the identified breeding lines conferring durable resistance to BB have been developed (Suh et al. 2013).

About 31 brown planthopper (BPH) resistance genes have been identified to date, most of which are biotype-specific. However, only a few resistance genes such as *Bph3*, *Bph18*, *Bph20*, *Bph21,* and *Bph31* have broad-spectrum resistance to different biotypes, but only three of these five genes (*Bph18*, *Bph20,* and *Bph21*) confer durable and stable resistance to different BPH biotypes (Rahman et al. 2009; Jena and Kim 2010; Suh et al. 2011; Ji et al. 2016). Near-isogenic lines (NILs) of 11 BPH resistance genes and their pyramids in the IR24 background have been developed, and bioassays of several BPH populations of the Philippines have been carried out. NILs with strong resistance to different BPH populations have also been identified (Jena et al. 2017). These BPH NILs are important resources to monitor resistance genes against BPH populations affecting rice cultivars, and breeders can incorporate target resistance genes into elite varieties.

Several genes having strong resistance to BB, leaf blast, and BPH have been pyramided in the japonica genetic background, and the resulting lines confer strong resistance to these diseases and insects (Suh et al. 2015). These breeding lines can be used as new sources of resistance to insects and diseases in resistance breeding programs (Fig. 4.4).

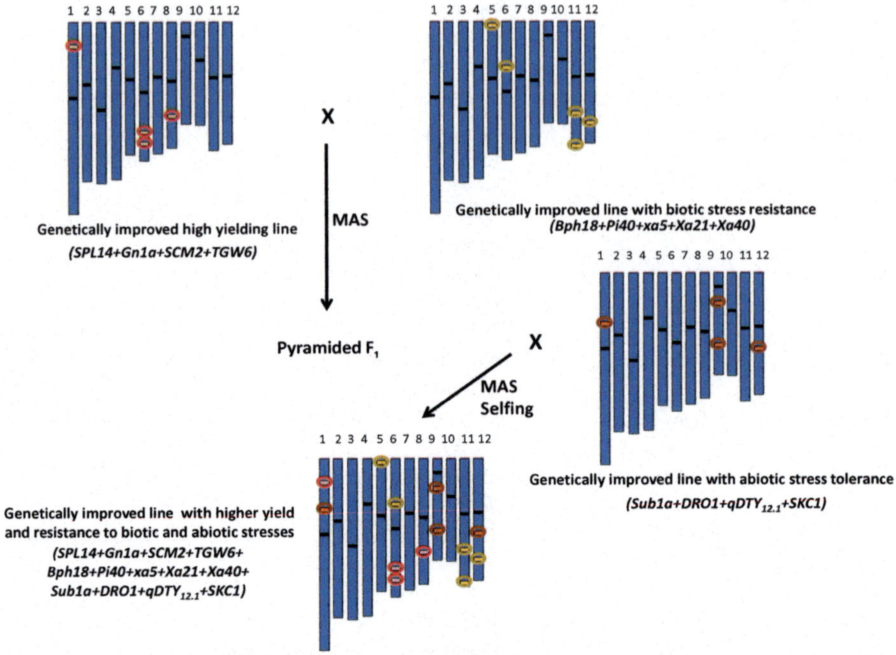

Fig. 4.4 Strategy for the genetic improvement of rice yield with pyramiding of multiple genes (*SPL14* + *Gn1a* + *SCM2* + *TGW6*+ *Bph18* + *Pi40* + *xa5* + *Xa21* + *Xa40*+ *Sub1a* + *DRO1* + *qDTY*$_{12.1}$ + *SKC1*) conferring resistance to biotic and tolerance to abiotic stresses. Color coding is as follows: *black rectangles* for centromere, *pink* for yield-related loci/genes, *yellow* for biotic stress resistance genes, and *orange* for abiotic stress tolerance genes. *MAS* marker-assisted selection

4.6 Genetic Improvement for Abiotic Stress Tolerance

In addition to problematic soil conditions, rice production is impaired in adverse ecosystems by drought, salinity, and submergence. Improving the genetic architecture of the rice plant will enable it to tolerate these abiotic stresses. Advances in molecular genetics and biotechnology have resulted in the identification of genes and gene-based markers for tolerance to drought, salinity, and submergence (Table 4.2). These genes can be introduced into different elite cultivars to develop stress-tolerant cultivars with higher rice production (Fig. 4.4).

Major QTLs with large genetic effects (32–33% genetic variation) for grain yield under drought stress are available (Bernier et al. 2007). QTL fine-mapping studies have identified multiple intra-QTL genes for the full impact of the QTL *qDTY12.1* on yield under drought stress (Dixit et al. 2015). These genes can be transferred efficiently by MAS into drought-sensitive cultivars that could produce higher yields under drought stress environments. Root trait is one of the important discriminating factors for the genetic improvement of drought resistance in rice. The gene *DRO1* which controls the alteration of the root system architecture to improve drought avoidance has been cloned and characterized (Uga et al. 2013). This gene can be introduced into drought-sensitive rice cultivars by MAS approach, allowing the development of new high-yielding cultivars with drought tolerance.

The mechanism of salt tolerance in crops is well established (Yamaguchi and Blumwald 2005). Lowering the Na^+ concentration in the plant tissue and decreasing Na^+/K^+ helps increase the plant's tolerance to salt stress. Salt tolerance is a complex trait and is controlled by QTLs, which have been identified and fine mapped in different land races of rice. A major QTL, *SKC1* derived from a land race cultivar (Nona Bokra) for shoot K^+ concentration in seedlings, has been identified. Fine mapping of the QTL locus revealed that the *SKC1* gene encodes an HKT-type transporter (Lin et al. 2004; Ren et al. 2005), suggesting that it contributes to the Na^+ and K^+ translocation probably by unloading Na^+ from the xylem tissue. It is necessary to make genetic changes in elite rice cultivars by introducing new genes or QTL pyramids to develop salt-tolerant lines for cultivation under salt stress environments. Besides q*Saltol*, other novel QTLs are reported from different sources of salinity tolerance that are quite effective against seedling stage salinity stress (Bizimana et al. 2017). Rice responds differentially to salinity at different stages of plant growth and development. The seedling and reproductive stages are the growth stages most sensitive to salinity stress with very weak association, suggesting that they are regulated by different processes and sets of genes/QTLs (Singh and Flowers 2010; Mohammadi et al., 2013; Hossain et al., 2015). Therefore, to breed a good rice variety, it needs to be screened for both sensitive growth stages. IRRI has standardized phenotyping technique exclusively for the reproductive stage salinity tolerance that is being followed by the researchers as reliable technique for QTL mapping studies (Ahmadizadeh et al., 2016).

Rice production in a rainfed ecosystem is often affected by flash flooding at the vegetative stage, damaging completely the whole rice crop due to lack of submergence tolerance especially in modern cultivars. Some traditional indica rice cultivars like FR13A can survive for 2 weeks under submergence. Using the QTL fine-mapping strategy, a major QTL, *Sub1,* which is linked to submergence tolerance, has been identified (Xu et al. 2000). Map-based cloning of this gene revealed that *Sub1* encodes putative ethylene response factors (ERFs), and the responsible gene has been named as *Sub1A* (Xu et al. 2006). Ectopic expression of the *Sub1A-1* gene in an intolerant variety increases submergence tolerance without yield penalty. Even though the genetic mechanism of submergence tolerance is still unclear, the introduction of the *Sub1A-1* gene into different elite cultivars can increase rice production under submergence stress. Currently, Sub1 varieties are developed and deployed in several countries such as India, Indonesia, Bangladesh, Nepal, and the Philippines. Another gene for anaerobic germination (AG1) has been identified and cloned (Kretzschmar et al. 2015). Promising breeding lines are also developed with tolerance to submergence and anaerobic germination.

4.7 Utilization of New Isogenic Lines/Populations for Breeding Applications

At IRRI, ideal populations such as NILs, chromosome segment substitution lines (CSSL), recombinant breeding lines (RILs), backcross inbred lines (BILs), and multi-parent advance genetic intercross (MAGIC) lines have been developed to precisely identify traits contributing to higher yield potential and resistance/tolerance to major biotic and abiotic stresses (Ramos et al. 2016; Ali et al. 2010; Bandilo et al. 2013). These modern technologies will aid precision breeding for traits/genes needed for the genetic improvement of rice.

4.8 Enhancing the Cultivated Rice Genome with Novel Genes from Wild *Oryza* Species

Rice has evolved as a model crop species, and the availability of whole-genome sequencing has accelerated research on genetic improvement for resistance against biotic and abiotic stresses. Additionally, the sequences of genomes of different wild species of *Oryza* have been completed to utilize novel genes/traits associated with rice improvement (Wing R, Amiraju J, Kudrana D, Jena K.K. personal communication). There are ten different genomes assigned to distantly related species of *Oryza* besides the AA genome of cultivated rice and the six wild species (Marathi et al. 2015). The wild species of *Oryza* are rich sources of genes that control traits of agronomic importance which are rare in cultivated rice. This makes it imperative to explore the genes from the wild species, which were lost during the early domestication process, and to transfer those value-added genes into cultivated rice. Some of these genes have been identified, genetically studied, characterized, and pyramided in the background of rice cultivars (Fig. 4.3). These include traits such as resistance to BPH, BB, and blast diseases (Ji et al. 2016: Du et al. 2009: Zhou et al. 2006; Jena et al. 2006; Jeung et al. 2007, Kim et al. 2015) and tolerance to major abiotic stresses such as drought, submergence, salinity, and acid sulfate soil (Jena 2010). Monosomic alien addition lines have also been developed, and value-added traits inherited from distantly related wild species genomes have been transferred into rice cultivars (Jena et al. 2016). A number of CSSLs having yield-enhancing loci such as early heading, panicle length, grain number per panicle, and grain sizes derived from the AA genome of the wild species *O. longistaminata* have been developed as well (Ramos et al. 2016). It is essential to transfer these genes in the genetic background of some elite cultivars to further improve rice production and productivity.

4.9 Future Prospects and Conclusions

Advances in molecular genetics and genomics, coupled with improved conventional breeding technologies based on DNA marker-assisted breeding, are becoming vital approaches in unraveling the intricacies of the rice genome and helping breeders to apply these technologies for rice genetic improvement, ultimately leading to increased production. Although several complex traits are associated with higher genetic yield potential of rice, it has become possible to dissect these traits such as number of grains per panicle, panicle branching, grain size, and resistance to lodging. These traits have been incorporated into less productive cultivars. The development of NILs and QTL cloning has increased the understanding on the function of yield-enhancing genes which can be incorporated into cultivars. Thus, the genetic improvement of rice could be achieved through MAS or genetic transformation for selected yield-enhancing traits such as grain number per panicle and panicle branching (Xing and Zhang 2010). As this chapter has reported, some indica rice cultivars have been identified with improved yield potential due to yield-enhancing genes (Kim et al. 2016; Jena KK, Kim SR, Ramos J personal communication). These breeding lines can be combined further with several biotic and abiotic stress resistance genes to develop genetically stable rice cultivars with high yield potential (Fig. 4.4).

In spite of advances in genomics, however, there is a knowledge gap on the use of appropriate genes for improving different traits. Many tools and novel genetic stocks underlying genes of functions are available (Zhang et al. 2008), but more understanding is needed on the source and sink relationship in rice, leading to the systematical manipulation of value-added traits to provide ideal sink and source traits to the biological productivity of the rice plant. It is imperative, therefore, to explore novel genetic variations from a diverse collection of germplasm for traits related to photosynthesis efficiency, including rice plant architecture, under different environmental conditions. It is also crucial to make genetic improvements on the rice plant by manipulating value-added yield component traits and designing the ideal plant architecture to make the plant stable under different ecosystems and efficiently utilize nitrogen and solar energy. These efforts are all geared toward developing the ability of the rice cultivars to resist multiple biotic stresses and to produce higher yield for global food security. To overcome the problem of nutrition deficiency in malnourished children in rice-growing countries of Asia, IRRI has developed transgenic lines with higher grain nutrition contents such as provitamin A, iron, and zinc. Additional experiments on field trials of these traits and their bioavailability are needed before making these products available for consumption.

Acknowledgment We are grateful to the Global Rice Science Partnership (GRiSP) program of IRRI for supporting the writing of this review article. We thank Dr. R. K. Singh, rainfed lowland rice breeder, IRRI, for providing information on breeding structure and salinity tolerance. We like to thank Ms. Joie Ramos, associate breeder, and Ricky Vinarao, assistant breeder of the Novel Gene Resources Team of the Plant Breeding Division for developing the graphics of the traits for genetic improvement of rice. We thank to the editorial team of IRRI Communications for their excellent editing the review article.

References

Ahmadizadeh M, Vispo NA, Calapit-Palao CDO, Pangaan ID, Dela Viña C, Singh RK (2016) Reproductive stage salinity tolerance in rice: a complex trait to phenotype. Ind J Plant Physio (Springer) 21:528. https://doi.org/10.1007/s40502-016-0268-6

Ali ML, Sanchez PL, SB Y, Lorieux M, Eizenga GC (2010) Chromosome segment substitution lines: a powerful tool for the introgression of valuable genes from *Oryza* wild species into cultivated rice (*O. sativa*). Rice 3:218–234

Ashikari M, Sakakibara H, Lin S, Yamamoto T, Takashi T, Nishimura A, Angeles ER, Qian Q, Kitano H, Matsuoka M (2005) Cytokinin oxidase regulates rice grain production. Science 309:741–745

Bandilo N, Raghavan C, Muyco PA, Sevilla MAL, Lobina IT, Dilla-Ermita CJ, Tung CW, McCouch S, Thomson M, Mauleon R, Singh RK, Gregorio G, Redona E, Leung H (2013) Multi-parent advanced generation inter-cross (MAGIC) populations in rice: progress and potential for genetics research and breeding. Rice 6:11

Bernier J, Kumar A, Ramaiah V, Spaner D, Atlin G (2007) A large-effect QTL for grain yield under reproductive-stage drought stress in upland rice. Crop Sci 47:507–516

Bizimana JB, Luzi Kihupi A, Murori RW, Singh RK (2017) Identification of quantitative trait loci for salinity tolerance in rice (*Oryza sativa L.*) using IR29/Hasawi mapping population. J Genet (accepted)

Chang TT (1976) The origin, evolution, cultivation, dissemination and diversification of Asian and African rices. Euphytica 25:435–441

Dixit S, Biswal AK, Min A, Henry A, Oane RH, Raorane ML, Longkumer T, Pabuayon IM, Mutte SK, Vardarajan AR, Miro B, Govindan G, Albano-Enriquez B, Pueffeld M, Sreenivasulu N, Slamet-Loedin I, Sundarvelpandian K, Tsai YC, Raghuvanshi S, Hsing YI, Kumar A, Kohli A (2015) Action of multiple intra-QTL genes concerted around a co-localized transcription factor underpins a large effect QTL. Nat Sci Rep 5:15183–15189

Du B, Zhang W, Liu B, Hu J, Wei Z, Shi Z et al (2009) Identification and characterization of *Bph14*, a gene conferring resistance to brown planthopper in rice. Proc Natl Acad Sci 106:22163–22168

Ehsan S, Eizenga GC (2014) Unraveling the secrets of rice wild species. In: Wengui Yan, Jinsong Bao (eds) Agricultural and biological sciences "rice germplasm, genetics and improvement", ISSN 978–953–51-1240-2. https://doi.org/10.5722/58393

Godfray HCJ, Beddington JR, Crute IR, Haddad L, Lawrence D, Muir JF, Pretty J, Robinson S, Thomas SM, Toulmin C (2010) Food security: the challenge of feeding 9 billion people. Science 327:812–818

Goto F, Yoshihara T, Shigemoto N, Toki S, Takaiwa F (1999) Iron nutrition of rice seed by the soybean ferritin gene. Nat Biotech 17:282–286

Hedden P (2003) The genes of the green revolution. Trends Genet 19:5–9

Hefferon KL (2015) Nutritionally enhanced food crops: progress and prospectives. Int J Mol Sci 16:3895–3914

Hossain H, Rahman MA, Alam MS, Singh RK (2015) Mapping of quantitative trait loci associated with reproductive-stage salt tolerance in rice. J Agron Crop Sci 201:17–31

Huang R, Jiang L, Zheng J, Wang T, Wang H, Huang Y, Hong Z (2013) Genetic bases of rice grain shape: so many genes, so little known. Trends Plant Sci 18:218–226

Ishimaru K, Hirotsu N, Madoka Y, Murakami N, Hara N, Onodera H, Kashiwagi T, Ujiie K, Shimizu B, Onishi A, Miyagawa H, Katoh E (2013) Loss of function of the IAA-glucose hydrolase gene TGW6 enhances rice grain weight and increases yield. Nat Genet 45(6):707–711

Iyer-Pascuzzi AS, McCouch SR (2007) Recessive resistance genes and the *Oryza sativa-Xanthomonas oryzae* pv. *oryzae* pathosystem. Mol Plant-Microbe Interact 20:731–739

Jena KK (2010) The species of the genus *Oryza* and transfer of useful genes from wild species into cultivated rice, *O. sativa*. Breed Sci 60:518–523

Jena KK, Kim SR (2010) Current status of brown planthopper (BPH) resistance and genetics. Rice 3:161–171

Jena KK, Jeung JU, Lee JH, Choi HC, Brar DS (2006) High-resolution mapping of a new brown planthopper (BPH) resistance gene *Bph18(t)*, and marker-assisted selection for BPH resistance in rice (*Oryza sativa* L.) Theor Appl Genet 112:288–297

Jena KK, Ballesfin ML, Vinarao RB (2016) Development of *Oryza sativa* L. by *Oryza punctata* Kotschy ex Steud. Monosomic addition lines with high value traits by interspecific hybridization. Theor Appl Genet 129:1873–1886

Jena KK, Hechanova SL, Verdeprado H, Prahalada GD, Kim SR (2017) Development of 25 near-isogenic lines (NILs) with ten BPH resistance genes in rice (*Oryza sativa* L.): production, resistance spectrum and molecular analysis. Theor Appl Genet (in press) https://doi.org/10.1007/s00122-017-2963-8

Jeung JU, Heu SG, Shin MS, Vera Cruz CM, Jena KK (2006) Dynamics of *Xanthomonas oryzae* pv. *oryzae* populations in Korea and their relationship to known bacterial blight resistance genes. Phytopathology 96:867–875

Jeung JU, Kim BR, Cho YC, Han SS, Moon HP, Lee YT, Jena KK (2007) A novel gene, *Pi40(t)*, linked to the DNA markers derived from NBS-LRR motifs confers broad spectrum of blast resistance in rice. Theor Appl Genet 115:1163–1177

Ji H, Kim SR, Kim YH, Suh JP, Park HM, Sreenivasulu N, Misra G, Kim SM, Hechanova SL, Kim H, Lee GS, Yoon UH, Kim TH, Lim H, Suh SC, Yang J, An G, Jena KK (2016) Map-based cloning and characterization of the *BPH18* gene from wild rice conferring resistance to brown planthopper (BPH) insect pest. Sci Rep 6:34376

Khush GS (2001) Challenges for meeting the global food and nutrient needs in the new millennium. Proc Nutr Sci 60:15–26

Khush GS (2005) What it will take to feed 5.0 billion rice consumers in 2030. Plant Mol Biol 59:1–6

Khush GS, Brar DS (1998a) The application of biotechnology to rice. In: Ives C, Bedford B (eds) Agricultural biotechnology in international development. CAB International, Walingford, pp 92–121

Khush GS, Brar DS (1998b) New tools to increase the efficiency of both phases. In: Biotechnology for rice breeding: progress and impact. Part III: Progress in rice genetic improvement for food security. Proceedings of the 20th session of International Rice Commission. FAO, Bangkok, Thailand. pp. 306

Kim SM, Suh JP, Qin Y, Noh TH, Reinke RF, Jena KK (2015) Identification and fine-mapping of a new resistance gene, *Xa40*, conferring resistance to bacterial blight races in rice (*Oryza sativa* L.) Theor Appl Genet 128:1933–1943

Kim SR, Ramos J, Ashikari M, Virk PS, Torres EA, Nissila E, Hechanova SL, Mauleon R, Jena KK (2016) Development and validation of allele-specific SNP/indel markers for eight yield-enhancing genes using whole-genome sequencing strategy to increase yield potential of rice, *Oryza sativa* L. Rice 9:12

Kretzschmar T, Pelayo MAF, Trijatmiko KR, Gabunada LFM, Alam R, Jimenez R, Mendioro MS, Slamet-Loedin IH, Sreenivasulu N, Bailey-Serres J, Ismail AM (2015) A trehalose-6-phosphate phosphatase enhances anaerobic germination tolerance in rice. Nat Plants 1:15124

Lin HX et al (2004) QTL for Na and K uptake of the shoot and root controlling rice salt tolerance. Theor Appl Genet 108:253–260

Liu L, Tong H, Xiao Y, Che R, Xu F, Hu B, Liang C, Chu J, Li J, Chu C (2015) Activation of big grain 1 significantly improves grain size by regulating auxin transport in rice. Proc Natl Acad Sci 112:11102–11107

Lucca P, Hurrell R, Potrykus I (2002) Genetic engineering approaches to improve the bioavailability and the level of iron in rice grains. Theor Appl Genet 102:392–397

Marathi B, Jena KK (2015) Floral traits to enhance out-crossing for higher hybrid seed production in rice: present status and future prospects. Euphytica 201:1–14

Marathi B, Ramos J, Hechanova SL, Oane RH, Jena KK (2015) SNP genotyping and characterization of pistil traits revealing a distinct phylogenetic relationship among the species of *Oryza*. Euphytica 201:131–148

Miura K, Ikeda M, Matsubara A, Song XJ, Ito M, Asano K, Matsuoka M, Kitano H, Ashikari M (2010) OsSPL14 promotes panicle branching and higher grain productivity in rice. Nat Genet 42(6):545–549

Miura K, Ashikari M, Matsuoka M (2011) The role of QTLs in the breeding of high-yielding rice. Trends Plant Sci 16(6):319–326

Mohammadi R, Mendioro MS, Diaz GQ, Gregorio GB, Singh RK (2013) Mapping quantitative trait loci associated to yield and yield components under reproductive stage salinity stress in rice (*Oryza sativa* L.) J Genet 92(3):433–443

Ookawa T, Hobo T, Yano M, Murata K, Ando T, Miura H, Asano K, Ochiai Y, Ikeda M, Nishitani R, Ebitani T, Ozaki H, Angeles ER, Hirasawa T, Matsuoka M (2010) New approach for rice improvement using a pleiotropic QTL gene for lodging resistance and yield. Nat Comms 1:132

Peng S, Huang J, Sheehy JE, Laza RC, Visperas RM, Zhong X, Centeno GS, Khush GS, Cassman KG (2004) Rice yields decline with higher night temperature from global warming. Proc Natl Acad Sci 101:9971–9975

Rahman ML, Jiang W, Chu SH, Qiao Y, Ham TH, Woo MO, Lee J, Khanam MS, Chin JH, Jeung JU, Brar DS, Jena KK, Koh HJ (2009) High-resolution mapping of two rice brown planthopper resistance genes *Bph20(t)* and *Bph21(t)*, originating from *Oryza minuta*. Theor Appl Genet 119:1237–1246

Ramos JM, Furuta T, Uehara K, Chihiro N, Angeles-Shim RB, Shim J, Brar DS, Ashikari M, Jena KK (2016) Development of chromosome segment substitution lines (CSSLs) of *Oryza longistaminata* A. Chev. & Rohr in the background of the elite japonica rice cultivar, Taichung 65 and their evaluation for yield traits. Euphytica 210:151–163

Ren ZH, Gao JP, Li LG, Cai XL, Huang W, Chao DY, Zhu MZ, Wang ZY, Luan S, Lim HX (2005) A rice quantitative trait locus for salt tolerance encodes a sodium transporter. Nat Genet 37:1141–1146

Ronald PC, Albano B, Tabien R, Abenes L, Wu K, McCouch S, Tanksley SD (1992) Genetic and physical analysis of rice bacterial blight resistance locus, *Xa21*. Mol Gen Genet 236:113–120

Seck PA, Diagne A, Mohanty S, Wooperis MCS (2012) Crops that feed the world. Food Sec 4:7–24

Singh RK, Flowers TJ (2010) The physiology and molecular biology of the effects of salinity on rice. In: Pessarakli M (ed) Handbook of plant and crop stress, 3rd edn. Taylor and Francis, Florida, pp 899–939

Song WY, Wang GL, Chen LL, Kim HS, Pi LY, Holsten T, Gardner J, Wang B, Zhao WX, Zhu LH, Fauquet C, Ronald P (1995) A receptor kinase-like protein encoded by the rice disease resistance gene, *Xa21*. Science 270:1804–1806

Song XJ, Huang W, Shi M, Zhu MZ, Lin HX (2007) A QTL for rice grain width and weight encodes a previously unknown RING-type E3 ubiquitin ligase. Nat Genet 39:623–630

Srinivasan B, Gnanamanickam S (2005) Identification of a new source of resistance in wild rice, *Oryza rufipogon*, to bacterial blight of rice caused by Indian strains of *Xanthomonas oryzae* pv. *oryzae*. Curr Sci 88:25

Suh JP, Noh TH, Kim KY, Kim JJ, Kim YG, Jena KK (2009) Expression levels of three bacterial blight resistance genes against K3a race of Korea by molecular and phenotype analysis in japonica rice (*O. sativa* L.) J Crop Sci Biotechnol 12(3):103–108

Suh JP, Yang SJ, Jeung JU, Pamplona A, Kim JJ, Jong-Hee Lee JH, Hong HC, Yang CI, Kim YG, Jena KK (2011) Development of elite breeding lines conferring Bph18 gene-derived resistance to brown planthopper (BPH) by marker-assisted selection and genome-wide background analysis in japonica rice (*Oryza sativa* L.) Field Crop Res 120:215–222

Suh JP, Jeung JU, Noh TH, Cho YC, Park SH, Park HS, Shin MS, Kim CK, Jena KK (2013) Development of breeding lines with three pyramided resistance genes that confer broad-spectrum bacterial blight resistance and their molecular analysis in rice. Rice 6:5

Suh JP, Cho YC, Won YJ, Ahn EK, Baek MK, Kim MK, Kim BK, Jena KK (2015) Development of resistant gene-pyramided *Japonica* rice for multiple biotic stresses using molecular marker-assisted selection. Plant Breed Biotech 3:333–345

Trijatmiko KR, Dueñas C, Tsakirpaloglou N, Torrizo L, Arines FM, Adeva C, Balindong J, Oliva N, Sapasap MV, Borrero J, Rey J, Francisco P, Nelson A, Nakanishi H, Lombi E, Tako E,

Glahn RP, Stangoulis J, Chadha-Mohanty P, Johnson AAT, Tohme J, Barry G, Slamet-Loedin IH (2016) Biofortified indica rice attains iron and zinc nutrition dietary targets in the field. Sci Rep 6:19792

Uga Y, Sugimoto K, Ogawa S, Rane J, Ishitani M, Hara N, Kitomi Y, Inukai Y, Ono K, Kanno N, Inoue H, Takehisa H, Motoyama R, Nagamura Y, Wu J, Matsumoto T, Takai T, Okuno K, Yano M (2013) Control of root system architecture by *DEEPER ROOTING 1* increases rice yield under drought conditions. Nat Genet 45:1097–1102

Wang S, Wu K, Yuan Q, Liu X, Zheng R, Zhu H, Dong G, Qian Q, Zhang Z, Fu X (2012) Control of grain size, shape, and quality by OsSpl16 in rice. Nat Genet 44:950–954

Xing Y, Zhang Q (2010) Genetic and molecular bases of rice yield. Annu Rev Plant Biol 61:421–442

Xu K, Xu X, Roland PC, Mackill DJ (2000) A high-resolution linkage map of the vicinity of the rice submergence tolerance locus Sub1. Mol Gen Genet 263:681–689

Xu K, Xu X, Fukao T, Canlas P, Maghirang-Rodriguez R, Heuer S, Ismail AM, Bailey-Serres J, Ronald PC, Mackill DJ (2006) 'Sub1A' is an ethylene-response factor-like gene that confers submergence tolerance to rice. Nature 442(10):705–708

Yamaguchi T, Blumwald E (2005) Developing salt tolerant crop plants: challenges and opportunities. Trends Plant Sci 10(12):615–620

Ye X, Al-Babili S, Kloti A, Zhang A, Lucca P, Beyer P, Potrykus I (2000) Engineering the provitamin A (b-carotene) biosynthetic pathway into (carotenoid-free) rice endosperm. Science 287:303–305

Yoshida S (1981) Fundamentals of rice crop science. The International Rice Research Institute, Los Baños, Laguna, p 251

Zeigler RS, Thome J, Nelson J, Levy M, Correa-Victoria FJ (1994) Lineage exclusion: a proposal for linking blast population analysis to resistance breeding. Rice blast disease. CAB International, Wallingford, pp 267–292

Zhang Q, Li J, Xue Y, Han B, Deng XW (2008) Rice 2020. A call for an international coordinated effort in rice functional genomics. Mol Plant 1:715–719

Zhang W, Wu L, Wu X, Ding Y, Li G, Li J, Weng F, Liu Z, Tang S, Ding C, Wang S (2016) Lodging resistance of japonica rice (*Oryza sat*iva L.): morphological and anatomical traits due to top-dressing nitrogen application rates. Rice 9:31

Zhou B, Qu S, Liu G, Dolan M, Sakai H, Lu G, Bellizzi M, Wang GL (2006) The eight amino acid differences within three leucine rich repeats between *Pi2* and *Piz-t* resistance proteins determine the resistance specificity to *Magnaporthe grisea*. Mol Plant-Microbe Interact 19:1216–1228

Chapter 5
Cassava (*Manihot esculenta* Crantz)

Hernan Ceballos and Clair H. Hershey

5.1 History of Cultivation

Cassava is a crop of Neotropical origin and significant economic relevance, particularly in the lowland tropics. Its main product is the starchy roots that are generally harvested about a year after planting. It has been suggested that the *Manihot* genus emerged in Mesoamerica and from there diversified to the north and south (Deputié et al. 2011). Endemic *Manihot* species can be found from southwestern North America to central Argentina (Rogers and Appan 1973; Nassar and Ortiz 2008; Deputié et al. 2011). The taxonomy of the genus has been generally understudied. Rogers and Appan (1973) in their comprehensive monograph described 98 *Manihot* species. More recent taxonomic updates were done by Allem (1999, 2002), Allem et al. (2001) and Second et al. (1997).

The botanical and geographic origin of cultivated cassava is still unclear. Early publications suggested that cassava had an unknown ancestry but was likely the by-product of indiscriminate introgression among some of the wild relatives in Mexico and Mesoamerica, probably including *M. aesculifolia* (H.B.K.) Pohl (Rogers and Appan 1973; Bertram 1993; Bertram and Schaal 1993). Renvoize (1973) suggested that cassava was domesticated in Mesoamerica and northern South America. However, today the prevailing hypothesis is that cultivated cassava originated in South America (Allem 2002; Olsen and Schaal 1999, 2001; Nassar and Ortiz 2008). Allem suggested in different articles (Allem 1999, 2002; Allem et al. 2001) that cultivated cassava was domesticated directly from an extant wild

H. Ceballos (✉) • C.H. Hershey
Centro Internacional de Agricultura Tropical, Cali, Colombia
e-mail: h.ceballos@cgiar.org; clair.hershey@gmail.com

H. Campos, P.D.S. Caligari, *Genetic Improvement of Tropical Crops*,
DOI 10.1007/978-3-319-59819-2_5

species (either *Manihot esculenta* Crantz ssp. flabellifolia (Pohl) Ciferri or *Manihot esculenta* Crantz ssp. peruviana (Mueller Argoviensis)). Another contrasting view of the South American origin of cultivated cassava was proposed by Nassar in 1978, suggesting that the origin and domestication of cultivated cassava occurred from a natural hybrid probably between *M. pilosa* and other species (Nassar 1978, 2000). It is possible that several independent domestication events have taken place in different sites and times. Upon domestication, cassava was disseminated through tribal migrations (particularly the Arawak people) in pre-Columbian times (Nassar 2000).

The timing of domestication has also not been determined. Archaeological evidence of vegetatively propagated crops is generally limited, and cassava is no exception. It is clear, however, that the Mochica culture flourishing in the northern coast of Peru (2200 through 1200 years ago) knew about cassava, as illustrated by a beautiful ceramic artwork (www.museolarco.org). Domestication of cassava, therefore, must have taken place at least 3000 years ago. Cassava was widely grown in pre-Columbian times. The Portuguese introduced the crop into West Africa in the 1500s where it spread quickly thanks to its resilience, flexibility of harvest and diversity of uses. From Africa, cassava spread eastwards and eventually into Asia, where it had also been introduced by the Spanish into the Philippines (Byrne 1984).

More than 20 million ha of cassava are harvested annually, most of it (72%) in Africa, followed by 18% in SE Asia and 12% in the Americas (FAOSTAT, averages from the last three available years – 2012–2014). Worldwide average yield (2012–2014 avg.) is 11.1 t of fresh roots per ha but varies widely (8.4; 12.3 and 21.3 t/ha, respectively, in Africa, America and Asia). Nigeria, Brazil and Thailand are interesting examples, for cassava production and use, in their respective continents. Nigeria is the largest producer of cassava worldwide, with more than 6.7 million ha planted and 51.1 million t of fresh roots harvested (average yield of 7.6 t/ha). Brazil plants about 1.6 million ha with an average yield of 14.2 t/ha. However, large differences in productivity can be observed, for example, between NE and Southern Brazil with average yields of 10.3 and 19.8 t/ha, respectively, for the 2001–2013 period (IBGE 2014). Thailand plants 1.4 million ha of cassava each year, and the average yield is above 22.0 t/ha. It can be postulated that productivity of cassava is closely linked to the strength and stability of markets. SE Asia and Southern Brazil have well-established and reliable markets for cassava which explains the motivation to achieve average productivity above 20 t/ha in those regions. In Africa, on the other hand, markets are not so well developed thus partially explaining the low average productivity of about 8 t/ha. These figures are helpful to understand that proper technologies have been developed for cassava to express its high yield potential, but their adoption by farmers depends on the market perspectives.

Among the key technologies available to farmers are improved varieties. One of the first reports on cassava variety assessment and selection was published in Brazil, in 1899 (Zehntner 1919), reported by Gonçalves Fukuda et al. (2002). Modern breeding programs were initiated during the first half of the last century (Byrne 1984; Jennings and Iglesias 2002) in Brazil (Graner 1935; Lozano et al. 1978; Normanha 1970), Ghana (Hahn et al. 1979), India (Abraham 1957), Indonesia, (Bolhuis 1953), Madagascar, (Cours 1951), and Tanzania (Nichols 1947; Jennings

1957). However, with the exception of Brazil, most of these early efforts were discontinued as a result of the dismantling of the colonial system. The creation of cassava improvement programs at the International Institute of Tropical Agriculture (IITA based in Ibadan, Nigeria) and Centro Internacional de Agricultura Tropical (CIAT in Cali, Colombia) broke the previous isolation that breeding efforts had had in the past (Byrne 1984) and contributed to the establishment of vigorous and successful programs in the most important cassava-growing countries. Most of the information presented in this chapter has been generated by the collaboration between the international centres and national agriculture research programs.

Markets in cassava are diverse. The crop was initially domesticated for the direct use of the roots, which contain little else in addition to starch (e.g. low concentration of proteins, fat/oil and traces of micronutrient mineral and vitamins). Low to very high levels of cyanogenic glucosides (CG) can be found in roots from sweet/cool or bitter cassava cultivars, respectively. CG are eliminated through alternative processing techniques. Different cultures developed diverse ethnic products from cassava roots, for example: gari, kokonte and fufu in Africa; farinha and cassabe in the Americas; and krupuk, gaplek and sago pearls in Asia (Cock 1985; Nweke 2004). Processing roots soon after the harvest is also important to prevent post-harvest physiological deterioration (PPD) that spoils roots 2–3 days after harvest. Dried chips or silage of cassava roots can be used for animal feeding. Cassava foliage is also used for human consumption, particularly in Africa (Diasolua Ngudi et al. 2003; Lancaster and Brooks 1983; Muoki and Maziya-Dixon 2010), animal feeding (Balagopalan 2002; Buitrago 1990; FAO 2013) and even insect production for human consumption (Caparros Megido et al. 2016).

Globally, in the period 1970–2003, the main uses of cassava roots were for food (54%), followed by feed (30%) and other uses including starch production (4%) (Prakash 2008). During this period, however, there were considerable changes, with a clear trend of increasing industrial processing. Global use of cassava for feed was affected by the reduction of imports from the European Union in the 1980s. Production of starch on the other hand increased considerably in the same period (by 17.5% annually according to Prakash 2008). Today, cassava is the second most important source of starch worldwide (Stapleton 2012). In the 2000s a considerable amount of cassava roots started to be used for the production of fuel ethanol (Anyanwu et al. 2016; Chen et al. 2016; Liang et al. 2016; Nguyen et al. 2007; Sriroth et al. 2010).

The diversity of uses for cassava roots and foliage illustrates the challenges that cassava breeders face. Each end use imposes a set of requirements, sometimes contrasting, that varieties need to meet for them to be acceptable to farmers and processors.

5.2 Biology

Cassava is a perennial species usually grown as an annual crop. The roots can be harvested from 6 up to 24 months after planting (MAP). More typically, however, farmers harvest cassava 10–12 MAP, at the end of the dry season (or before the cold season in subtropical regions), when root quality would be optimum, and store the stems only for a few weeks until the arrival of the rains (or several weeks during winter). Commercial multiplication of cassava is achieved by planting stem cuttings. Farmers grow clonal hybrids, so after having access to planting material of a good variety they only have to make sure that health, nutrition and physiological status of the planting material is maintained at optimum levels. There is a genetic component, however, of the capacity of planting material to withstand storage and to sprout quickly and vigorously (Ceballos et al. 2011). This characteristic has become more critical in recent years due to the more erratic arrival of the rains as a result of climate change.

Farmers collect the main stems before harvesting the roots. Depending on the variety and growing conditions, length of harvested stems can range from 50 to 200 cm. There is certain variation in sprouting capacity/vigour, depending on the sections of the stem from which the cuttings come. In a 10–12-month-old cassava, the best cuttings correspond to the middle half section of the stems (Ceballos and Calle 2010). This generates, unavoidably, certain phenotypic variation among plants in the same plot which may even increase the experimental error in evaluation trials. Cuttings are generally 20–30 cm long and have five to seven nodes. Cuttings can be planted horizontally, vertically or at a certain angle. If not planted horizontally, the cuttings are introduced into the ground at least half their length and taking into consideration the orientation of the stem (proximal end down and distal end up) as apical dominance occurs during sprouting. Adventitious roots develop from the underground portion of the cutting, either from the base of the nodal units or from the cut base of the stem. About 2 to 3 MAP, some roots start to swell and become storage roots. Above ground, buds sprout to produce leaves, and below the apical meristems, the stem starts elongating. Usually one or two buds will sprout to generate the respective stems. If cuttings are planted horizontally, many potential stems may emerge, and then some thinning may be advisable. Ideally one or two main stems should develop from each cutting. Planting density is typically 1 × 1 m for a 10,000 plants/ha density. Higher densities are acceptable (up to 20,000 pl/ha, depending on varieties and environmental conditions).

The multiplication rate through stem cuttings in cassava is low. Because of differences in plant architecture, there is large variation in the number of cuttings that a plant can produce. Plants from some genotypes can produce as many as 30 cuttings per plant, whereas in other genotypes plants may barely produce five cuttings. When hundreds or thousands of genotypes need to be screened in uniform conditions, a maximum of eight (at best ten) cuttings per plant can safely be assumed from each and every genotype. This low reproductive rate implies that several years are required until enough planting material is available for multilocation trials. Breeders are chronically short of planting material, which imposes many limitations on the evaluation process. For example, little progress has been made in identifying early bulking germplasm because it would require sacrificing some plants whose stems cannot be used as source of planting material (as they would have been harvested off season). Similarly, it may be desirable to store stems for longer periods of time (to select for capacity to sprout quickly and vigorously, even after long storage) but that would also expose losing some plants when cuttings fail to sprout. In turn, this would complicate the logistics of the following stages in the selection process. Although rapid multiplication schemes based on micro-cuttings or tissue culture approaches can be implemented, they are expensive and have not been routinely used.

Cassava is monoecious, i.e. with separate female and male flowers, occurring in the same inflorescence (raceme or panicle). Male flowers are more numerous and develop in the upper section of the inflorescence (Gonçalves Fukuda et al. 2002; Perera et al. 2012). Female flowers are fewer and are found in the proximal branches of the inflorescence (Fig. 5.1a). Anthesis of female flowers occurs about 10–14 days before that of male flowers (protogynia). Cassava naturally outcrosses (mostly by insects). Self-pollination can occur when male and female flowers on different branches of the same plant are open, or different plants of the same genotypes simultaneously produce male and female flowers.

Inflorescences always develop at the apex of the developing stem. Sprouting of the buds below the inflorescence allows further growth of the plant. Therefore, the plant first flowers and then develops branches (Gonçalves Fukuda et al. 2002).

Fig. 5.1 (**a**) Illustration of an inflorescence with female flowers in the basal section already open and male flowers in the *top* still not open. (**b**) Erect plant architecture. (**c**) Defoliated plant illustrating the four consecutive flowering (and branching) events. (**d**) Highly branching plant architecture of a clone planted by farmers in Central Ghana

Every flowering event, therefore, results in branching (Fig. 5.1c). Some genotypes flower early and several times (starting from 3 MAP and up to five times) during a growth cycle and others flower little or late (e.g. once at eight MAP). Erect, non-branching types are frequently preferred by farmers because this plant architecture facilitates cultural practices and results in good production of vegetative planting material, and its transport and storage are easier (Fig. 5.1b). Longer stems can withstand a longer storage period and may be a key trait to deal with climate change (Ceballos et al. 2012). In Africa, on the other hand, early branching clones are sometimes preferred by farmers (Fig. 5.1d); an advantage may be early canopy closure to help control weeds.

The general advantages of erect clones which do not branch or branch late in the season result in a dilemma for the breeder because the production of botanical seed from these genotypes is sparse and slow and, ultimately, more expensive. There are indeed many cassava clones with profuse, early and frequent flowering. These materials would facilitate greatly the production of segregating seed. However, as stated above, this flowering behaviour is closely related to an early (i.e. low to the ground) and frequent branching (Fig. 5.1c, d), which is generally undesirable. Since initiation, frequency and prolificacy of flowering are under genetic control, the progenies from early flowering types tend to inherit the trait and, along with it, the undesirable branching plant architecture. Perhaps one of the most important areas of research to accelerate and improve the impact of breeding, therefore, would be the development of a protocol for temporarily inducing flowering in cassava. Conventional breeding would benefit by reducing the costs and time currently required to make crosses and obtain seeds. Induction of flowering would also facilitate the development of inbred progenitors through successive self-pollinations. Accelerated and synchronized flowering in cassava crossing blocks would also be useful for the implementation of genomic selection. There are ongoing approaches to induce flowering through grafting, photoperiod modulation and the use of plant growth regulators.

The development and growth of cassava are rather simplistic; it does not mature from the phenological point of view. After planting the buds sprout to produce leaves and stems. As the leaf area index increases, excess photosynthates are produced by the source (canopy) and are then stored in the sink (roots). The only function of the storage root of cassava is as an energy reserve organ for the mother plant. As the conditions are conducive for photosynthesis (e.g. adequate light, water and temperature), the plant continues growing and partitioning photosynthates into the roots. The root does not reach physiological maturity at any given time. If environmental conditions become limiting for further growth (e.g. a dry season begins or temperatures fall below optimum), the plant ceases growing; starch accumulation will cease or become very slow. In most cassava-growing regions of the world, there will be a "dormant" period in which the plant stops accumulating energy in the root. It is usually at the end of that time when farmers harvest the crop. The roots will have optimum quality (particularly in relation to dry matter content – DMC), and harvested stems will have to be stored for a short period of time before growing conditions improve (e.g. rains arrive). If the plants are left in the ground, growth will restart using the energy stored in the roots. DMC in the root, therefore, falls drastically if plants are allowed to restart growth before they are harvested. Some genotypes quickly recover DMC but not others. The capacity of recovering DMC after a second growth stage is an important characteristic in those systems when harvest is extended to take place during a second year of growth as is done in Southern Brazil and Paraguay.

The fact that cassava roots can be harvested at any time or age of the plant (provided that they are old enough to have had adequate time to accumulate starch) is a great advantage for cassava and explains the remarkable food security role of the crop. Farmers can leave the crop in the field until needed. Cassava does not have a phenological stage during its growth when it is particularly vulnerable to environmental constraints (such as water stress during flowering of cereals), thus its recognized drought tolerance.

5.3 Genetics

Cassava is frequently considered an historical polyploid species based on the basic chromosome number of species in the Euphorbiaceae family (Westwood 1990). However, cytogenetic analyses during meiosis consistently have found the presence of 18 bivalents which are small and similar in size (Hahn et al. 1990; Umanah and Hartmann 1973; Wang et al. 2011). In some cases, occurrence of univalents/trivalents and late bivalent pairing has been reported. Cassava is therefore a functional diploid ($2n = 2x = 36$) (Jennings 1963; Westwood 1990; De Carvalho and Guerra-M 2002; Nassar and Ortiz 2008). Magoon and co-workers suggested in 1969 that certain portions of the genome may be duplicated, and, therefore, cassava may be a segmental allotetraploid.

Very few Mendelian genetic studies have been conducted and reported in cassava. Gonçalves Fukuda and co-workers summarized in 2002 a few cases in which inheritance of different traits had been reported. Narrow-lobed leaves are a dominant trait controlled by a single gene, and darker colour of the external root peel is dominant over the light coloration (Graner 1942; Jos and Hrishi 1976). Male sterility is a recessive monogenic trait (Jos and Bai 1981; Jos and Nair 1984). Pale green in the stem collenchyma is dominant over dark green and controlled by a single gene; yellow root parenchyma is partially dominant over white and controlled by two genes (Hershey and Ocampo-N 1989; Iglesias et al. 1997; Morillo-C et al. 2012). The stem in zigzag is a recessive trait, and it has been used as a marker gene in cross identification. However, since some nutrient deficiencies can also cause the zigzag trait, the genetic component may not always be clear. Red is dominant to green in leaf nerves and has also been used to distinguish clones derived from crosses from those of self-pollinations (Kawano et al. 1978). Unfortunately, none of these traits (except parenchyma pigmentation which is linked to higher carotenoids content) have economic relevance. More recently, the inheritance due to a single recessive mutation was confirmed for waxy (amylose-free) starch (Aiemnaka et al. 2012). Resistance to cassava mosaic disease-CMD (a devastating disease present in Africa, India and Sri Lanka) seems to be controlled in some cases by a single dominant gene (Rabbi et al. 2014a).

Quantitative genetic information, on the other hand, has been more widely studied. The vegetative reproduction of cassava allows the quantification of within-family genetic variation, which, in turn, allows a test for epistasis in diallel crosses (Cach et al. 2005; Pérez et al. 2005a, b). These diallel studies covered progenitors adapted to three different agroecological regions of Colombia. The most relevant results from these studies have been summarized in Table 5.1 for two key traits (fresh root yield, FRY, and DMC). Two features are worth highlighting. The use of heterozygous progenitors results in large within-family genetic variation. Nonadditive genetic effects (dominance and epistasis) are clearly relevant for the expression of these two traits but particularly for FRY. This type of information has been useful in the prediction that genomic selection would not be effective for traits such as FRY.

Additional quantitative genetic information has been produced in Africa and Asia. These studies can be grouped as diallel analyses and North Carolina II designs (Bueno 1991; Chipeta et al. 2013; Easwari Amma et al. 1995; Kamau et al. 2010; Lokko et al. 2006a; Njenga et al. 2014; Owolade et al. 2006; Parkes et al. 2013; Were et al. 2012; Zacarias and Labuschagne 2010). In general, all these papers support the information presented in Table 5.1 regarding the importance of nonadditive genetic effects for FRY.

Cassava, as most outcrossing species, shows considerable inbreeding depression for FRY (Gonçalves Fukuda et al. 2002; Kawuki et al. 2011b; Rojas-C et al. 2009). When inbreeding depression is important in a crop, it is often the case that heterosis will also be prevalent. The relative importance of nonadditive genetic effects for FRY suggested by diallel and North Carolina II Designs agree with the inbreeding depression observed for this trait by several studies.

Table 5.1 Variance estimates (standard errors within parenthesis) for FRY and DMC in three different diallel sets evaluated in the three environments for cassava production in Colombia

	Fresh root yield (t/ha^{-1})			Dry matter content (%)		
Genetic parameter	Acid soil	Subhumid	Mid-altitude	Acid soil	Subhumid	Mid-altitude
σ^2_G (between)	1.65	13.09	42.78	1.60	0.77	0.35
	(2.95)	(4.74)	(13.27)	(0.66)	(0.29)	(0.12)
σ^2_G (within)	21.08	127.21	288.93	3.22	5.56	0.12
	(2.30)	(7.65)	(1918)	(0.17)	(0.31)	(0.12)
σ^2_A (additive)	−1.49	17.82	11.88	3.38	1.45	0.99
	(6.32)	(13.75)	(24.67)	(2.40)	(0.99)	(0.47)
σ^2_D (dominant)	9.03	23.87	152.11	0.87	0.77	−0.21
	(7.93)	(11.15)	(49.08)	(0.67)	(0.50)	(0.13)
Epistasis test	15.05	100.40	168.91	0.87	4.26	−0.32
	(6.74)	(12.74)	(39.72)	(1.29)	(0.67)	(0.92)

5.4 Germplasm: Intra- and Interspecific[1]

One of the challenges in *Manihot* taxonomy is that interspecific hybridization between cultivated cassava and wild relatives, and among some of the wild relatives, occurs readily. It is feasible therefore that many wild species may have experienced important introgression of cultivated cassava alleles and vice versa. This has been suggested, for example, in the case of *M. glaziovii* (Nassar 2000; Bredeson et al. 2016). Ideally, wild species should be collected in isolated areas where no cassava is cultivated. This is critically important. The first source of resistance to cassava mosaic disease (CMD), a devastating virus disease in Africa, was first reported in *M. glaziovii* by the East African Agriculture and Forestry Research Organization, based at Amani (then Tanganyika). That program then developed interspecific hybrids to generate the first cassava clones with resistance to CMD, sources which are still widely used in breeding in both East and West Africa.

As for any crop, the future potential of cassava to contribute to the sustainable benefit of humankind will rely fundamentally on safe long-term conservation of broad-based genetic resources and their use in effective breeding programs. These resources are basically the landrace varieties that evolved for centuries under farmer and natural selection and about 100 wild species of the genus *Manihot*. The genus is native to the Americas, and most of the genetic diversification has occurred here. Traders first introduced cassava into Africa and later into Asia. Both continents have become important secondary centres of genetic diversity, especially Africa.

Cassava is a vegetatively propagated crop, while all the wild *Manihot* species are seed-propagated in their natural environments. In order to preserve the genetic integrity of a landrace, cassava must be conserved in vegetative form. The most common forms of conservation are as field-grown plants or as plantlets started from meristem tips, cultured on sterile artificial media, under light, temperature and media conditions that induce slow growth. For either field or in vitro conservation, expensive periodic regeneration is required, at a much higher frequency (typically every 12–24 months) than is typical for seed conservation.

Many cassava-growing countries have established a gene bank of local landraces, managed by government organizations. Most of these collections have been established since the 1970s, but some much more recently. Up until the 1970s, and through the 1980s, relatively few landraces had been lost due to broad-scale replacement by new, bred varieties. However, the risks are now much higher for genetic erosion of landraces, due to multiple factors, including success of new varieties, replacement of cassava by other crops in some regions (e.g. southern cone of South America), crop intensification and the associated trends towards less on-farm variety diversity and, possibly, climate change.

---

[1] The following section draws heavily from an extensive survey and review of cassava genetic resources supported by the Crop Trust and published as part of their series on Crop Conservation Strategies (Hershey 2008).

Most countries note collection gaps (less so for Asia), due to lack of funding, losses of diversity due to natural disasters and social conflict, difficult access to areas for collecting and inadequate collecting techniques of the past. Nearly all programs rely primarily on field-grown plants but may have part of their collection in vitro as well. In vitro collections have had a mixed success except in the few institutions where they are well funded for the long-term such as Embrapa, Brazil; CIAT, Colombia; and IITA, Nigeria. Globally, only about one-quarter of accessions held by national programs appear to be conserved in vitro. Two international centres (CIAT and IITA) maintain regional collections for the Americas and Asia (CIAT) and for Africa (IITA). There are very few national gene banks that have the capacity to carry out safe international exchange in situations where viruses, of quarantine significance, are present. Most international exchange is facilitated via the international centres.

About two-thirds of cassava is currently grown in Africa, but probably well over half the landraces occur in the Americas. This is to be expected in view of origin of the species in the Americas. A study in 2008 (Table 5.2) estimates that some 27,000 distinct landraces of cassava are conserved in situ and about 10,000 maintained in gene banks. Hershey (2008) proposed that a total of about 15,000 landrace varieties should be conserved ex situ in order to represent the complete genetic diversity of the species. However, these estimates are based on very tentative results from genetic diversity studies, and much more information from molecular diversity studies will be needed to develop more precise information.

Currently CIAT curates the gene bank with the largest number of cassava landrace varieties (about 5500 accessions), sourced from throughout the crop's origin in the Americas and from Asia. IITA holds an extensive collection from West Africa and is developing plans to incorporate a larger representation from East Africa. A conservation strategy should consider security, cost and efficiency in its design (Epperson et al. 1997). Security is a function of both the number of replications of a gene bank (in different sites or in different forms) and the management level of each. Field gene banks are the least secure, followed by in vitro slow growth and, finally, cryo-conservation. Although cryo-conservation has been researched for more than two decades for cassava, and with relatively good success, there are as yet no cassava gene banks which rely on it as a main form of conservation.

Table 5.2 Estimates of cassava landraces in situ and ex situ in major growing regions

Region	Est. total landrace varieties	Est. total ex situ accessions[a]	Est. in CGIAR gene banks	Est. no of landraces missing from CGIAR gene banks
Africa	7480	3743	2112	5368
Asia	2965	1132	257	2708
Americas	15,925	5148	4851	11,074
Global	26,986	10,068	7205	19,954

Source: Hershey (2008)

[a]Landraces held in gene banks, excluding breeding lines and estimated duplicates within and across collections

Hershey (2008), based on survey and workshop results, proposed a comprehensive conservation strategy consisting of the following elements:

- Collecting in priority areas is carried out to fill gaps, with the aid of genetic diversity studies and GIS.
- National program gene banks and international centre gene banks are systematically compared for matching and non-matching accessions, based on passport, morphological and molecular information. This would evolve into a common cassava registry at a global level.
- CIAT and IITA duplicate all the landraces of national program collections, in their respective regions of responsibility (CIAT: Americas and Asia; IITA: Africa). Currently they appear to maintain about 50–60% of these accessions.
- National programs commit to at least one working gene bank that serves the purposes both of conservation at a moderate level of security and evaluation.
- CIAT and IITA maintain at least two forms of each accession. Currently this may be an in vitro active gene bank plus a *black box* duplicate kept in another centre. In the future, cryopreserved accessions will be either the main or the backup gene bank.
- CIAT and IITA commit to making the material they maintain available to national program gene banks, when requested.
- CIAT and IITA commit to meeting the demands and phytosanitary requirements for international exchange of cassava landrace varieties under terms of the International Treaty. Along with this, it is urgent to develop protocols for the safe movement of vegetative germplasm between the Americas and Africa.
- There is a mechanism developed for periodic interaction among stakeholders. Most notably this will be between the international centres and the national programs. Each will have a formal responsibility to periodically inform the other of the status of collections.

Duplicate identification, further improvements for in vitro slow growth techniques, improving cryo-conservation, and flower induction for seed conservation are all research areas outside the funding stream for routine conservation, but which will contribute to greater conservation and use efficiencies in the long term. Cryo-conservation is clearly an option for effective, inexpensive, secure long-term conservation, but work remains to be done on achieving an adequate recovery level for about one-third of accessions (based on results from CIAT's core collection). Research should continue on improving recovery of these recalcitrant types before committing to large-scale cryo-conservation of any gene bank.

As a future alternative to vegetative cassava gene banks, the seed from self-pollinated accessions could be a less expensive and more efficient conservation method and would be equally or more effective for breeding programs. Since many cassava accessions do not readily flower, there is a need for research on the induction of flowering in order to have a broadly applicable strategy for seed conservation. Long-term, a conservation strategy that consists of a combination of cryo-conserved meristem shoots and seed maintained in conventional cold storage might be envisioned. This would combine the advantages of both seed and vegetative conservation in a low-cost, secure system.

The wild *Manihot* species present a situation that is simpler in some aspects but more complex in others, compared to cassava. It is simpler in that only a handful of institutions are involved in conservation – mainly Embrapa and the University of Brasilia in Brazil, CIAT and IITA. It is more complex in that:

- The taxonomy of species is still poorly defined.
- The highest concentration of species is native to threatened habitats. This is especially true in the Campo Cerrado of South-Central Brazil, where the expansion of agriculture and urbanization are rapidly encroaching on the wild species habitats.
- A secondary centre of diversity, with a distinct set of species, exists in Mesoamerica. Here, and especially in Mexico, cassava is a relatively unimportant crop, and it is difficult for these governments to justify investment in *Manihot* conservation in terms of value addition to the crop per se.
- Fewer than half the species are conserved in vitro, and very few are protected in national or regional reserves, in their native habitat.
- Wild species conservation presents many challenges, especially with regard to regeneration. Progress is being made both in seed and in vitro propagation, but much remains to be done.
- The value of the wild species is continually becoming more evident as new characters are identified with potential for transfer to cassava, and the techniques for efficient transfer and selection of specific genes are developed.
- Genes from cultivated cassava may have been introgressed into wild relatives.

Cassava is unusual among major crops in that there is relatively little differentiation between landrace varieties and "modern"-bred varieties. This is due in large part to the late and relatively low investment that cassava breeding has had and to the fact that the large majority of clients for new varieties are the small-holder farmers who grow the crop with minimal inputs for traditional markets (especially in Africa). This relatively low differentiation between landraces and bred materials has allowed breeders to continue to make wide use of gene banks to introduce new traits into breeding populations and to derive superior individuals without the need for extensive further crossing to eliminate unfavourable genes. However, this is changing, and breeders will ultimately broaden the gap between what is available in gene banks or in situ, compared to new varieties for modern management and markets.

Cassava gene banks have been moderately to well evaluated for basic agronomic traits like plant architecture, yield in multiple agroecosystems, DMC, cyanogenic glucosides (CG) and reaction to a range of pests and diseases. In addition, significant numbers of accessions have been evaluated for nutrient use efficiency, multiple root quality traits such as amylose/amylopectin ratio, sugar, and carotenoid content and tolerance to PPD. Systematic efforts to screen for useful recessive traits, such as amylose-free starch (Morante et al. 2016), need to be conducted. As we face new challenges such as climate change and the spread of pests and diseases, and new opportunities with new markets, novel genetic diversity will become ever more important, such as capacity to withstand long storage periods of the stems. From the standpoint of ease of use by breeders, the first step should be the full exploration of the *M. esculenta* gene pools. There is little advantage – and many disadvantages – to the extraction of genes from wild species if the same genes are available in cassava landraces. For example, the use of *M. glaziovii* as a source of resistance to cassava mosaic disease required decades of breeding in order to restore agronomic performance in genotypes carrying the CMD resistance. Initially it was believed that there were no sources of resistance within the cultivated species, but more recently it has been shown that variation for resistance does exist. Molecular markers will ultimately allow more efficient identification and extraction of genes from wild species (Duitama 2017). There is an urgent need to collect, conserve and evaluate these species as a resource for future breeding.

5.5 Crossing Approaches

Controlled pollinations in cassava result in the production of full-sib families (Fig. 5.2a). Several publications illustrate crossing techniques (Kawano 1980; Byrne 1984; Jennings and Iglesias 2002; Gonçalves Fukuda et al. 2002). There is no evidence of incompatibility, but there is variation in the success of seed set among different crosses (Njoku et al. 2015a). A maximum of three seeds per pollination can be obtained, although averages in controlled pollinations are considerably lower (ranging from one to two seeds). Making controlled pollinations in cassava is not difficult. Male flowers, ready for anthesis, are collected in the morning hours before they open. At the same time, on the day of anthesis, female flowers are covered (before stigmas are exposed). Field workers can easily recognize those flowers that will open each day, and they do these operations efficiently. Collecting or covering flowers before anthesis prevents contamination with undesired pollen.

Open pollinations can also be used as source of botanical seed. Polycross nurseries are planted following a special field design to favour pollinations of different genotypes (Wright 1965). At CIAT all the seeds collected from a given female progenitor in a polycross define a unique half-sib family. The female parent is known, since the seeds are collected from identified genotypes. The source of pollen that generated each seed is unknown, but a set of potential male progenitors is restricted to the genotypes that make up the polycross nursery each year.

The major bottleneck in the production of seed in cassava breeding programs is the scarcity of flowering and/or problems of synchronizing the flowering of particular pairs of genotypes that the breeder wants to cross. Efforts to develop a protocol for pollen conservation several decades ago were unfortunately fruitless (Orrego and Hershey 1984). The preference in many regions of the world for erect plant architecture aggravates matters further because this phenotype implies late flowering and few flowering events that the breeder can exploit. Certain environmental conditions may favour flowering. Longer photoperiods and cooler temperatures have been reported to stimulate flowering (De Bruijn 1977; Keating 1982). Lack of balance in the number of crosses representing each progenitor has implications for breeding. Indirect information suggests that at least 200–300 genotypes

Fig. 5.2 (**a**) Illustration of female (*top left*) and male (*mid-left*) flowers and procedure used in controlled pollinations (*bottom left*) which take place in the afternoon. Female flowers are covered with bags before they open and the bags are removed 2–3 days after pollination (*centre top*). Bags are placed again over near-mature fruits to collect seeds after dehiscence. Male flowers are collected in the morning and kept in plastic containers until needed (*centre bottom*). (**b**) Example of "asparagus" phenotype (*top*) and aerial view of a trial where it was compared with cassava genotypes with normal phenotype

would be required to properly assess the breeding value of a given progenitor (Ceballos et al. 2016a). This figure is often difficult to attain and breeders may have to wait for 2–3 years until enough crosses from a given set of progenitors have been made and the resulting seed harvested.

The late flowering habit of erect plant types implies that breeders typically need 18–24 months to produce an acceptable number of seeds (e.g. at least 20–30) from any planned cross. This time frame was not a problem for ordinary breeding projects where the main objective was developing high-yielding materials adapted to a particular agroecological zone. If seed from a given cross could not be evaluated in one year, it would be included in the following year. However, as breeding projects sought to develop clones with special attributes (e.g. high carotenoids or special starches), the slow rate in the production of botanical seed became a logistic problem as crosses were usually concentrated during 12–18 months. Special breeding projects are usually supported by time-constrained research grants that impose a restriction in the time allocated to make crosses. The Next-Generation Cassava Breeding Project (www.nextgencassava.org) is evaluating the potential of genomic selection in cassava (de Oliveira et al. 2012; Ly et al. 2013; Rabbi et al. 2014b; Wolfe et al. 2016a, b). It recognized, early on, that the induction of flowering was a key requirement because it would allow the achievement of a more balanced number of progenies from each progenitor and shorten the length of each recurrent selection cycle.

Another example of difficulties in making crosses in cassava can be given with the so-called "asparagus cassava" (Fig. 5.2b). This particular phenotype is characterized by leaves without petiole and absence of flowering within the first 8–10 MAP (no or very late branching). This very particular phenotype could be a new plant type as it is well adapted for high planting densities (e.g. 40,000 pl ha^{-1} instead of the normal 10,000). Preliminary results have demonstrated that "asparagus cassava" responds better to high densities than ordinary phenotypes (CIAT, unpublished results). However, breeding this type of cassava would be nearly impossible due to the current difficulties producing segregating seeds.

Because of the reasons described above, the induction of flowering in cassava has become an important research objective. Grafting of stems from non-flowering genotypes into a rootstock from an early and frequently flowering genotype has induced flowering in some genotypes, but not in others (Ceballos et al. 2017). There are ongoing efforts to induce earlier flowering and stimulate number of flowers produced and seed set through modulation of photoperiod or application of plant growth regulators. These efforts have been successful in some genotypes but not in others. Preliminary results indicate that the application of certain growth regulators has been successful for inducing flowering in the "asparagus cassava" (CIAT, unpublished data). Genetic transformation modulating the FT locus has also been successfully achieved (Adeyemo et al. 2008).

5.6 Breeding Schemes

Several reviews on cassava breeding have been made over the years (Byrne 1984; CIAT 1991; Gonçalves Fukuda et al. 2002; Jennings and Hershey 1985; Jennings and Iglesias 2002; Ceballos et al. 2004, 2007a, 2010, 2012; Kawano and Cock 2005; Kawuki et al. 2011a). These reviews provide new information regarding discoveries of new sources for relevant traits, consolidated information regarding the relative importance of genotype, environment and their interaction for productivity and, more recently, on the potential of molecular markers. However, basically the breeding scheme has remained unchanged since the inception of modern cassava breeding in the late 1960s. Below is a brief description of the scheme used at CIAT. All programs apply some variation of a scheme which begins with selection of parents; crossing to produce a segregating population; seedling nurseries where each individual is genetically distinct; then a series of clonal trials of increasing plot size; number of replications and number of testing sites, ending in regional and on-farm trials; and finally official release. Typically, the time frame from selection of parents to release is at least 10 years and often longer. Under highly accelerated schemes, release could take place in as few as 6 or 7 years.

Figure 5.3 illustrates a general scheme of the different trials used in cassava breeding. In the case of cassava, botanical seeds from a cross between two heterozygous parents are normally referred to as the F_1 generation, in contrast to common use of the term to refer to progeny from two homozygous parents. Botanical seeds from full- or half-sib families are germinated and the resulting seedlings grown in a screenhouse for 1–2 months. Some programs, however, germinate the seeds directly in the field, a possibility especially under high soil temperature conditions. Vigorous and healthy seedlings are then transplanted to the field and grown for 10–11 months. The size of seedling nurseries varies considerably depending on seed availability, regions, objectives and program resources to manage nurseries. If selection can be effectively made on a one-plant plot, i.e. for traits with high heritability, then the F_1 nurseries tend to be relatively large (>15,000 plants). Strong selection pressure is used at the seedling plant stage in few specific cases where the breeder aims at selecting for a high-heritability trait: resistance to CMD (Rabbi et al. 2014a; Jennings and Iglesias 2002), bio-fortification for enhanced carotenoids content (Belalcazar et al. 2016; Ceballos et al. 2013), waxy starch (Aiemnaka et al. 2012) or white vs. brown root surface.

When the goal is to generate clones with good agronomic performance and high productivity, in the absence of a limiting factor such as CMD, selection based on a single plant plot is not reliable. In these cases, therefore, selection pressure cannot be high at the F_1 stage, and the nursery size tends to be smaller, such that the large majority of genotypes can be cloned for more complete evaluation in larger plots.

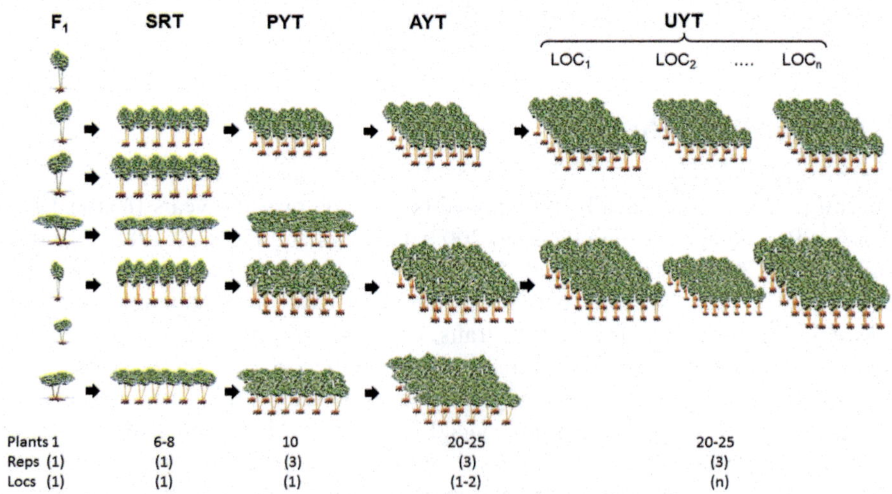

Fig. 5.3 Illustration of the different stages of evaluation used at CIAT in the selection of clones eventually released for commercial growth by farmers. The number of plants per plot, replications and locations for each stage is indicated at the *bottom*

Seedling nurseries at CIAT for a particular target environment are typically <10,000 plants. Genotypes with undesirable phenotypes (e.g. poor vigour, susceptibility to thrips, chlorosis, very low or high harvest index, etc.) are also eliminated by default (Fig. 5.4a). The seedling plants (F_1 in Figs. 5.3 and 5.4a) are used as source of planting material for the next stage in the selection process and must, therefore, provide six to eight good quality cuttings. At CIAT, the seedling nursery is planted at Palmira Experimental Station, which offers nearly ideal growing conditions (including irrigation) to guarantee excellent nutritional and sanitary conditions of the resulting planting material. The next stage in the selection process is the single-row trial (SRT) which is planted in the target environment. In Colombia, CIAT has four main targets: subhumid, acid-soil savannas, mid-altitude valleys and the highland environments.

In SRT each genotype is planted in a single row with six to eight plants per row (the number may change from one year to another, but it is uniform for each trial). These are large experiments with 1500–2500 genotypes (1–2 ha) and therefore prone to large experimental errors (Fig. 5.4b). Target environments and evaluation plots lack uniformity which reduces the precision of evaluations and reliability of data. At CIAT, therefore, the SRTs are split in three to four blocks (strata), and selection is made within each stratum (Ceballos et al. 2007a). This strategy reduces the environmental variation among strata in the selection process (Gardner 1961).

Fig. 5.4 Illustration of different types of trials in the cassava breeding scheme. (**a**) Seedling nursery (F_1) in which adequate plant architecture, vigour, health and harvest index are key selection criteria; (**b**) single-row trials (*SRT*) are usually very large (up to 2 ha); (**c**) preliminary yield trials (PYT) follow a special design that leaves one empty row between plots to reduce plant competition among different genotypes. Notice the difference in plant height in two neighbouring clones; (**d**) plots in advanced yield trials (AYT) and uniform yield trials (UYT) are larger, and only the six or nine central plants are harvested, to eliminate effects of inter-genotypic competition. The surrounding plants in the plot periphery are left in the field and used as source of planting material

A selection index (SIN, described in next section) is used to facilitate selection of genotypes evaluated at SRT.

The following stage in the selection process is the preliminary yield trails (*PYT*) in which each genotype is planted in three replications with ten-plant plots (two rows of five plants). The six to eight plants from SRT are used as source of planting material for the PYTs. Because of limitation in the amount of planting material available, a special design is often used for PYTs. In cassava there is still considerable variation in the farmer-preferred plant architecture, ranging from short and bushy to tall and erect (Fig. 5.1). To minimize inter-genotypic competition arising from differences in plant height and architecture, an empty row is left between neighbouring plots (Fig. 5.4c). Plant spacing is set to 0.8×0.8 m within plots and 1.6 m between plots, for a final plant density of 9600 pl ha^{-1}. The number of genotypes evaluated in PYTs is still relatively large (200–500), and therefore they are split into three to five separate trials with 50–100 clones each. Selection is conducted within each trial but is not as stringent as for SRT. About 50–120 genotypes are selected for the following stage (advanced yield trials or AYT).

Plots in AYT have four (or five) rows with five plants each. The six (or nine) central plants are harvested and the resulting data used for selection. The surrounding plants in the periphery are left in the field and used as source of planting material when needed (Fig. 5.4d). Occasionally AYT may be planted for two consecutive years or at two or more locations. The same selection index is used throughout the different stages of the selection process, but as the number of genotypes is gradually reduced, more information is taken in later stages (e.g. boiling time, cooking quality, CG etc.).

The last stage in the breeding scheme is the uniform yield trials (UYT) which are conducted in several locations and at least two consecutive years. About 20–25 experimental clones and 5–8 commercial checks/controls are involved in these trials, which have three replications and plots similar to those used in AYT. Genotypes reaching UYT are usually incorporated as progenitors in crossing blocks and the best two to three may be considered for release as official varieties.

There are many variations that can be introduced to this generalized scheme, to achieve specific goals for any given breeding program or to adapt to local conditions and resources. For example, CIAT uses a disease "hotspot" in the eastern plains region (Llanos Orientales) to screen all material that has reached intermediate selection stage. For example, at the same time a set of genotypes is evaluated in the PYT in the principal target environment of the Atlantic coast region (subhumid tropics), a set of the same genotypes is evaluated in a SRT in the Llanos for resistance to bacterial blight and super-elongation disease.

5.7 Traits

For many years, the main objective of cassava breeding programs was to produce high-yielding clones that were adapted to the target environments and tolerant to their main biotic and abiotic stresses. However, several traits, in addition to high and stable productivity, are required in order for clones to be adopted by farmers. Some of these additional characteristics are common across regions (e.g. capacity of stems to withstand long storage period or fast and vigorous sprouting of the cuttings). But others may be contrasting and region specific (e.g. erect plant architecture is preferred in many Asian and Latin American countries, whereas bushy types may be preferred in certain regions of Africa).

Depending on the target region, different biotic stresses have to be considered by breeders. Bacterial blight (CBB) is found widely across the three continents. Across Africa, in India and Sri Lanka, different strains of CMV are a major threat to cassava. Cassava brown streak virus (CBSV) originated in East Africa but is spreading westward. In Latin America, cassava frogskin disease (CFSD) (whose aetiology is not clearly established) is a problem that can be managed with simple cultural practices (e.g. confirm that roots are symptomless before collecting the stems as source of planting material an approach that is also useful for CBSV) but may also be a target for resistance breeding. Thrips can have a devastating effect on cassava. Fortunately, pubescence of newly emerging leaves at the shoot tips offers an excellent and stable source of resistance. Whiteflies can cause direct damage to cassava, but their most serious impact is transmitting viruses such as CMV and CBSV. An excellent source of resistance has been found in landraces from Ecuador and Peru (Bellotti and Arias 2001). Mites can have detrimental effects on cassava, particularly during the dry seasons. There is an interesting example of interaction between genotype and biological control in relation to mites. In Africa, the management of the cassava green mite problem is achieved through biological control by exotic predatory mites. However, cassava morphological traits can be improved to provide better shelter and ensure continuous survival of the natural enemies of the cassava green mite (Chalwe et al. 2015; Molo et al. 2016).

End uses of cassava drastically define the traits that breeders have to take into consideration. For starch, ethanol or dried chips for animal feeding the key traits will be DMC and FRY. White parenchyma is preferred by the starch industry, but the enhanced nutritional quality of yellow roots would be preferable for animal feeding. The great diversity of food uses of cassava was thoroughly described by Balagopalan (2002). Some regional and ethnic uses of cassava such as farinha and casabe (Amazon basin), kokonte (Ghana), gaplek and krupuk (Indonesia) or tapioca pearls (India) would also benefit from adequate DMC and FRY. However other ethnic uses require additional traits. Boiled cassava roots require low levels of CG, reduced boiling time and consumer-preferred texture. On the other hand, African products such as fufu and gari require proper poundability or mealiness. It is not clear which are the anatomical or biochemical characteristics that define cooking quality, mealiness or poundability, and there is ongoing research to elucidate these characteristics to facilitate the selection process made by breeders.

As breeding techniques and cassava utilization evolved, however, there has been a growing tendency to shift from breeding to develop *general-purpose* cultivars towards more specialized, market-oriented products. This new trend is a result of the confluence of different circumstances briefly described below.

The first modern cassava varieties started to be released in the early 1980s through early 1990s. These varieties fulfilled the basic requirements of adequate levels and stability of productivity, acceptable levels of DMC and, depending on the end uses, root quality traits. This early group of varieties includes, for example, the very successful variety KU50 released in Thailand in 1992 (FAO 2013; Fisher et al. 2014; Kawano 2003; Kawano and Cock 2005; Kawano et al. 1998). A second wave of improved clones was released during the 2000s, and a third wave is being released in the 2010s. The varieties released were very successful in increasing productivity (particularly in SE Asia where it basically doubled in the last 30 years). However, genetic gains in the second and third waves of new varieties seem to have plateaued (Ceballos and Hershey 2016). The breeders' hopes of producing new clones with higher productivity than that of already-released varieties face increasing challenges. It has been postulated that the difficulties of increasing productivity in cassava arise from the heterozygous nature of the progenitors used and the strong influence of nonadditive genetic effects already described in Table 5.1 (Ceballos et al. 2015, 2016a). These problems require change(s) in the current breeding methods, which will be discussed later in this chapter.

At the same time that breeders recognized the need for alternative breeding approaches for increasing productivity, their attention has also shifted towards other traits with commercial relevance and with higher heritabilities. This is the case of breeding for increased carotenoids content (Ceballos et al. 2013; Esuma et al. 2016; Morillo-C et al. 2012; Njenga et al. 2014; Njoku et al. 2015b; Ssemakula and Dixon 2007) or cassava with special starch functional properties (Carvalho et al. 2004; Ceballos et al. 2007b, 2008; Morante et al. 2016).

The identification of traits with high commercial value (such as starch mutants) or the development of clones with desirable nutritional traits (e.g. high carotenoids or low levels of CG) highlight the importance of a thorough and complete screening of gene banks as indicated in the section of germplasm resources. The development of reverse genetic molecular approaches (such as Eco-TILLING) or similar technologies (such as pooled targeted resequencing of DNA to detect rare SNPs in specific genes) would make the identification of useful traits much easier (Duitama et al. 2017). As we expand our understanding of the specific needs for the ever-widening uses of cassava, these molecular tools offer a very appealing alternative for the identification of useful (recessive) traits.

Cassava roots spoil quickly due to PPD. The short shelf life of roots imposes many limitations to post-harvest handing, transport and processing of cassava roots (Beeching et al. 1993; Reilly et al. 2007; Vlaar et al. 2007). Sources of tolerance to PPD have been reported (Morante et al. 2010; Moyib et al. 2015), but changes during root storage, including starch losses, limit the shelf life after harvest to less than 2 weeks (Sánchez et al. 2013). Little progress can be achieved through conventional breeding to reduce the negative impact of PPD until an adequate protocol for proper screening is developed. The experimental errors associated with PPD are huge and the effect of genotype-by-environment interaction unacceptably large (CIAT unpublished data).

Increase in levels of atmospheric CO_2 is one of the most certain expectations of climate change and the environmental conditions for the next few decades. It is expected that increased CO_2 will have a positive effect on cassava productivity by making photosynthesis (specifically CO_2 fixation) more efficient. This is an interesting situation because it would result in reducing the physiological advantages that C4 crops such as maize, sugar cane and sorghum have over C3 plants such as cassava, wheat or soybean (Long et al. 2004, 2006). Free-Air CO_2 Enrichment (FACE) methods allow field evaluation of crops under elevated CO_2 concentrations that simulate the predicted levels for the decades to come (El-Sharkawy 2009). Modelling and FACE results could be very useful guides in the molecular optimization of the photosynthetic apparatus to maximize carbon gains without increasing crop inputs (Zhu et al. 2007, 2010). Preliminary results indicate that increases in productivity under elevated CO_2 concentrations failed to meet theoretical expectations. If this is confirmed, it can be hypothesized that some bottlenecks in the complex metabolic processes following CO_2 fixation during photosynthesis prevent the theoretical expectations. Overcoming these potential bottlenecks may allow breeding to maximize the enhanced productivity that elevated CO_2 concentrations will have on cassava in years to come.

5.8 Field Trialling and Selection Approaches

The experimental design used for the different stages of breeding trials is basically a randomized block. Excel spreadsheets satisfy the needs and allow for the use of tablets or small portable computers for direct data uploading in the field. Statistical analyses (e.g. ANOVA) are not used routinely as statistical significances are not as relevant as properly ranking the different genotypes according to their performance – although error variances do give a good indication of what reliance one can place on the trial results in general. A method for the adjustment for missing plants has been developed (Pérez et al. 2010). However, correcting for missing plants may have the negative effect of rewarding genotypes with weak sprouting capacity after a normal storage period of the stems. A critical concept in the implementation of evaluation trials is to stratify them when they are large and the field is variable (Gardner 1961). Alternatively, large trials can be split into few smaller ones.

A few traits are broadly accepted across breeding programs as common key goals for improvement: high FRY, high and stable DMC, suitable plant architecture, resistance to locally important pests and diseases and harvest index (HIN). At CIAT, in addition to individual ratings, breeders integrate plant architecture and resistance to biotic/abiotic stresses into a single score indicating overall desirability of the above-ground plant appearance (plant type score or PTS) where 1 is very good and 5 is very poor. This is essentially a subjective answer to the question: "How much do I like the overall appearance of this genotype, based on the above-ground parts of the plant?" It is a sort of subjective ideotype, based on the breeder's experience and skill, and may be included among the more objective "select" or "reject" criteria.

Because of the low heritability of FRY in early stages of selection, cassava breeders for many years have applied indirect selection for yield by using correlated traits with higher heritabilities, such as HIN (Kawano et al. 1998). Since 2000, CIAT has used a selection index (SIN) that integrates four high-priority variables, assigning them best judgement weight (in italics in the formula below). These weights have been established by the breeder's experience (Ceballos et al. 2012) and can be modified with time:

$$\text{SIN} = \left(\text{FRY}^{*}\,\mathit{10}\right) + \left(\text{DMC}^{*}\,\mathit{10}\right) - \left(\text{PTS}^{*}\,\mathit{5}\right) + \left(\text{HIN}^{*}\,\mathit{3}\right)$$

In the case of PTS, the desired target is a lower score. Therefore, a negative sign is assigned to the respective term in the SIN equation. The variables used in SIN are measured in units that differ drastically in magnitude. To overcome this problem, standardized values are used ($x_i - X/\sigma$, where x_i is the individual observation, X is the average across genotypes and σ is the standard deviation) in the estimation of SIN. About 15% of clones evaluated in SRT are selected. Other traits, such as cooking quality, may be incorporated in the selection process but in later stages (AYT or UYT). Selection pressure is gradually relaxed through the scheme (e.g. 30–40% of clones evaluated in AYT may be selected for evaluation in UYT).

The selection process and criteria described above are ideally suited for a goal of high productivity. However, as stated earlier, other traits may be required for varietal adoption, particularly where cassava plays an important role in food security. Participatory breeding approaches (Gonçalvez Fukuda and Saad 2001; Gonçalves Fukuda et al. 2002; Grüneberg et al. 2009; Kamau et al. 2011; Manu-Aduening et al. 2006) are ideally suited for these conditions. Participatory breeding allows for much broader selection criteria, e.g. taking into account traits such as "maturity" period, suitability for intercropping, leaf production, taste, bitterness, processing amenability and cooking quality and even some traits that may just have a role as morphological markers such as petiole or shoot colour, leaf lobule shape or branching characteristics (Benesi et al. 2010). Participatory breeding emerged as a response to failures observed in some breeding efforts that limited the benefits of the green revolution in many developing countries for different crops. Many of the participatory breeding principles were initially developed in cassava (Ashby et al. 1987; Prudencio et al. 1992). Because of the size of the early trials (SRT and PYT), participation by farmers in the selection process may start in AYT or UYT.

Participatory approaches are based on the farmers' ability to select what is best for his/her conditions, a process which proved its power most significantly through the domestication of crops. However, it must also take advantage of the scientific knowledge generated in the past century. A major issue that is particularly acute in participatory schemes relates to the negative impact of genotype-by-environment effects (Grüneberg et al. 2009). Local selection of germplasm may fail to take advantage of selection based on multilocation evaluations where stability of performance is a key objective. If possible, the same set of genotypes should be grown in different environments by farmers that have common interests. Farmers may select the materials of their preference, but breeders can then select for the best across-environment genotypes which should offer a more stable performance.

A critical decision that breeders constantly take and debate about is the progenitors to be included in the crossing blocks. More often than not, progenitors used in cassava breeding are just clones with outstanding performance or carrying a desirable trait (e.g. resistance to CMD or CBB, amylose-free starch or high carotenoids, etc.). Alternatively, the use of breeding value or general combining ability (Falconer 1981) as a criterion for choosing progenitors in cassava breeding has been proposed (Ceballos et al. 2004). This initial idea was renewed (in a more sophisticated way) with the implementation of genomic selection (de Oliveira et al. 2012; Rabbi et al. 2014a, b). The usefulness of selection of progenitors based on breeding values is reduced by their heterozygous nature and the resulting within-family genetic variation (Ceballos et al. 2016a). Additionally, at least for fresh root yield, there is a strong influence of nonadditive genetic effects, further reducing the predictive value of general combining ability (Table 5.1, Ceballos et al. 2015, 2016a).

Another important factor affecting the selection process is the relationship between different characteristics that can be positively or negatively correlated. For many years, for example, selection in early stages such as SRT was based preferably on HIN rather than FRY (Kawano 2003). The rationale was that these two variables are closely associated, but HIN has higher heritability than FRY (at least in unreplicated trials). An analysis of the selection process after 14 years of continuous evaluations in the subhumid environment of Colombia was recently published (Joaqui et al. 2016). In this study, the benefits of using HIN as an indirect selection criterion for enhanced productivity in SRT were questioned. HIN will still be a key criterion for selection at the seedling plant stage (F1). Another interesting observation was that the relationship between FRY and DMC was weakly but positively associated in SRT ($r = 0.21$) a result similar to the one reported by Kawano et al. in 1987. However, that relationship gradually becomes negative in successive stages of the selection process. In UYT the correlation was $r = -0.42$. This clear and consistent trend suggests that cassava genotypes attain high dry matter productivity by maximizing DMC or FRY, or else through a compromise for "acceptable" levels in both variables simultaneously. But it is difficult for a given genotype to express simultaneously maximum levels of DMC and FRY. The nature of the association between DMC and FRY cannot be properly stated without a clarification in which stage of the selection process this association is being considered. Similar conclusions can be drawn from Kawano et al. (1998).

New technologies are likely to have a positive impact on cassava breeding, and one of these is high-throughput phenotyping. The way that near infrared spectroscopy (NIR) benefited breeding for high carotenoid content has already been reported (Belalcazar et al. 2016; Davrieux et al. 2016; Sánchez et al. 2014). The use of drones has become a standard strategy to analyse canopy size and health. However, little progress has been made for monitoring root growth in cassava until now. The use of ground penetrating radar offers promising results that would allow non-destructive monitoring of root growth through the growing season (Delgado et al. 2017). Early bulking has been among the traits sought after by cassava breeders, particularly in Africa (Kamau et al. 2011; Okechukwu and Dixon 2009; Olasanmi et al. 2014;

Tumuhimbise et al. 2014). But the resources required to managing multiple harvests for many genotypes, and chronic lack of planting material during the selection process, prevent breeders from sacrificing some plants to assess FRY before the usual harvesting time (11–12 MAP). This new technology offers promising advantages not only for selection of genotypes but also adequate cultural practices, for example, in the type and timing of fertilizers.

Cassava production in many regions of the world is seasonal. It is very common to have a large peak of availability of roots for periods of only 3–6 months. This is a major problem for large processing facilities that can operate only a few months of the year. An alternative solution to this problem is to delay harvest for a few months. This is how the cassava starch sector operates in Southern Brazil and Paraguay. Cassava is planted in early spring (e.g. August) and harvest can be split in two batches. Farmers may harvest part of the field early the following year in April–June and leave the rest for a delayed harvest in November–February. This approach benefits not only processors that can have their facilities operational for most of the year but also for farmers. Delayed harvests imply a sharp increase in the productivity (almost doubled, without a major increase in the costs). A key requirement for this approach, however, is that DMC at harvest time is at an optimum. In the system described above, DMC drops when plants reinitiate growth in August. This is a well-known phenomenon related to starch hydrolysis in the root (van Oirschot et al. 2000). When a similar strategy was attempted in tropical regions, however, DMC in available clones did not recover after 1 or 2 months after re-initiation of growth. There are ongoing efforts to select for a rapid recovery of DMC in delayed harvests. However, to do so, special trial arrangements are required: in the same locations, plots need to be harvested at the standard age (10–11 MAP) and then in delayed harvests (15–16 and 18–19 MAP).

5.9 Tissue Culture and Clonal Production

Tissue culture plays several fundamental roles related indirectly or directly to cassava breeding. The applications of tissue culture techniques are similar in cassava to those in other vegetatively propagated crops. Tissue culture has been used in cassava for rapid multiplication of clean planting material (Wasswa et al. 2010; Wongtiem et al. 2011), conservation/exchange of germplasm (Angel et al. 1996; Escobar et al. 1997; Roca 1984), embryo rescue approaches in interspecific crosses (Akinbo et al. 2010; Biggs et al. 1986; Fregene et al. 1999) and as a required system for genetic transformation (Bull et al. 2009; Ibrahim et al. 2008; Liu et al. 2011; Mongomake et al. 2015; Raemakers et al. 2001; Taylor et al. 2001, 2012). These are just representative reports that can be found in the literature for this broad range of applications of tissue culture. The conservation and exchange of germplasm have direct impact on breeding activities. Rapid multiplication and cleaning of planting material from diseases (particularly viruses) are important for the spreading and fast adoption of new varieties.

Tissue culture systems have been developed and used in cassava since the 1970s and are quite refined for routine use, such as for slow growth plantlet culture in sterile media for germplasm conservation and international exchange. For these systems, essentially any genotype can be successfully cultured in vitro with a single media formulation and light/temperature conditions. The wild species are more complex, and different media and growth conditions are required for different species. The development of friable embryogenic calli for transformation or gene editing is also rather genotype-sensitive, though many genotypes have now been successfully cultured for transformation.

Cassava is one of the few vegetatively propagated crops where important tissue culture efforts have been made for the production of doubled haploids. Ceballos et al. (2015) have listed the advantages that the use of inbred progenitors would have for the genetic enhancement of cassava:

- *Implementing the back-cross scheme and trait introgression.* The deployment and impact of desirable traits (resistance to diseases and pests, special starch quality traits, herbicide tolerance, etc.) is slow and limited because their introgression requires breeding for a new variety de novo. Back-crossing is a highly successful breeding scheme used in many crops (Xu and Crouch 2008), but it cannot be applied to cassava because of the heterozygous nature of the progenitors currently used. The relevance of the homozygote advantage has been recently highlighted by the efforts to deploy commercial cassava varieties with the waxy (amylose-free) starch. Introgressing a single recessive gene implied developing a new variety. The best varieties from the first batch of waxy clones had a productivity level similar to normal (non-waxy) clones released 30 years ago. Although the second generation of waxy varieties are expected to quickly catch up (Karlström et al. 2016), the cost of introgressing a simple trait is unacceptably high.
- *Doubling of breeding value*: Self-pollinating a heterozygous source of resistance to a given pest or disease and selecting a homozygous descendent would double the breeding value of the material when used as progenitor. Whereas 50% of the gametes produced by the heterozygous source carry the resistance gene, 100% of gametes from the homozygous version do (Ceballos et al. 2016a; Kaweesi et al. 2016).
- *Reduction of genetic load*: Inbreeding exposes undesirable recessive alleles and allows a rapid reduction in their frequency (which tends to be relatively high in heterozygous populations).
- *Discovery of useful recessive traits*: There are many examples in the literature of useful recessive traits including in cassava (Ceballos et al. 2007b, 2008).
- *Facilitated germplasm exchange and conservation*: When inbred progenitors are available, their conservation and exchange could be through true-breeding botanical seed.
- *Development of superior hybrids by design, not by trial and error*: Hybrid vigour (e.g. nonadditive genetic effects) can be progressively improved, but only through reciprocal recurrent selection methods (Bernardo 2014; Hallauer

and Miranda Fo 1981) or through inbred line development within heterotic groups. Improving heterosis would be slow if no inbreeding were employed. The impact that the use of inbred progenitors has had in maize is unquestionable (Troyer 2006). The use of inbred progenitors offers the chance to maintain favourable gene combinations at different loci controlling the small, but critically relevant, nonadditive genetic effects as demonstrated in the case of maize (Crow 2000).

- *Facilitated conventional and molecular genetic studies*: The availability of homozygous progenitors would facilitate greatly the logistics of conventional and molecular genetic studies (Gallais and Bordes 2007; Tuvesson et al. 2007).
- *Shortening the length of breeding cycles*: The starting point for the selection process in Fig. 5.3 could be, for example, 30 plants (as currently required for PYT).
- *True collaboration and synergies among the few cassava breeding programs*: Development and sharing of inbred lines with known characteristics and combining abilities by land granted US universities were the backbone of maize research during a significant part of the last century. Today, the collaboration between cassava programs is limited because of problems related to sharing germplasm (in vitro) and because the materials shared are finished products whose main characteristic is an outstanding performance in certain environments. Ideally, however, breeding programs should exchange good progenitors that can be used by other programs in crosses with local germplasm.

Inbred lines could be produced after successive self-pollinations. A few programs, especially CTCRI in India and IITA in Nigeria, have had medium- or long-term inbreeding programs. The first reported sequenced genome in cassava was on an S_3 line developed at CIAT (Prochnik et al. 2012), which demonstrates that this is a feasible approach for cassava. However, successive self-pollinations are not practical because unavoidably it favours the production of early flowering types which are not desirable for breeding purposes and would require too many years (12–15 years). Instant homozygosity can be achieved through the production of doubled haploids, an advantage that was recognized many years ago (Woodward and Puonti-Kaerlas 2001). There are important ongoing efforts to develop a protocol for the production of doubled haploids through different strategies based on anther and microspore culture (androgenesis), ovary and ovule culture (gynogenesis) and wide crosses with *Ricinus communis* or irradiated pollen (parthenogenesis). There has been considerable progress towards inducing cell division in gametic tissue during the past few years (Perera et al. 2012, 2013), and research is now focusing on regenerating plants. Protocols for early embryo rescue (7–14 days after pollination) have also been developed, and plants were successfully regenerated. Early embryo rescue is fundamental in gynogenesis and parthenogenesis because the absence or abnormal fertilization in which these technologies are based results in weak embryos, poor or absent endosperm development and a strong tendency for fruit abortion.

5.10 Seed Production and Marketing

This chapter will mostly refer to *seed* as the botanical seed resulting from hybridization and used in the initial stages of selection of breeding programs. In this section, however, *seed* refers to the stem pieces or other clonal propagules used for planting the crop in other experimental or commercial production conditions. For the large majority of cassava production around the world, seed systems used by farmers have changed little over the past centuries. Because it is a vegetatively propagated crop, farmers are able to save planting material from their own production fields or trade informally with their neighbours, without any perceived need for a formal seed system to provide certified disease-free, high-quality and true-to-type seed. Farmers are generally underinformed about the potential improvements that can be made to seed quality, and scientists generally have inadequate research evidence to convince farmers of better options. This is due to the inadequate research on the subject, to the poor dissemination of results in easily accessed publications and to the generally poor extension systems to disseminate available information.

The basic feature of this traditional seed system is that farmers either store stems over a short period of time (from several days to a few months but typically a few weeks) or plant immediately after harvest of the previous crop without any significant storage period. Usually there is little or no discrimination between plants that are used for seed and those that are used strictly for root production. There are, however, efforts to promote the idea that part of the field should be targeted as the source of planting material for the following season. This system is described below.

This traditional system will adequately supply grower needs under stable production conditions and where no new varieties are being introduced. But obviously the goal of breeders is to see dissemination and impact from new varieties as efficiently and as quickly as possible. Multiplication and dissemination through traditional systems are very slow processes. This has the advantage of minimizing risk, in that any new variety will be well-proven over many years before it is grown on a large area, making dramatic failure unlikely. Cassava's traditional slow multiplication rate is thus a built-in risk-management system to give varieties a chance to progressively prove themselves in real-life farm and market situations. On the other hand, for varieties that are truly superior, the impacts of their advantages and benefits may only be felt slowly and for a limited number of end users.

This informal system has obvious advantages and a proven track record of success. It has allowed cassava to succeed and expand as a crop over many centuries. Nonetheless, there are a number of changes impacting current and future cassava production which will drive changes in the way seed is produced and marketed. This section will review the drivers of change in cassava seed production and the opportunities afforded by new technologies and new systems.

Traditional seed systems work just fine where there are few production constraints and where farmers have no need or motivation to adopt new varieties at a higher pace. But these conditions are increasingly rare in modern times. Cassava production has continued to expand rapidly in the past 50 years, with much of this expansion in areas of stress such as poor soils and periodic drought, conditions which affect seed quality. In addition, pests and diseases are spreading, especially in Asia and Africa, and one of their primary modes of dispersion is through infested or infected stems used as planting material. Climate change includes enhanced uncertainty in the initiation of the rainy season. Delayed arrival of the rains implies extended storage period of the stems and (eventually) the possibility of sharp reduction in their sprouting capacity.

Consequently, improved seed systems have two main functions, which may be combined or independent in any given situation. First, seed systems are needed to improve access to high-quality seed even where current varieties are grown and used on-farm. Second, seed systems are needed to accelerate access to new varieties by farmers. The CGIAR Research Program on Roots, Tubers and Bananas (CRP-RTB) has developed a framework for intervening in seed systems, specifically aimed at the major vegetatively propagated crops (RTB 2016).

5.10.1 Improving On-Farm Seed Production of Current Varieties

Typically, under stable production systems, about 10% of the plants in a cassava field need to be used for seed for the next planting season, i.e. a reproduction rate of about 1:10. This varies with variety and growing conditions. Typically, farmers do not make a distinction between managing plants that will be used for seed and those that will not. In other words, the planting material may be chosen on the basis of various factors, such as harvest date (i.e. material harvested closest to next planting season will be used for seed) or convenience for transportation of seed to next field for planting. There is generally little perception that different management practices can have significant effect on the quality of seed and consequently yield of the subsequent crop. Nonetheless, some broad guidelines can be provided that will allow farmers to maximize seed quality and the resulting contributions to yield.

A basic tenet of producing high-quality seed on-farm is that production management specifically oriented towards seed quality will be different from management aimed at maximizing returns from harvesting and selling roots. At the same time, since the commercial product (the roots) are not propagative material, there may be no, or only a small, trade-off between maximizing income from roots and producing highest-quality seed. A concept promoted at CIAT and with the Colombian national research agency, Corpoica, is the "corner of prosperity" for seed management. For maintaining the same area of the same varieties from year to year, a farmer will need to use about 10% of the plants from each variety to supply the next season's planting material. CIAT and Corpoica suggest that

10% of the production area (the "corner of prosperity") be dedicated to management that will optimize quality of stem cuttings. This strategy will involve, in broad terms:

- Selection of the best part of the field (well-drained; most fertile soil; protected against mammalian invaders such as deer, wild pigs, etc.; easiest access to irrigation if available).
- Fertilizer to optimize plant health (which may be at levels that are higher than those that maximize net returns for root production).
- Irrigation where extreme drought stress will significantly impact plant growth and development.
- Excellent weed management.
- Periodic inspection for preventive management of any negative impacts.
- Control of pests and diseases, including roguing if virus symptoms appear.
- Harvest of stems near to next season planting, to minimize seed deterioration during storage.
- Harvest stems from plants whose roots have been inspected to be symptom-free from diseases such as CFSD or CBSD.

Unfortunately, there is little quantitative information to indicate specific returns on these various management practices, and there is a high level of need for well-designed research to obtain such data. While there are many publications recommending practices to produce and select high-quality planting material, the quantification of benefits that farmers should expect by specific management of a small part of their production field for high-quality seed is still not well established.

Even if a farmer does not manage a "corner of prosperity" specifically for quality seed production, he or she can adopt several practices to improve seed quality. These have been well-documented in several publications (Ceballos and Calle 2010); Howeler and Maung Aye 2014). One of the key practices will be "positive selection" at harvest time. This concept involves inspection of both tops and roots in order to identify best sources of seed. In many cases, a healthy-looking plant above ground will correspond to healthy roots, high yield, and high-quality seed. However, there are some key seed-disseminated pathogens that may not have visible leaf or stem symptoms. For example, cassava frogskin disease (CFSD) and cassava brown streak disease (CBSD) may both be symptomless above ground while having moderate or even severe root symptoms. In these cases, the stems would carry the pathogen into the next production cycle and initiate a series of cycles of degeneration of seed quality. While there are sophisticated tools such as real-time PCR to detect the virus or phytoplasma, farmers can make good progress in reducing inoculum pressure by harvesting plants before cutting stems for seed and only selecting those plants without root symptoms of CFSD or CBSD. Empirical results have demonstrated that positive selection can efficiently keep CFSD under control in farmers' fields.

5.10.2 Sustainable Systems for Supplying High-Quality Seed of New Varieties

The goal of most breeding programs is to periodically introduce new varieties to farmers – quickly and at scale. The traditional farmer-to-farmer distribution system will not adequately achieve this, and other, more intensive systems of intervention will need to be developed. Unlike the situation for seed crops, and especially hybrid varieties, the private seed sector for cassava is poorly developed and, in fact, non-existent in most countries due to generally low and variable demand. Most commonly, national research and extension agencies work together (sometimes in collaboration with processing enterprises such as starch factories) before and after varietal release with a system of multiplication and distribution to farmers. However, beyond the initial release and small-scale distribution, most countries do not have in place an effective continuing system to provide quality seed.

The majority of the research on developing high-quality seed at mass scale has focused on tissue culture systems and other forms of intensive rapid propagation. However, to date, these intensive systems have had little practical success outside the laboratory and experiment station. Exceptions are primarily where tissue culture has been applied at the very earliest stages of variety multiplication, followed by conventional multiplication through stem cuttings in later stages, up to farmer acquisition and use. Current thinking among most donors and the CGIAR is to promote public/private sector alliances which rely on some public support for start-up costs but ultimately are driven mainly by profits in the private sector to sustain seed systems that provide additional profits to farmers and have zero or minimal ongoing public sector costs. Two prominent examples are seed systems in Uganda and Nigeria. In both cases, the emphasis is on developing viable private sectors. This is a major challenge which has never been achieved on a significant scale. While seed traders are common in major producing countries, there is typically little management for seed quality. In SE Asia, for example, large-scale seed exchange occurs across the Thai-Vietnam-Cambodia borders but with little inspection or control for varietal purity, physiological quality or phytosanitary status. As cassava production intensifies and new superior varieties are more in demand, likewise the incentives for good seed systems will increase and should provide opportunities for sustainable public/private initiatives.

In 2016, a major cassava seed systems' initiative was launched in Nigeria, funded by the Bill and Melinda Gates Foundation and implemented by CRP-RTB and IITA. This initiative pulls together the key lessons from many years of (often-failed) seed systems work, to try to positively affect productivity and profitability in the world's top cassava producer. It is based, first and foremost, on the principles of sustainability, i.e. the mutual long-term profit advantages that will accrue to both private and public/private sector seed producers and farmers. In Nigeria, demand for seed is expected to be driven both by the need to provide seed of varieties degenerated from accumulated effects of CMD and by the demand for new, higher-yielding varieties to supply dynamic cassava markets.

5.10.3 Rapid Multiplication

5.10.3.1 In Vitro Systems to Support Seed Production

When a cassava plant is infected with a virus or phytoplasma, there may be few options to eliminate that pathogen other than tissue culture systems that involve some combination of small apical meristem culturing and thermo- or chemotherapy. Such systems have proven effective for viruses such as CMD and CBSV and for the virus-/phytoplasma-induced CFSD. It is also effective for most bacterial and fungal pathogens and for all insects and mites.

A system of maintaining breeder's seed in an in vitro, pathogen-free environment is fundamental to assuring long-term access to highest-quality, true-to-type seed. Such systems typically need to be managed by only one or a very few laboratories in a given country, due to costs and expertise involved. Theoretically, tissue culture systems can also achieve very high multiplication rates – almost unlimited. At the high end of these possibilities is somatic embryogenesis, where individual cells can be induced to develop into somatic embryos and subsequently, through "artificial seeds," into plants. In this case, millions of plants could be produced from a small amount of leaf tissue and the hardening of plantlets quickly become a major bottleneck. There are intermediate systems, such as micropropagation from in vitro plantlets, where nodal pieces are cut and propagated into new media, successively until the desired number of plants is derived to harden and taken to the field. In vitro systems have even been proposed and tested for use at the village and farm level in order to cut costs and improve efficiency, but these systems have been largely unsuccessful.

5.10.4 Rapid Multiplication from Sprouted Shoots and from Mini-Stakes

Intermediate-level technologies (between the traditional stem cuttings and sophisticated tissue culture systems) have been developed and used to some degree in applied, on-farm systems. If stems are planted horizontally in beds, the growing shoots can be harvested periodically over several months, rooted in water, hardened in containers (e.g. pots or bags) and transplanted to the field to produce conventional seed (mature stems) for further seed production or for commercial root production. In this system, a single plant in the field can theoretically produce several thousand plants within a year's time. However, as with tissue culture systems, the practical use has been quite limited due to costs and the feasibility of covering those costs through added value compared to more conventional multiplication.

A very low-level technology to increase rate of reproduction for cassava is simply to reduce the length of the stem used for propagation. For example, if an average plant can produce ten 20 cm stakes, it could produce forty 5 cm stakes and quadruple the standard multiplication rate. The constraint to this system is that the plants resulting from the shorter stem pieces are likely to be less vigorous and robust, at least initially, which will require additional management inputs such as more fertilizer, water and weed control. There are numerous successful examples of the mini-stake system applied at the experimental level and some cases where it is adopted by progressive farmers.

5.10.4.1 The Way Forward for Cassava Seed Management

There have probably been more failures than successes in investments aimed at developing sustainable improved seed systems for cassava. While there is considerable research describing optimum seed management, there have been relatively few cases where this knowledge has translated to consistent farmer demand for improved seed. This demand is essential to sustain private sector investment. It now seems clear that in the past, donor or publicly funded investments in seed systems have overemphasized the more sophisticated approaches of tissue culture and other technologies that have been difficult to implement in ways that provide clear profit both for seed producers and for farmer-customers. The most difficult challenge for profit-based systems is in situations where the demand for new varieties is low. This can be the case where high-yielding, high-quality varieties are already available, and there is little pressure for farmers to change varieties.

This may be the situation in large areas of SE Asia, where varieties such as KU50 are already widespread and new varieties with clear superior performance are not common. The same is true of areas of Colombia (North Coast and Llanos) and Brazil (South) where new varieties are already widely used. While there is clearly continuing interest in testing and adopting new varieties in these areas, there are not compelling forces for change. On the other hand, in much of Africa, disease pressure, especially CMD and CBSD, is forcing farmers either to look for varieties that are resistant or to purchase clean seed of current varieties.

One of the major lessons from seed systems research in cassava is that clean seed by itself is not sufficient to control seed-borne diseases. For example, the Great Lakes Cassava Initiative (CRS 2010) attempted to produce and distribute seed free of CMV as a means of controlling the disease. However, the reinfection rate was too high, even during the seed multiplication process, such that the farmer had only a small or a short-term advantage to clean seed if the varieties were susceptible. It now seems clear that, in the case of some of the key pathogens, a strategy combining resistance and clean seed systems is essential to success.

5.11 Biotechnology Tools and Their Use in Cassava

A wide array of biotechnology tools has been developed during the past three decades. To facilitate a description of their impact and potential in cassava breeding, they will be grouped into five types and discussed separately below.

The applications and huge potential of tissue culture protocols have already been described earlier.

5.11.1 Molecular Markers for Diversity and Identity Studies

Isozymes were the first type of molecular markers used in cassava. Typically, the first applications focused on analysing genetic diversity or for identification purposes (Ramírez et al. 1987; Hussain et al. 1987; Lefèvre and Charrier 1993). During the 1990s, different types of (DNA or RNA) molecular markers were gradually developed and used for these same initial purposes (Asante and Offei 2003; Carmo et al. 2015; Carvalho and Schaal 2001; Chepkoech et al. 2015; Kawuki et al. 2009; Maredia et al. 2016; Marmey et al. 1993; Moyib et al. 2007; Rabbi et al. 2015; Zacarias et al. 2004). Markers have also been used to distinguish hybrids from self-pollinations in breeding nurseries (Otti et al. 2011). These applications for molecular marker technologies offer clear advantages that become even more evident with their constant reduction of costs and enhancement of discriminating capacity.

The use of markers to identify clones grown by farmers is a powerful tool for impact assessment studies which are always relevant for cassava, considering the informal or weak seed systems. More often than not, there is no reliable way to know the area planted to different cassava varieties (with the exception of Thailand). Diversity studies based on molecular markers have been particularly useful to assess the relationship among different *Manihot* species and the evolution of this genus (Deputié et al. 2011; Olsen and Schaal 2001; Roa et al. 1997; Second et al. 1997).

There is a growing interest and need to conduct diversity studies aiming at identifying potential heterotic groups in cassava. One of the proposed strategies to be able to resume strong genetic gains for productivity would be the implementation of reciprocal recurrent selection (Ceballos et al. 2015) as already done in sweetpotato. This approach relies on two (perhaps three) breeding populations which exhibit high heterosis when crossed with each other (Hallauer and Miranda Fo 1981; Bernardo 2014). Unfortunately, genetic distances, based on molecular markers, do not seem to be good predictors for identifying potential heterotic groups in cassava (Ceballos et al. 2016b). A strategic effort needs to be made in cassava to develop a population structure that would facilitate the creation or identification of heterotic groups. One way to achieve this would be to focus on diverse gene pools that have evolved isolated from each other over a long period of time. CIAT has been working on the definition of diverse gene pools from its large germplasm collection using SNPs markers. Eight subpopulations have emerged from this diversity study (Becerra Lopez-Lavalle 2015). Representatives of each pool could be used initially as a proto-heterotic grouping.

5.11.2 Molecular Markers for Diagnostic Tools and Plant Health

Molecular diagnostic tools can be used for diseases of complex aetiology, such as frogskin disease (Alvarez et al. 2009; Calvert et al. 2008), detection and quantification of viral diseases (Monger et al. 2001; Kaweesi et al. 2014) and analysis of their genetic diversity (Calvert et al. 2008; Legg et al. 2011; Monger et al. 2001), in the characterization and diversity studies of fungal and bacterial diseases (Restrepo and Verdier 1997; Álvarez et al. 2003; Wydra et al. 2004) as well as in gene expression studies in host-pathogen interactions (Kemp et al. 2004, 2005; Fregene et al. 2004; Maruthi et al. 2014). An interesting application of molecular markers has been for the dissection of the pathway leading to post-harvest physiological deterioration (PPD) in cassava roots (Reilly et al. 2007). This is not a comprehensive list of publications on the subject.

The availability of tools for understanding genetic diversity of pests and diseases is an important asset for developing strategies for durable and efficient resistance in cassava. Reliable and affordable diagnostic tools are fundamental for the safe exchange of cassava germplasm.

5.11.3 Marker-Assisted Selection (MAS)

The first molecular map for cassava was published nearly two decades ago (Fregene et al. 1997). Since these pioneering days, a large number of research articles have been published and will not be exhaustively listed here. QTL maps for a broad range of traits (reaction to anthracnose, CMD, CBB, whiteflies, HCN or CG, DMC, early bulking, PPD, plant architecture, carotenoids and protein contents in roots, FRY, etc.) and based on different type of markers (RFLP, AFLP, SSR, DArT, RAPD, etc.) are available (Akinbo et al. 2007, 2011, 2012; Blair et al., 2007; Chen et al. 2012; Ferguson et al. 2012; Kizito et al. 2007; Kunkeaw et al. 2011; Okogbenin and Fregene 2002, 2003; Mkumbira et al. 2003; Whankaew et al. 2011 to list just a few). A comprehensive summary of applications of molecular markers in cassava and progress so far achieved was published by Ferguson and co-workers in 2011. In the last two decades, the cost of genotyping has been drastically reduced. The latest technology of genotyping by sequencing based on SNPs has also been applied to cassava (Rabbi et al. 2014b) and opens up the possibility of implementing genomic selection and marker-assisted recurrent selection as described in the following section.

CMD is not present in Latin America, and therefore nearly all cassava germplasm from this region appears to be susceptible to the disease (Okogbenin et al. 1998). Resistance was reported to be controlled by a single dominant gene, designated as CMD2 (Akano et al. 2002; Fregene et al. 2000) although evidence for different sources has also been reported (Lokko et al. 2006a, b; Okogbenin et al. 2012). CIAT, in collaboration with cassava breeding programs in the African National Agriculture Research System and International Institute of Tropical Agriculture

(IITA), has exploited CMD2 for the development of CMD-resistant Latin American germplasm in the centre of origin using MAS (Akano et al. 2002; Okogbenin et al. 2007, 2012). This is the first report of MAS in cassava. The advantage of the application is obvious: because CMD is absent in the Americas, the availability of molecular markers was the only alternative for selection.

To this day, the selection for resistance to CMD is the only example of MAS applied to cassava (Ferguson et al. 2011). Although markers are available for other traits (e.g. waxy starch, yellow root parenchyma, etc.), their use is economically not competitive. For these high-heritability traits, selection could be made as early as 6 months after planting a seedling nursery (Belalcazar et al. 2016), with the advantage that along with the key trait, the breeder can select for other characteristics (vigour, plant architecture, resistance to thrips, etc.). Typically, a seedling nursery for these types of traits has around 15,000–20,000 genotypes. Growing 1–2 ha of seedling plants is still much less expensive than extracting and analysing DNA samples for so many genotypes, but costs for the latter continue to decline.

The use of markers for selection purposes in a large segregating population has so far failed to deliver the expected benefits. MAS, however, could have interesting applications for the selection of progenitors (Ceballos et al. 2016a). For example, progenitors that are homozygous for CMD2 would have twice the breeding value (for this trait) compared with heterozygous progenitors. Markers could be used to identify genotypes that are homozygous for CMD2. In the case of carotenoids, it may be desirable to reduce the activity of carotenoid β-hydroxylase, controlling the conversion of β-carotene into other molecules. Molecular markers could be used to identify genotypes heterozygous for this gene, self-pollinate them and select progenies that are homozygous recessive (Ceballos et al. 2013; Morillo-C et al. 2012). Again, by this process, the breeding value of such self-pollinated genotypes would be better than that of its progenitor. As is often the case, application of new technologies to cassava requires special adaptation. Most efforts to apply MAS in cassava have been for selection in segregating progenies, which has until now had negligible impact. However, no effort has been made so far, in the selection of better progenitors.

The genome of cassava has been sequenced (Bredeson et al. 2016; Prochnik et al. 2012). The information generated by these studies and the public access to it is very relevant. For example, it has been used for determining the sequence in cassava of key genes related to starch biosynthesis or herbicide tolerance. With that information, reverse genetic approaches to screen germplasm collections in search of sources of useful alleles is an alternative that has been already been initiated (Duitama et al. 2017) as described earlier.

5.11.4 Genomic Selection

Genomic selection (GS) simultaneously tags many loci across the entire genome to estimate genomic estimated breeding values or GEBV. It offers several advantages and overcomes key problems of MAS based on QTLs. Key among the different advantages is that several different traits can be improved simultaneously through a selection index, similar to those based on phenotypic traits and economic values. de Olviera and co-workers suggested the potential of GS for cassava in 2012. GS was also proposed for other crops (Heslot et al. 2012) characterized by the fact that breeding is based on the use of inbred progenitors. Different articles emphasize that GS would maximize genetic gains by unit of time (De Oliveira et al. 2012; Heffner et al. 2009). Like conventional breeding, GS has serious limitations for the selection of low narrow-sense heritability traits because of their low additive genetic effects in relation to the phenotypic variance. Genotype-by-environment interactions also affect the precision of GS estimates as well as the relatedness among genotypes. Ceballos et al. (2015) predicted that GS would be effective in high-heritability traits such as plant architecture, pest and disease resistance or DMC. However, GS was not expected to be efficient in improving FRY because of the relatively high influence of nonadditive genetic effects and the large within-family genetic variation generated by the heterozygous progenitors used in cassava (Table 5.1). Other factors may limit further the usefulness of breeding value in cassava (Ceballos et al. 2016a; Joaqui et al. 2016).

There is an ongoing project (NextGen Cassava Project) to test GS in cassava led by Cornell University with field work in Uganda and Nigeria (Rabbi et al. 2015). As expected, GS is proving to be efficient improving traits such as resistance to CMD (Wolfe et al. 2016b; Rabbi et al. 2014a). Whereas DMC was also increased considerably, progress increasing FRY was, as predicted, not satisfactory (Wolfe et al. 2016a). An honest assessment of cost/benefit for technologies such as GS will have to be made in the years to come. It should be pointed out that improving resistance to CMD or increasing DMC is not enough for such an expensive approach that has the additional restriction of the need of "closed" breeding populations. The use of inbred progenitors would overcome some of the problems that implementing GS in cassava has. The induction of flowering would also benefit greatly the success of GS. Alternatives of GS combined with reciprocal recurrent selection would also be an interesting approach to overcome, at least partially, the problems of nonadditive genetic effects influencing FRY, which is ultimately the most important trait to improve.

5.11.5 Genetic Transformation and Gene Editing

The first published reports of genetic transformation in cassava date back from two decades ago (Li et al. 1996; Raemakers et al. 1996; Schöpke et al. 1996; Sarria et al. 1995). The technology offers a great potential considering the problems related to trait introgression discussed earlier, challenges to improve the productivity after successful clones such as KU50 were released and limited knowledge of genetic variability available in cassava germplasm. As it is often the case, genetic transformation initially suffered from genotypic dependency: initially a single genotype could be efficiently transformed (60444). However, protocols were improved quickly, and many genotypes have now been transformed (Chauhan et al. 2015; Liu et al. 2011; Raemakers et al. 2001; Taylor et al. 2004, 2012; Zainuddin et al. 2012). The commonly used methods for the genetic transformation of cassava include *Agrobacterium*-mediated gene delivery and particle bombardment. The explants used for transformation include somatic cotyledons and friable embryogenic calli.

A broad range of traits have been considered for genetic transformation: (a) Resistance to CMD and CBSD (Bi et al. 2010; Chellappan et al. 2004; Ntui et al. 2015; Patil et al. 2011; Vanderschuren et al. 2007; Zhang et al. 2005; Yadav et al. 2011); (b) Enhanced nutritional quality of the roots including high carotenoids, Fe, Zn, proteins as well as reduction in cyanogenic glucosides (Jørgensen et al. 2005; Leyva-Guerrero et al. 2012; Sayre et al. 2011; Welsch et al. 2010); (c) Quantity and quality of starches (Ihemere et al. 2006; Koehorst-van Putten et al. 2012; Zhao et al. 2011); (d) Reduction of PPD (Zidenga et al. 2012) and physiological traits such as leaf retention (Zhang et al. 2010); (e) Herbicide tolerance (Sarria et al. 2000, initially reported in 1995); (f) Induction of flowering (Adeyemo et al. 2008).

Genetic transformation for new starch types (e.g. waxy starch) has been successful and the phenotype of transgenic material fulfilled expectations regarding amylose content in the starch (Koehorst-van Putten et al. 2012; Zhao et al. 2011). The discovery of a spontaneous mutation for waxy starch may have reduced the appeal of waxy transgenic cassava. However, the complications of introgressing a single recessive gene into successful commercial varieties should be taken into consideration, at least for countries where the growth and commercialization of transgenic crops/products are more relaxed. Improving AGPase activity can contribute to the conversion of sugar to starch and subsequently increase the starch quantity. Early work (Ihemere et al. 2006) suggested that this is indeed the case. However, no further research on the subject has been published since then. Smith (2008) pointed out some of the potential stumbling blocks that may explain why no further efforts in this area have been published.

With the support of the Bill and Melinda Gates Foundation and several years of global cooperation, the BioCassava Plus project team has developed a number of transgenic cassava lines with value-added traits, such as improved protein content and increased vitamin A, iron and zinc contents. In spite of the promising results forecasted by Sayre and co-workers in 2011 - genetic transformation regarding enhanced nutritional quality, not a single product has shown promising results in the field. In the case of high carotenoids, transgenic materials show a drastic reduction of DMC. Early claims for enhanced protein content in the root had to be retracted. In addition, attempts to elevate protein levels in roots had impacts on overall nitrogen allocation between leaves and roots reflecting the strong leaf sink strength for reduced nitrogen (Leyva-Guerrero et al. 2012). Recent efforts by BioCassava Plus initiative using a new construct for enhanced Fe and Zn in the roots have yielded promising results (Gaitán-Solís et al. 2015).

The alternative to develop virus resistant transgenic cassava is also a justified and relevant objective. Although there is genetic resistance to CMD, apparently it is based on a single source. Chance of the resistance breaking down is a feasible threat, thus justifying the development of a new source of resistance. In the case of CBSD, available natural resistance is not as effective as in the case of CMD2, and, therefore, efforts have concentrated on resistance to CBSD. Transformed materials have shown excellent levels of resistance to CBSD (Chauhan et al. 2015; Odipio et al. 2014; Vanderschuren et al. 2012). However, transgenic material resistant to CBSD would not be useful if they are susceptible to CMD. Therefore, farmers preferred varieties carrying the CMD2 resistance were transformed to incorporate the proven resistance to CBSD. Unfortunately, the process of somatic embryogenesis used to regenerate cassava caused the resulting plants to become susceptible to CMD, even though the original clones carried the CMD2 resistance (Beyene et al. 2016). This is a systematic process observed by two independent research groups.

Genetic transformation remains a promising technology. Early experiences have been humbling and have helped expose the complexities in plant physiology and metabolism. Hopes to increase productivity or enhanced levels of proteins in the roots may need a long time to materialize. But for simpler goals such as starch variants, virus resistance (in spite of the current drawback of loss of CMD2 resistance), herbicide tolerance and (perhaps) tolerance to PPD, the technology may have a positive impact. Advances have also been made regarding the legal framework for field evaluation of transgenic materials in different African countries, Colombia, Indonesia and China. No commercial exploitation of transgenic cassava has been requested or authorized yet.

Associated with the potential of genetic transformation is the emerging alternative of gene editing, for example, based on CRISPR (Odipio et al. 2016). Gene editing can overcome some of the regulatory problems that genetic transformation has. Regeneration of protoplasts is feasible in cassava (Sofiari et al. 1998). This is important because regulatory issues are less limiting when regeneration is from edited protoplasts. Four copies of the GUS gene have already been edited in cassava (P. Chavarriaga, personal communication), and it is expected that soon field evaluations of edited cassava will be reported. Editing inbred progenitors would also allow overcoming the regulatory issues, and this is yet another reason justifying the need to develop inbred cassava.

5.12 Future Prospects and Outlook

Cassava is today a key food security staple and a competitive commodity for different industries. It will be even more important in the next few decades. Population growth will be particularly high in Africa, where cassava is a fundamental source of calories for the diet of millions of people. Increases in the concentration of CO_2 in the air will improve its productivity, thus reducing the physiological advantage that C4 crops such as maize have as a feedstock for industrial processes. There is, however, an urgent need to fundamentally change the way cassava is bred, so that major advances in productivity can be attained again and new technologies can finally have a positive impact on the crop.

Cassava is the only diploid species, among the major staple crops, whose breeding is based on the use of heterozygous progenitors. Breeding should shift from the current system in which elite progenitors are crossed – hoping to find an outstanding hybrid by a process that is largely trial and error – to improving progenitors that would more predictably produce better clones. Initially, this approach is more complex because a pre-breeding step (e.g. parental development) needs to be added. However, the added complexity is only temporary. Once a group of a few elite inbred progenitors, producing outstanding hybrids, is identified, the system becomes much simpler and more efficient. There are several advantages that justify the use of inbred progenitors: genetic variability would be conveniently partitioned in ways that breeders can more easily exploit; dominance and epistatic effects can be efficiently and predictably exploited when inbred progenitors are used; heterosis can gradually and systematically be enhanced; trait introgression would not result in yield penalties as is happening today and back-crossing could be implemented; storage and exchange of germplasm (as botanical seed that breeds true) would be facilitated; and collaboration among the few cassava breeding programs will be greatly facilitated allowing true synergies among them.

Genomic selection offers a great potential. It allows identifying genotypes carrying a set of desirable alleles that can reliably be identified as such. However, genomic selection only allows "seeing" the good and bad alleles that each genotype has. Combining desirable alleles and maintaining them together are nearly impossible if heterozygous progenitors are used. On the other hand, if inbred progenitors are used, desirable alleles can gradually and systematically be assembled in a given genotype. Every step accumulating desirable alleles is maintained. In the current system, every meiotic event recombines randomly the alleles, essentially erasing a considerable proportion of the genetic progress made in generating the genotype where it takes place. Current applications of GS in maize breeding could be readily implemented for cassava, if inbred progenitors were available.

Other applications of molecular markers can have an impact in cassava breeding as well. However, cost/benefits should be taken into account. MAS for the sake of using it should be avoided. There are examples where molecular markers offer appealing advantages that have not been exploited. Improving the breeding value of progenitors could have an immediate and obvious impact. Molecular markers could be used to screen germplasm collections in search of sources of useful mutations. The discovery of the waxy starch mutation in cassava illustrates the wealth of genetic diversity that has remained hidden in germplasm collections and needs to be exposed and exploited. Screening germplasm collections through conventional phenotyping methods is cumbersome, time consuming and expensive. Molecular tools offer clear advantages that have not yet been considered. These applications of molecular markers are helpful to support the idea that new technologies in cassava need to be applied but in ways that differ from the conventional uses for other crops.

Genetic transformation in cassava has evolved considerably in the last two decades. Protocols of genetic transformation in cassava have improved considerably in the last two decades. Protocols are efficient and many cultivars can be transformed. A broad set of traits have been addressed by several laboratories that have succeeded transforming cassava in Africa, Asia, Latin America, Europe and the USA. Although there have been unexpected drawbacks, they have contributed to a better understanding of the complexities of gene regulation and expression, and some of them should be quickly overcome. It would be very useful to have at least one example of commercial exploitation of transgenic technologies in cassava. Surprisingly no effort has been made to release herbicide-tolerant transgenic cassava. Weed management is labour-intensive (requiring one to two person-months each year per hectare) and is often carried out by women. Herbicide tolerance would also allow direct planting, with further protection of the environment, reducing costs and allowing more mechanized operations. This would also be fundamental for the future of cassava as a competitive feedstock for industrial uses.

Finally, as different emerging technologies are developed and applied, there is a tension between the old and the new approaches. Young scientists, trained in novel technologies, have the legitimate interest in applying them to cassava. Cassava research now is a multipronged effort with limited coordination. Scientists unavoidably promote the use of the technologies with which they work. This is a reasonable attitude. However, farmers' needs require that the best and most efficient technologies are used to maximize the chances of developing superior hybrids. There is also a tension between centres that specialize in a particular technology and scan for opportunities to apply them. The commitment of these centres is for these technologies, not to cassava. Ideally, resources should be allocated to institutions and programs that have clear and long-standing commitment to cassava research.

Acknowledgments This research was undertaken as part of the CGIAR Research Program on Roots, Tubers and Banana (RTB) and HarvestPlus, part of the CGIAR Research Program on Agriculture for Nutrition and Health (A4NH). It has also received financial support from the Bill and Melinda Gates Foundation and USAID.

References

Abraham A (1957) Breeding of tuber crops in India. Indian J Gen Plant Breed 17:212–217

Adeyemo OS, Davis SJ, Chavarriaga P, Tohme J, Ceballos H, Fregene M (2008) Strategies for enhancing flowering in cassava using molecular tools: towards a more efficient breeding program. First Scientific Meeting of the Global Cassava Partnership GCP-1. Ghent. p 51

Aiemnaka P, Wongkaew A, Chanthaworn J et al (2012) Molecular characterization of a spontaneous waxy starch mutation in cassava (*Manihot esculenta* Crantz). Crop Sci 52:2121–2130

Akano AO, Dixon AGO, Mba C, Barrera E, Fregene M (2002) Genetic mapping of a dominant gene conferring resistance to cassava mosaic disease. Theor Appl Genet 105:521–525

Akinbo O, Gedil M, Ekpo EJA, Oladele J, Dixon AGO (2007) Detection of RAPD markers-linked to resistance to cassava anthracnose disease. Afr J Biotechnol 6:677–682

Akinbo O, Labuschagne M, Fregene M (2010) Embryo rescue as a method to develop and multiply a backcross population of cassava (*Manihot esculenta* Crantz) from an interspecific cross of *Manihot esculenta* ssp. *flabellifolia*. Afr J Biotechnol 9:7058–7062

Akinbo O, Labuschagne M, Fregene M (2011) Introgression of whitefly (Aleurotrachelus socialis) resistance gene from F1 inter-specific hybrids into commercial cassava. Euphytica 183:19–26

Akinbo O, Labuschagne M, Marin J et al (2012) QTL analysis for root protein in a backcross family of cassava derived from *Manihot esculenta* ssp *flabellifolia*. Trop Plant Biol 5:1–12

Allem AC (1999) The closest wild relatives of cassava (*Manihot esculenta* Crantz). Euphytica 107:123–133
Allem AC (2002) The origins and taxonomy of cassava. In: Hillocks RJ, Tresh JM, Bellotti AC (eds) Cassava: biology, production and utilization. CABI Publishing, Wallingford, pp 1–16
Allem AC, Mendes RA, Salomão AN, Burle ML (2001) The primary gene pool of cassava (*Manihot esculenta* Crantz subspecies *escultenta*, Euphorbiaceae). Euphytica 120:127–132
Alvarez E, Mejía JF, Llano GA, Loke JB, Calari A, Duduk B, Bertaccini A (2009) Characterization of a phytoplasma associated with frogskin disease in cassava. Plant Dis 93:1139–1145
Álvarez E, Mejía JF, Valle T (2003) Molecular and pathogenicity characterization of *Sphaceloma manihoticola* isolates from south-central Brazil. Plant Dis 87:1322–1328
Angel F, Barney VE, Tohme J, Roca WM (1996) Stability of cassava plants at the DNA level after retrieval from 10 years of in vitro storage. Euphytica 90:307–313
Anyanwu CN, Ibeto CN, Ezeoha SL, Ogbuagu NJ (2016) Sustainability of cassava (*Manihot esculenta* Crantz) as industrial feedstock, energy and food crop in Nigeria. Renew Energy 81:745–752
Asante IK, Offei SK (2003) RAPD-based genetic diversity study of fifty cassava (*Manihot esculenta* Crantz) genotypes. Euphytica 131:113–119
Ashby JA, Quirós CA, Rivera YM (1987) Farmer participation in on-farm variety trials. Agricultural Administration (Research and Extention) Network Discussion Paper No. 22. London. Overseas Development Institute
Balagopalan C (2002) Cassava utilization in food, feed and industry. In: Hillocks RJ, Tresh JM, Bellotti AC (eds) Cassava: biology, production and utilization. CABI Publishing, Wallingford, pp 301–318
Becerra Lopez-Lavalle LA (2015) Revisiting cassava genetic diversity reveals eco-geographic signature of the crop's domestication. Plant and Animal Genome XXIII, San Diego
Beeching JR, Marmey P, Gavalda MC, Noirot M, Haysom HR, Hughes MA, Charrier A (1993) An assessment of genetic diversity within a collection of cassava (Manihot esculenta Cranz) germplasm using molecular markers. Ann Bot 72:515–520
Belalcazar J, Dufour D, Pizarro M et al (2016) High-throughput phenotyping and improvement in breeding cassava for increased carotenoids in the roots. Crop Sci 56:2916–2925
Bellotti AC, Arias B (2001) Host plant resistance to whiteflies with emphasis on cassava as a case study. Crop Prot 20:813–823
Benesi IRM, Labuschagne MT, Herselman L, Mahungu N (2010) Ethnobotany, morphology and genotyping of cassava germplasm from Malawi. J Biol Sci 10:616–623
Bernardo R (2014) Essentials of plant breeding. Stemma Press, Woodbury
Bertram RB (1993) Application of molecular techniques to genetic resources of cassava (Manihot esculenta Crantz, Euphorbiaceae): interspecific evolutionary relationships and intraspecific characterization. Ph.D. dissertation. University of Mariland. 465 pp
Bertram RB, Schaal BA (1993) Phylogeny of *Manihot* and the evolution of cassava. CBN Newsl 1:4–6
Beyene G, Chauhan RJ, Wagaba H et al (2016) Loss of CMD2-mediated resistance to cassava mosaic disease in plants regenerated through somatic embryogenesis. Mol Plant Pathol 17:1095–1110
Bi H, Aileni M, Zhang P (2010) Evaluation of cassava varieties for cassava mosaic disease resistance jointly by agro-inoculation screening and molecular markers. African J Plant Sci 4(9):330–338
Biggs BJ, Smith MK, Scott KJ (1986) The use of embryo culture for the recovery of plants from cassava (Manihot esculenta Crantz) seeds. Plant Cell Tissue Organ Cult 6:229–234
Blair MW, Fregene MA, Beebe SE, Ceballos H (2007) Marker assisted selection in common Beans and Cassava. In: Guimaraes EP, Ruane J, Scherf BD, Sonnino A, Dargie JD (eds) Marker-Assisted Selection (MAS) in crops, livestock, forestry and fish: current status and the way forward. Food and Agriculture Organization of the United Nations (FAO), Rome, pp 81–115
Bolhuis GG (1953) A survey of some attempts to breed cassava varieties with a high content of proteins in the roots. Euphytica 2:107–112
Bredeson JV, Lyons JB, Prochnik SE et al (2016) Sequencing wild and cultivated cassava and related species reveals extensive interspecific hybridization and genetic diversity. Nat Biotechnol. doi:10.1038/nbt.3535

Bueno A (1991) Estimación de los parámetros genéticos en la yuca. In: Hershey CH (ed) Mejoramiento genético de la yuca en América Latina. CIAT Publication, Cali, pp 197–220
Buitrago AJA (1990) La yuca en la alimentación animal. Centro Internacional de Agricultura Tropical. CIAT, Cali. 446 p
Bull SE, Owiti JA, Niklaus M, Beeching JR, Gruissem W, Vanderschuren H (2009) Agrobacterium-mediated transformation of friable embryogenic calli and regeneration of transgenic cassava. Nat Protoc 4:1845–1854
Byrne D (1984) Breeding cassava. Plant Breed Rev 2:73–134
Cach NT, Perez JC, Lenis JI, Calle F, Morante N, Ceballos H (2005) Epistasis in the expression of relevant traits in cassava (Manihot esculenta Crantz) for subhumid conditions. J Hered 96(5):586–592
Calvert LA, Cuervo M, Lozano I, Villareal N, Arroyave J (2008) Identification of three strains of a virus associated with cassava plants affected by frogskin disease. J Phytopathol 156:647–653
Caparros Megido R, Alabi T, Nieus C et al (2016) Optimisation of a cheap and residential small-scale production of edible crickets with local by products as an alternative protein-rich human food source in Ratanakiri Province, Cambodia. J Sci Food Agric 96:627–632
Carmo CD, Santos DB, Alves LB, Oliveira GAF, Oliveira EJ (2015) Development of TRAP (Target Region Amplification Polymorphism) as new tool for molecular genetic analysis in cassava. Plant Mol Biol Rep 33:1953–1966
Carvalho C-BLJ, Schaal BA (2001) Assessing genetic diversity in the cassava (Manihot esculenta Crantz) germplasm collection in Brazil using PCR-based markers. Euphytica 120:133–142
Carvalho LJCB, de Souza CRB, Cascardo JCM, Junior CB, Campos L (2004) Identification and characterization of a novel cassava (Manihot esculenta Crantz) clone with high free sugar content and novel starch. Plant Mol Biol 56:643–659
Ceballos H, Becerra LA, Calle F, Morante N, Ovalle T, Pérez JC, Hershey C (2016) Genetic distance and heterosis in cassava. Euphytica 210:79–92. doi:10.1007/s10681-016-1701-7
Ceballos H, Calle F (2010) Cassava. In: Fajardo J, NeBambi L, Larinde MI, Rosell C, Barker I, Roca W, Chujoy E (eds) Quality declared planting material. Protocols and standards for vegetatively propagated crops. FAO Plant Production and Protection Paper 195. p 33–40. Food and Agriculture Organization (FAO) of the United Nations, Rome
Ceballos H, Fregene M, Pérez JC, Morante N, Calle F (2007) Cassava genetic improvement. In: Kang MS, Priyadarshan PM (eds) Breeding major food Staples. Blackwell Publishing, Ames, pp 365–391
Ceballos, Hershey (2016) Road map for cassava genetic improvement. Proceedings of the World Congress on Root and Tuber Crops and Third Scientific Conference of the Global Cassava Partnership for the 21st Century. Nanning, China. January 2016
Ceballos H, Hershey C, Becerra-López-Lavalle LA (2012) New approaches to cassava breeding. Plant Breed Rev 36:427–504. Hoboken, Wiley-Blackwell
Ceballos H, Iglesias CA, Pérez JC, Dixon AGO (2004) Cassava breeding: opportunities and challenges. Plant Mol Biol 56:503–515
Ceballos H, Jaramillo JJ, Salazar S, Pineda LM, Calle F, Setter T (2017) Induction of flowering in cassava through grafting. J Plant Breed Crop Sci 9:19–29
Ceballos H, Kawuki RS, Gracen VE, Yencho GC, Hershey CH (2015) Conventional breeding, marker assisted selection, genomic selection and inbreeding in clonally propagated crops: a case study for cassava. Theor Appl Genet 128:1647–1667
Ceballos H, Morante N, Sánchez T, Ortiz D, Aragón I, Chávez AL, Pizarro M, Calle F, Dufour D (2013) Rapid cycling recurrent selection for increased carotenoids content in cassava roots. Crop Sci 53:2342–2351
Ceballos H, Okogbenin E, Pérez JC, Becerra LA, Debouck D (2010) Cassava. In: Bradshaw J (ed) Root and tuber crops. Springer Publishers, New York, pp 53–96
Ceballos H, Pérez JC, Joaqui-B O, Lenis JI, Morante N, Calle F, Hershey H (2016) Cassava breeding I: the value of breeding value. Front Plant Sci. doi:10.3389/fpls.2016.01227
Ceballos H, Ramirez J, Bellotti AC, Jarvis A, Alvarez E (2011) Adaptation of cassava to changing climates. In: Yadav S, Redden B, Hatfield JL, Lotze-Campen H (eds) Crop adaptation to climate change. Wiley-Blackwell Publishers. pp 411–425

Ceballos H, Sánchez T, Morante N et al (2007) Discovery of an amylose-free starch mutant in cassava (Manihot esculenta Crantz). J Agric Food Chem 55:7469–7476
Ceballos H, Sánchez T, Tofiño AP et al (2008) Induction and identification of a small-granule, high-amylose mutant in cassava (Manihot esculenta Crantz). J Agric Food Chem 56:7215–7222
Chalwe A, Melis R, Shanahan P, Chiona M (2015) Inheritance of resistance to cassava green mite and other useful agronomic traits in cassava grown in Zambia. Euphytica 205:103–119
Chauhan RD, Beyene G, Kalyaeva M, Fauquet CM, Taylor N (2015) Improvements in Agrobacterium-mediated transformation of cassava (*Manihot esculenta* Crantz) for large-scale production of transgenic plants. Plant Cell Tiss Org Cult 121:591–603
Chellappan P, Masona MV, Vanitharani R, Taylor NJ, Fauquet CM (2004) Broad spectrum resistance to ssDNA viruses associated with transgene-induced gene silencing in cassava. Plant Mol Biol 56:601–611
Chen X, Fu Y, Xia Z, Jie L, Wang H, Lu C, Wang W (2012) Analysis of QTL for yield-related traits in cassava using an F_1 population from non-inbred parents. Euphytica 187:227–234
Chen H, Xu M-l, Guo Q, Yang L, Ma Y (2016) A review on present situation and development of biofuels in China. J Energy Inst 89:248–255
Chepkoech E, Kinyua M, Kiplagat O, Arunga EE, Kimno S (2015) Genetic diversity of cassava mutants, hybrids and landraces using simple sequence repeat markers. Am J Exp Agric 5(4):287–294
Chipeta MM, Bokosi JM, Saka VW, Benesi IRM (2013) Combining ability and mode of gene action in cassava for resistance to cassava green mite and cassava mealy bug in Malawi. J Plant Breed Crop Sci 5:195–202
CIAT (1991) In: Hershey CH (ed). Mejoramiento genético de la yuca en América Latina. CIAT, Cali p 426. (ISBN 958-9183-16-6)
Cock J (1985) Cassava. New potential for a neglected crop. Westview Press, Boulder
Cours G (1951) Le manioc a Madagascar. Memoires de L'Institut Scientifique de Madagascar Serie B 3:203–400
Crow JF (2000) The rise and fall of overdominance. Plant Breed Rev 17:225–257
CRS (2010) Final report on the Great Lakes Cassava initiative. 137p. http://www.crs.org/sites/default/files/tools-research/final-report-great-lakes-cassava-initiative.pdf
Davrieux F, Dufour D, Dardenne P et al (2016) LOCAL regression algorithm improves NIRS predictions when the target constituent evolves in breeding populations. JNIRS 24:109–117
De Bruijn GH (1977) Influence of day length on the flowering of cassava. Trop Root Tuber Crops Newsl 10:1–3
De Carvalho RD, Guerra-M M (2002) Cytogenetics of Manihot esculenta Crantz (cassava) and eight related species. Hereditas 136:159–168
De Oliveira EJ, de Resende V, da Silva Santos V et al (2012) Genome-wide selection in cassava. Euphytica 187:263–276
Delgado A, Hays DB, Bruton RK, Ceballos H, Novo A, Boi E, Gomez Selvaraj M (2017) Ground penetrating radar: a case study for estimating root bulking rate in cassava (*Manihot esculenta* Crantz). Plant Methods 13:65. https://doi.org/10.1186/s13007-017-0216-0
Deputié A, Salick J, McKey D (2011) Evolutionary biogeography of Manihot (Euphorbiaceae), a rapidly radiating Neotropical genus restricted to dry environments. J Biogeogr 38(6):1033–1043
Diasolua Ngudi D, Kuo Y-H, Lambein F (2003) Amino acid profiles and protein quality of cooked cassava leaves or 'saka-saka'. J Sci Food Agric 83:529–534
Duitama J, Kafuri L, Tello D, Leiva AM, Hofinger B, Datta S, Lentini Z, Aranzales E, Till B, Ceballos H (2017) Deep assessment of genomic diversity in cassava for herbicide tolerance and starch biosynthesis. Comput Struct Biotechnol J 15:185–194
Easwari Amma CS, Sheela N, Thankamma Pillai PK (1995) Combining ability analysis in cassava. J Root Crops 21(2):65–71
El-Sharkawy MA (2009) Pioneering research on C4 leaf anatomical, physiological, and agronomic characteristics of tropical monocot and dicot plant species: implications for crop water relations and productivity in comparison with C3 cropping systems. Photosynthetica 47:163–183
Epperson JE, Pachico DH, Guevara CL (1997) A cost analysis of maintaining cassava plant genetic resources. Crop Sci 37:1641–1649

Escobar RH, Mafla G, Roca WM (1997) A methodology for recovering cassava plants from shoot tips maintained in liquid nitrogen. Plant Cell Rep 16:474–478

Esuma W, Herselman L, Labuschagne MT et al (2016) Genome-wide association mapping of provitamin A carotenoid content in cassava. Euphytica. doi:10.1007/s10681-016-1772-5

Falconer DS (1981) Introduction to quantitative genetics, 2nd edn. Longman, New York

FAO (2013) In: Howeler R, NeBambi L, Thomas G (eds) Save and grow: cassava. A guide to sustainable production intensification. Food and Agriculture Organization (FAO) of the United Nations, Rome, p 129 p. ISBN 978-92-5-107641-5

Ferguson ME, Hearne SJ, Close TJ et al (2012) Identification, validation and high-throughput genotyping of transcribed gene SNPs in cassava. Theor Appl Genet 124(4):685–695

Ferguson ME, Rabbi IY, Kim DJ, Gedil M, Becerra Lopez-Lavalle LA, Okogbenin E (2011) Molecular markers and their application to cassava breeding: past, present and future. Trop Plant Biol 5:95–109

Fisher T, Buerlee D, Edmeades G (2014) Crop yields and global food security. Will yield increase continue to feed the world? ACIAR Monograph, vol 158. Australian Centre for International Agricultural Research, Canberra, pp 285–289

Fregene M, Angel F, Gomez R, Rodríguez F, Chavarriaga P, Roca W, Tohme J (1997) A molecular genetic map of cassava (Manihot esculenta Crantz). Theor Appl Genet 95:431–441

Fregene M, Bernal A, Duque M, Dixon A, Tohme J (2000) AFLP analysis of African cassava (Manihot esculenta Crantz) germplasm resistant to the cassava mosaic disease (CMD). Theor Appl Genet 100:678–685

Fregene M, Matsumura H, Akano A, Dixon A, Terauchi R (2004) Serial analysis of gene expression (SAGE) of host–plant resistance to the cassava mosaic disease (CMD). Plant Mol Biol 56(4):563–571

Fregene M, Ospina JA, Roca W (1999) Recovery of cassava (Manihot esculenta Crantz) plants from culture of immature zygotic embryos. Plant Cell Tissue Organ Cult 55:39–43

Gaitán-Solís E, Taylor NJ, Siritunga D, Stevens W, Schachtman DP (2015) Overexpression of the transporters AtZIP1 and AtMTP1 in cassava changes zinc accumulation and partitioning. Front Plant Sci 6:492. https://doi.org/10.3389/fpls.2015.00492

Gallais A, Bordes J (2007) The use of doubled haploids in recurrent selection and hybrid development in maize. Crop Sci 47:S191–S200

Gardner CO (1961) An evaluation of effects of mass selection and seed irradiation with thermal neutrons on yields of corn. Crop Sci 1:241–245

Gonçalves Fukuda WM, de Oliveira, Silva S, Iglesias C (2002) Cassava breeding. Crop Breed Appl Biotech 2(4):617–638

Gonçalvez Fukuda WM, Saad N (2001) Participatory research in cassava breeding with farmers in Northeastern Brazil. Document CNPMF 99. EMBRAPA, Cruz das Almas. Bahia

Graner EA (1935) Contribuição para o estudo citológico da mandioca. Escola Superior de Agricultura Luiz de Queiroz, Piracicaba, pp 1–28

Graner EA (1942) Genética da *Manihot*. I Hereditariedade da forma da folha e da coloração da película externa das Raízes em *Manihot utilíssima*. Pohl Bragantia 2:13–22

Grüneberg W, Mwanga R, Andrade M, Espinoza J (2009) Breeding clonally propagated crops. In: Ceccarelli S, Guimarães EP, Weltzien E (eds) Plant breeding and farmer participation. Food and Agriculture Organization of the United Nations, Rome, pp 175–322

Hahn SK, Bai KV, Asiedu R (1990) Tetraploids, triploids, and 2n pollen from diploid interspecific crosses with cassava. Theor Appl Genet 79:433–439

Hahn SK, Terry ER, Leuschner K, Akobundu IO, Okali C, Lal R (1979) Cassava improvement in Africa. Field Crops Res 2:193–226

Hallauer AR, Miranda Fo JB (1981) Quantitative genetics in maize breeding, 2nd edn. Iowa State University Press, Ames

Heffner EL, Sorrells ME, Jannink J-L (2009) Genomic selection for crop improvement. Crop Sci 49:1–12

Hershey CH. 2008. A global conservation strategy for cassava (Manihot esculenta) and wild Manihot species (https://www.croptrust.org/wp-content/uploads/2014/12/cassava-strategy.pdf)

Hershey CH, Ocampo-N CH (1989) New marker genes found in cassava. Cassava Newsl 13(1):1–5

Heslot N, Yang H-P, Sorrells ME, Jannink J-L (2012) Genomic selection in plant breeding: a comparison of models. Crop Sci 52:146–160
Howeler RH, Maung Aye T (2014) Sustainable management of cassava in Asia: from research to practice. International Centre for Tropical Agriculture (CIAT); The Nippon Foundation, Vietnam. 168 p. (http://hdl.handle.net/10568/65172)
Hussain A, Brushuk W, Ramírez H, Roca W (1987) Identification of cassava (Manihot esculenta Crantz) cultivars by eletrophoretic patterns of esterase isozymes. Seed Sci Technol 15:19–22
IBGE (2014) Instituto Brasileiro de Geografia e Estatística (http://www.ibge.gov.br/). Produção Agrícola Municipal
Ibrahim AB, Heredia FF, Pinheiro CB, Aragao FJL, Campos FAP (2008) Optimization of somatic embryogenesis and selection regimes for particle bombardment of friable embryogenic callus and somatic cotyledons of cassava (Manihot esculenta Crantz). Afr J Biotechnol 7:2790–2797
Iglesias C, Mayer J, Chávez AL, Calle F (1997) Genetic potential and stability of carotene content in cassava roots. Euphytica 94:367–373
Ihemere U, Arias-Garzon D, Lawrence S, Sayre R (2006) Genetic modification of cassava for enhanced starch production. Plant Biotechnol J 4:453–465
Jennings DL (1957) Further studies in breeding cassava for virus resistance. E Afr Agric J 22:213–219
Jennings DL (1963) Variation in pollen and ovule fertility in varieties of cassava, and the effect of interspecific crossing on fertility. Euphytica 12:69–76
Jennings DL, Hershey CH (1985) Cassava breeding: a decade of progress from international programmes. In: Russell GE (ed) Progress in plant breeding-1. Butterworths, London, pp 89–116
Jennings DL, Iglesias CA (2002) Breeding for crop improvement. In: Hillocks RJ, Tresh JM, Bellotti AC (eds) Cassava: biology, production and utilization. CABI Publishing, Wallingford, pp 149–166
Joaqui BO, Lenis JI, Calle F et al (2016) Cassava breeding II: phenotypic correlations through the different stages of selection. Front Plant Sci 7:1649. (in press)
Jørgensen K, Bak S, Busk PK, Sørensen C, Olsen CE, Puonti- Kaerlas J, Møller BL (2005) Cassava plants with a depleted cyanogenic glucoside content in leaves and tubers. Distribution of cyanogenic glucosides, their site of synthesis and transport, and blockage of the biosynthesis by RNA interference technology. Plant Physiol 139:363–374
Jos JS, Bai KV (1981) Functional male sterility in cassava. Curr Sci (Banglare) 50:1035–1036
Jos JS, Hrishi N (1976) Inheritance of leaf shape in cassava. J Root Crop 2(2):10–12
Jos, Nair (1984) Genetics of male sterility in a genotype of cassava. Curr Sci (Bangalore) 53:494–496
Kamau J, Melis R, Laing M, Derera J, Shanahan P, Ngugi E (2010) Combining the yield ability and secondary traits of selected cassava genotypes in the semi-arid areas of Eastern Kenya. J Plant Breed Crop Sci 2(7):181–191
Kamau J, Melis R, Laing M, Derera J, Shanahan P, Ngugi ECK (2011) Farmers' participatory selection for early bulking cassava genotypes in semi-arid Eastern Kenya. J Plant Breed Crop Sci 3:44–52
Karlström A, Calle F, Salazar S, Morante N, Dufour D, Ceballos H (2016) Biological implications in cassava for the production of amylose-free starch: impact on root yield and related traits. Front Plant Sci 7:604. doi:10.3389/fpls.2016.00604
Kawano K (1980) Cassava. In: Fehr WR, Hadley HH (eds) Hybridization of crop plants. ASA, CSSA, Madison, pp 225–233
Kawano K (2003) Thirty years of cassava breeding for productivity – biological and social factors for success. Crop Sci 43:1325–1335
Kawano K, Amaya A, Daza P, Ríos M (1978) Factors affecting efficiency of hybridization and selection in cassava (Manihot esculenta Crantz). Crop Sci 18:373–376
Kawano K, Cock JH (2005) Breeding cassava for underprivileged: institutional, socio-economic and biological factors for success. J Crop Improv 14:197–219
Kawano K, Goncalves Fukuda WM, Cenpukdee U (1987) Genetic and environmental effects on dry matter content of cassava root. Crop Sci 27:69–74

Kawano K, Narintaraporn K, Narintaraporn P et al (1998) Yield improvement in a multistage breeding program for cassava. Crop Sci 38:325–332
Kaweesi T, Kawuki R, Kyaligonza V, Baguma Y, Tusiime G, Ferguson ME (2014) Field evaluation of selected cassava genotypes for cassava brown streak disease based on symptom expression and virus load. Virol J 11:216. doi:10.1186/s12985-014-0216-x
Kaweesi T, Kyaligonza V, Baguma Y, Kawuki R, Ferguson M (2016) Inbreeding enhances field resistance to cassava brown streak viruses. J Plant Breed Crop Sci 8:138–149
Kawuki R, Ferguson M, Labuschagne M, Herselman L, Kim DJ (2009) Identification, characterisation and application of single nucleotide polymorphisms for diversity assessment in cassava (Manihot esculenta Crantz). Mol Breed 23:669–684
Kawuki RS, Nuwamanya E, Labuschagne MT, Herselman L, Ferguson ME (2011) Segregation of selected agronomic traits in six S1 cassava families. J Plant Breed Crop Sci 3:154–160
Kawuki RS, Pariyo A, Amuge T et al (2011) A breeding scheme for local adoption of cassava (*Manihot esculenta* Crantz). J Plant Breed Crop Sci 3:120–130
Keating B (1982) Environmental effects on growth and development of cassava (*Manihot esculenta* Crantz) with special reference to photoperiod and temperature. Cassava News 1(10):10–12
Kemp BP, Beeching JR, Cooper RM (2005) cDNA-AFLP reveals genes differentially expressed during the hypersensitive response of cassava. Mol Plant Pathol 6:113–123
Kemp BP, Horne J, Bryant A, Cooper RM (2004) *Xanthomonas axonopodis* gumD gene is essential for EPS production and pathogenicity and enhances epiphytic survival on cassava (*Manihot esculenta*). Physiol Mol Plant Pathol 64:209–218
Kizito E, Rönnberg-Wästljung AC, Egwang T, Gullberg U, Fregene M, Westerbergh A (2007) Quantitative trait loci controlling cyanogenic glucoside and dry matter content in cassava (*Manihot esculenta* Crantz) roots. Hereditas 144:129–136
Koehorst-van Putten HJJ, Sudarmonowati E, Herman M, Pereira-Bertram IJ, Wolters AMA, Meima H et al (2012) Field testing and exploitation of genetically modified cassava with low-amylose or amylose-free starch in Indonesia. Transgenic Res 21:39e50
Kunkeaw S, Yoocha T, Sraphet S et al (2011) Construction of a genetic linkage map using simple sequence repeat markers from expressed sequence tags for cassava (*Manihot esculenta* Crantz). Mol Breed 27:67–75
Lancaster PA, Brooks JE (1983) Cassava leaves as human food. Econ Bot 37:331–348
Lefèvre E, Charrier A (1993) Heredity of seventeen isozyme loci in cassava (*Manihot esculenta* Crantz). Euphytica 66:171–178
Legg JP, Jeremiah SC, Obiero HM et al (2011) Comparing the regional epidemiology of the cassava mosaic and cassava brown streak virus pandemics in Africa. Virus Res 159:161–170
Leyva-Guerrero E, Narayanan NN, Ihemere U, Sayre RT (2012) Iron and protein biofortification of cassava: lessons learned. Curr Opin Biotechnol 23:257–264
Li HQ, Sautter C, Potrykus I, Puonti-Kaerlas J (1996) Genetic transformation of cassava (*Manihot esculenta* Crantz). Nat Biotechnol 14:736–740
Liang H, Ren J, Gao Z, Gao S, Luo X, Dong L, Scipioni A (2016) Identification of critical success factors for sustainable development of biofuel industry in China based on grey decision-making trial and evaluation laboratory (DEMATEL). J Clean Prod 131:500–508
Liu J, Zheng Q, Ma Q, Gadidasu KK, Zhang P (2011) Cassava genetic transformation and its application in breeding. J Integr Plant Biol 53:552–569
Lokko Y, Dixon AGO, Offei SK, Danquah EY (2006) Combining ability analysis of resistance to mosaic virus disease in cassava. Afr Crop Sci J 14(3):221–230
Lokko Y, Dixon A, Offei S, Danquah E, Fregene M (2006) Assessment of genetic diversity among African cassava *Manihot esculenta* Crantz accessions resistant to the cassava mosaic virus disease using SSR markers. Genet Resour Crop Evol 53:1441–1453
Long SP, Ainsworth EA, Leakey ADB, Nösberger J, Ort DR (2006) Food for thought: lower-than-expected crop yield stimulation with rising CO2 concentrations. Science 312:1918–1921
Long SP, Ainsworth EA, Rogers A, Ort DR (2004) Rising atmospheric carbon dioxide: plants face the future. Annu Rev Plant Biol 55:591–628

Lozano JC, Cock JH, Castaño J (1978) New developments in cassava storage. In: Brekelbaum T, Bellotti A, Lozano JC (eds) Proceedings of cassava protection workshop. CIAT, Cali, pp 135–141

Ly D, Hamblin M, Rabbi I et al (2013) Relatedness and genotype x environment interaction affect prediction accuracies in genomic selection: a study in cassava. Crop Sci 53:1312–1325

Magoon ML, Krishnan R, Bai KV (1969) Morphology of the pachytene chromosomes and meiosis in *Manihot esculenta* Crantz. Cytologia 34:612–626

Manu-Aduening JA, Lamboll RI, Ampong Mensah G, Lamptey JN, Moses E, Dankyi AA, Gibson RW (2006) Development of superior cassava cultivars in Ghana by farmers and scientists: the process adopted, outcomes and contributions and changed roles of different stakeholders. Euphytica 150:47–61

Maredia MK, Reyes BA, Manu-Aduening J et al (2016) Testing alternative methods of varietal identification using DNA fingerprinting: results of pilot studies in Ghana and Zambia. MSU International Development Working Paper No. 149. 36 p

Marmey P, Beeching J, Hamon S, Charrier A (1993) Evaluation of cassava (*Manihot esculenta* Crantz.) germplasm using RAPD markers. Euphytica 74:203–209

Maruthi MN, Bouvaine S, Tufan HA, Mohammed IU, Hillocks RJ (2014) Transcriptional response of virus-infected cassava and identification of putative sources of resistance for cassava brown streak disease. PLoS One. doi:10.1371/journal.pone.0096642

Mkumbira J, Chiwona-Karltun L, Lagercrantz U et al (2003) Classification of cassava into 'bitter' and 'cool' in Malawi: from farmers' perception to characterisation by molecular markers. Euphytica 132:7–22

Molo R, Aool W, Adumo S, Mutisya DL (2016) Integrating cassava varieties and *Typhlodramulus aripo* to sustain biological control of cassava green mite. Afr Crop Sci J 24(s1):117–126

Monger WA, Seal S, Cotton S, Foster GD (2001) Identification of different isolates of Cassava brown

Mongomake K, Doungous O, Khatabi B, Fondong VN (2015) Somatic embryogenesis and plant regeneration of cassava (*Manihot esculenta* Crantz) landraces from Cameroon. SpringerPlus 4:477. doi:10.1186/s40064-015-1272-4

Morante N, Ceballos C, Sánchez T et al (2016) Discovery of new spontaneous sources of amylose-free cassava starch and analysis of their structure and techno-functional properties. Food Hydrocoll 56:383–395

Morante N, Sánchez T, Ceballos H et al (2010) Tolerance to post-harvest physiological deterioration in cassava roots. Crop Sci 50:1333–1338

Morillo-C Y, Sánchez T, Morante N, Chávez AL, Morillo-C AC, Bolaños A, Ceballos H (2012) Estudio preliminar de herencia del contenido de carotenoides en raíces de poblaciones segregantes de yuca (*Manihot esculenta* Crantz). Acta Agron 61(3):253–264

Moyib KO, Mkumbira J, Odunola OA, Dixon AG, Akoroda MO, Kulakow P (2015) Genetic variation of postharvest physiological deterioration susceptibility in a cassava germplasm. Crop Sci 55:2701–2711

Moyib OK, Odunola OA, Dixon AGO (2007) SSR markers reveal genetic variation between improved cassava cultivars and landraces within a collection of Nigerian cassava germplasm. Afr J Biotechnol 6:2666–2674

Muoki PN, Maziya-Dixon B (2010) Household utilization of manioc (*Manihot esculenta* Crantz) in northern Mozambique. Ecol Food Nutr 49:337–356

Nassar NMA (1978) Conservation of the genetic resources of cassava (*Manihot esculenta*): determination of wild species localitities with emphasis on probably origin. Econ Bot 32:311–320

Nassar NMA (2000) Cytogenetics and evolution of cassava (*Manihot esculenta* Crantz). Genet Mol Biol 23:1003–1014

Nassar NMA, Ortiz R (2008) Cassava genetic resources: manipulation for crop improvement. Plant Breed Rev 31:247–275

Nguyen TLT, Gheewala SH, Garivait S (2007) Full chain energy analysis of fuel ethanol from cassava in Thailand. Environ Sci Technol 41:4135–4142

Nichols RFW (1947) Breeding cassava for virus resistance. East Afr Agric J 12:184–194

Njenga P, Edema R, Kamay J (2014) Combining ability for beta-carotene and important quantitative traits in a cassava F1 population. J Plant Breed Crop Sci 6(2):24–30

Njoku DN, Gracen VE, Offei SK, Asante IK, Egesi C, Kulakow P, Ceballos H (2015) Parent-offspring regression analysis for total carotenoids and some agronomic traits in cassava. Euphytica. doi:10.1007/s10681-015-1482-4

Njoku DN, Ikeogu UN, Ewa F, Egesi C (2015) Crossability and germinability potentials of some cassava (*Manihot esculenta* Crantz) progenitors for selection. J Plant Breed Crop Sci 7(3):61–66

Normanha ES (1970) Cassava breeding work at the São Paulo State. Agronomic Institute, Campinas, Brazil. In: Trabalhos do Encontro de Engenheiros Agrônomos do Estado de São Paulo, Brazil. pp 40–47

Ntui VO, Kong K, Khan RS, Igawa T, Janavi GJ, Rabindran R et al (2015) Resistance to Sri Lankan cassava mosaic virus (SLCMV) in genetically engineered cassava cv. KU50 through RNA silencing. PLoS One 10(4):e0120551. doi:10.1371/journal.pone.0120551

Nweke F (2004) New challenges in the cassava transformation in Nigeria and Ghana, EPT Discussion Paper, vol 118. International Food Policy Research Institute (IFPRI), Washington, DC

Odipio J, Alicai T, Nusinow D, Bart R, Ingelbrecht I, Taylor N (2016) CRISPR/CAS9 genome editing of putative cassava flowering genes. Proceedings of the World Congress on Root and Tuber Crops and Third Scientific Conference of the Global Cassava Partnership for the 21st Century. Nanning, China. Jan 2016

Odipio J, Ogwok E, Taylor NJ, Halsey M, Bua A, Fauquet CM, Alicai T (2014) RNAi-derived field resistance to cassava brown streak disease persists across the vegetative cropping cycle. GM Crops Food 5:16–19

Okechukwu RU, Dixon AGO (2009) Performance of improved cassava genotypes for early bulking, disease resistance, and culinary qualities in an Inland Valley ecosystem. Agron J 101:1258–1265

Okogbenin E, Egesi CN, Olasanmi B et al (2012) Molecular marker analysis and validation of resistance to cassava mosaic disease in elite cassava genotypes in Nigeria. Crop Sci 52:2576–2586

Okogbenin E, Fregene M (2002) Genetic analysis and QTL mapping of early bulking in an F1 segregating population from non-inbred parents in cassava (*Manihot esculenta* Crantz). Theor Appl Genet 106:58–66

Okogbenin E, Fregene M (2003) Genetic mapping of QTL affecting productivity and plant architecture in a full-sib cross from non-inbred parents in cassava (*Manihot esculenta* Crantz). Theor Appl Genet 107:1452–1462

Okogbenin E, Porto MCM, Dixon AGO (1998) Influence of planting season on incidence and severity of African cassava mosaic disease in the subhumic zone of Nigeria. In: Akoroda MO, Ekanayake IJ (eds) Root crops and poverty alleviation. International Institute of Tropical Agriculture (IITA), Ibadan, pp 388–392

Okogbenin E, Porto MCM, Egesi C et al (2007) Marker aided introgression of CMD resistance in Latin American germplasm for genetic improvement of cassava in Africa. Crop Sci 47:1895–1904

Olasanmi B, Akoroda MO, Okogbenin E et al (2014) Bulked segregant analysis identifies molecular markers associated with early bulking in cassava (*Manihot esculenta* Crantz). Euphytica 195(2):235–244

Olsen KM, Schaal BA (1999) Evidence on the origin of cassava: phylogeography of *Manihot esculenta*. Proc Natl Acad Sci USA 96:5586–5591

Olsen KM, Schaal BA (2001) Microsatellite variation in cassava (*Manihot esculenta*, Euphorbiaceae) and its wild relatives: further evidence for a southern Amazonian origin of domestication. Am J Bot 88(1):131–142

Orrego JI, Hershey CH (1984) Almacenamiento del pólen de yuca (*Manihot esculenta* Crantz) por medio de liofilización y varios regímenes de humedad y temperatura. Acta Agron 34(1):21–25

Otti G, Fakoya A, Andrew I, Gedil M (2011) Development of genomic tools for verification of hybrids and selfed progenies in cassava (*Manihot esculenta*). Afr J Biotechnol 10:17400–17408

Owolade OF, Dixon AGO, Adeoti AYA (2006) Diallel analysis of cassava genotypes to anthracnose disease. World J Agric Sci 2:98–104

Parkes EY, Fregene M, Dixon A, Peprah BB, Labuschagne MT (2013) Combining ability of cassava genotypes for cassava mosaic disease and cassava bacterial blight, yield and its related components in two ecological zones in Ghana. Euphytica 194:13–24

Patil BL, Ogwok E, Wagaba H et al (2011) RNAi-mediated resistance to diverse isolates belonging to two virus species involved in cassava brown streak disease. Mol Plant Pathol 12(1):31–41

Perera PIP, Ordoñez CA, Becerra López-Lavalle LA, Dedicova B (2013) A milestone in the doubled haploid pathway of cassava. Protoplasma 251:233–246

Perera PIP, Dedicova B, Ordoñez C, Kularatne JDJS, Quintero M, Ceballos H (2012) Recent advances in androgenesis induction of cassava (*Manihot esculenta* Crantz). Acta Hortic 961:319–325

Pérez JC, Ceballos H, Calle F, Morante N, Gaitán W, Llano G, Alvarez E (2005) Within-family genetic variation and epistasis in cassava (*Manihot esculenta* Crantz) adapted to the acid-soils environment. Euphytica 145(1–2):77–85

Pérez JC, Ceballos H, Jaramillo G, Morante N, Calle F, Arias B, Bellotti AC (2005) Epistasis in cassava adapted to mid-altitude valley environments. Crop Sci 45:1491–1496

Pérez JC, Ceballos H, Ramirez IC, Lenis JI, Calle F, Morante N, Lentini M. del C. (2010) Adjustment for missing plants in cassava evaluation trials. Euphytica 172(1):59–65

Prakash A (2008) Cassava: international market profile. Background paper for the Competitive Commercial Agriculture in Sub–Saharan Africa (CCAA) Study. Trade and Markets Division. Food and Agriculture Organisation of the United Nations, Rome

Prochnik S, Marri PR, Desany B et al (2012) The cassava genome: current progress, future directions. Trop Plant Biol 5:88–94

Prudencio CY, Orkwor GC, Kissiedu AFK (1992) The relationships between cassava variety set characteristics, farmers' food security objectives, environmental and socio-economic conditions in Africa. Agric Syst 39:387–408

Rabbi I, Hamblin M, Gedil M, Kulakow P, Ferguson M, Ikpan AS, Ly D, Jannink J-L (2014) Genetic mapping using genotyping-by-sequencing in the clonally propagated cassava. Crop Sci 54:1384–1396

Rabbi IY, Hamblin MT, Lava Kumar P, Gedil MA, Ikpan AS, Jannink J-L, Kulakow PA (2014) High-resolution mapping of resistance to cassava mosaic geminiviruses in cassava using genotyping by sequencing and its implications for breeding. Virus Res 186:87–96

Rabbi I, Kulakow P, Manu-Aduening J et al (2015) Tracking crop varieties using genotyping-by-sequencing markers: a case study using cassava (*Manihot esculenta* Crantz). BMC Genet 16:115. doi:10.1186/s12863-015-0273-1

Raemakers K, Schreuder M, Pereira I, Munyikwa T, Jacobsen E, Visser R (2001) Progress made in FEC transformation of cassava. Euphytica 120:15–24

Raemakers CJJM, Sofiari E, Taylor NJ, Henshaw GG, Jacobsen E, Visser RGF (1996) Production of transgenic cassava plants by particle bombardment using luciferase activity as the selection marker. Mol Breed 2:339–349

Ramírez H, Hussain A, Roca W, Bushuk W (1987) Isozyme electrophoregrams of sixteen enzynes in five tissues of cassava (*Manihot esculenta* Crantz) varieties. Euphytica 36:39–48

Reilly K, Bernal D, Cortes DF, Gomez-Vasquez R, Tohme J, Beeching JR (2007) Towards identifying the full set of genes expressed during cassava post-harvest physiological deterioration. Plant Mol Biol 64:187–203

Renvioze BS (1973) The area of origin of *Manihot esculenta* as a crop plant: a review of the evidence. Econ Bot 26:352–360

Restrepo S, Verdier V (1997) Geographical differentiation of the population of *Xanthomonas axonopodis* pv. *manihotis* in Colombia. Appl Environ Microbiol 63:4427–4434

Roa AC, Maya MM, Duque MC, Tohme J, Allem AC, Bonierbale MW (1997) AFLP analysis of relationships among cassava and other *Manihot* species. Theor Appl Genet 95:741–750

Roca WM (1984) Cassava. In: Sharp WR, Evans DA, Ammirato PV, Yamada Y (eds) Handbook of plant cell culture. Crop species. Macmillan, New York, pp 269–301

Rogers DJ, Appan SG (1973) *Manihot* and *Manihotoides* (Euphorbiaceae), a computer-assisted study, Flora Neotropica, monograph, vol 13. Hafner Press, New York. 272 p

Rojas-C M, Pérez JC, Ceballos H, Baena D, Morante N, Calle F (2009) Introduction of inbreeding and analysis of inbreeding depression in eight S1 cassava families. Crop Sci 49:543–548
RTB (2016) The CGIAR Research Program on Roots, Tubers and Bananas (RTB) 2016 Annual Report. www.cgiar-rtb-2016-annual-report
Sánchez T, Ceballos H, Dufour D et al (2014) Carotenoids and dry matter prediction by NIRS and hunter color in fresh cassava roots. Food Chem 151:444–451
Sánchez T, Dufour D, Moreno JL, Pizarro M, Arango I, Domínquez M, Ceballos H (2013) Changes in extended shelf life of cassava roots during storage in ambient conditions. Postharvest Biol Technol 86:520–528
Sarria R, Torres E, Angel F, Chavarriaga P, Roca WM (2000) Transgenic plants of cassava (*Manihot esculenta*) with resistance to Basta obtained by *Agrobacterium*-mediated transformation. Plant Cell Rep 19:339–344
Sarria R, Torres E, Balcazar, M, Destafano-Beltran L, Roca WM (1995) Progress in Agrobacterium-mediated transformation of cassava (*Manihot esculenta* Crantz). In: Proceedings of the Second International Scientific Meeting Cassava Biotechnology Network (22–26 August 1994, Bogor, Indonesia) (Working document 150), CIAT, Cali. pp 241–244
Sayre R, Beeching J, Cahoon E et al (2011) The BioCassava plus program: Biofortification of cassava for sub-Saharan Africa. Annu Rev Plant Biol 62:251–272
Schöpke C, Taylor NJ, Carcamo R, Konan NK, Marmey P, Henshaw GG, Beachy RN, Fauquet CM (1996) Regeneration of transgenic cassava plants (*Manihot esculenta* Crantz) from microbombarded embryogenic suspension cultures. Nat Biotech 14:731–735
Second G, Allem A, Emperaire L et al (1997) AFLP based Manihot and cassava numerical taxanomy and genetic structure analysis in progress: implications for dynamic conservation and genetic mapping. Afr J Root Tuber Crops 2:140–147
Smith AM (2008) Prospects for increasing starch and sucrose yields for bioethanol production. Plant J 54:546–558
Sofiari E, Raemakers CJJM, Bergervoet JEM, Jacobsen E, RGF V (1998) Plant regeneration from protoplasts isolated from friable embryogenic callus of cassava. Plant Cell Rep 18:159–165
Sriroth K, Piyachomkwan K, Wanlapatit S, Nivitchanyong S (2010) The promise of a technology revolution in cassava bioethanol: from Thai practice to the world practice. Fuel 89:1333–1338
Ssemakula G, Dixon A (2007) Genotype x environment interaction, stability and agronomic performance of carotenoid-rich cassava clones. Sci Res Essay 2(9):390–399
Stapleton G (2012) Global starch market outlook and competing starch raw materials for starches by product segment and region. Cassava Starch World 2010. Centre for Management Technology (CMT), Cambodia
Taylor N, Chavarriaga P, Raemakers K, Siritunga D, Zhang P (2004) Development and application of transgenic technologies in cassava. Plant Mol Biol 56:671–688
Taylor N, Gaitán-Solís E, Moll T et al (2012) A high-throughput platform for the production and analysis of transgenic cassava (*Manihot esculenta*) plants. Trop Plant Biol 5:127–139
Taylor NJ, Masona MV, Carcamo R, Ho T, Schöpke C, Fauquet CM (2001) Production of embryogenic tissues and regeneration of transgenic plants in cassava (*Manihot esculenta* Crantz). Euphytica 120:25–34
Troyer AF (2006) Adaptedness and heterosis in corn and mule hybrids. Crop Sci 46:528–543
Tumuhimbise R, Shanahan P, Melis R, Kawuki R (2014) Genetic variation and association among factors influencing storage root bulking in cassava. J Agric Sci 153:1267–1280
Tuvesson S, Dayteg C, Hagberg P et al (2007) Molecular markers and doubled haploids in European plant breeding programs. Euphytica 158:305–312
Umanah EE, Hartmann CRW (1973) Chromosome numbers and karyotypes of some Manihot species. J Am Soc Hortic Sci 98(3):272–274
van Oirschot QEA, O'Brien GM, Dufour D, El-Sharkawy MA, Mesa E (2000) The effect of preharvest pruning of cassava upon root deterioration and quality characteristics. J Sci Food Agr 80:1866–1873

Vanderschuren H, Moreno I, Anjanappa RB, Zainuddin IM, Gruissem W (2012) Exploiting the combination of natural and genetically engineered resistance to cassava mosaic and cassava brown streak viruses impacting cassava production in Africa. PLoS One 7:e45277

Vanderschuren H, Stupak M, Fütterer J, Gruissem W, Zhang P (2007) Engineering resistance to geminiviruses—review and perspectives. Plant Biotechnol J 5:207–220

Vlaar PWL, van Beek P, RGF V (2007) Genetic modification and its impact on industry structure and performance: post-harvest deterioration of cassava in Thailand. J Chain Netw Sci 7:133–142

Wang C, Lentini Z, Tabares E et al (2011) Microsporogenesis and pollen formation in cassava (*Manihot esculenta* Crantz). Biol Plant 55(3):469–478

Wasswa P, Alicai T, Mukasa SB (2010) Optimisation of in vitro techniques for cassava brown streak virus elimination from infected cassava clones. Afr Crop Sci J 18:235–241

Welsch R, Arango J, Bär C et al (2010) Provitamin a – accumulation in cassava (*Manihot esculenta*) roots driven by a single nucleotide polymorphism in a phytoene synthase gene. Plant Cell 22:3348–3356

Were WV, Shanahan P, Melis R, Omari OO (2012) Gene action controlling farmer preferred traits in cassava varieties adapted to mid-altitude tropical climatic conditions of western Kenya. Field Crop Res 133:113–118

Westwood NN (1990) Maintenance and storage: clonal germplasm. Plant Breed Rev 7:111–128

Whankaew S, Poopear S, Kanjanawattanawong S, Tangphatsornruang S, Boonseng O, Lightfoot DA, Triwitayakorn K (2011) A genome scan for quantitative trait loci affecting cyanogenic potential of cassava root in an outbred population. BMC Genomics 12:266. http://www.biomedcentral.com/1471-2164/12/266

Wolfe MD, Egesi C, Kawuki R, Kulakow P, Rabbi IY, Jannink J-L (2016) Next generation cassava breeding: progress implementing genomic selection" in Proceedings of the World Congress on Root and Tuber Crops and Third Scientific Conference of the Global Cassava Partnership for the 21st Century (Nanning). Available online at: http://www.gcp21.org/wcrtc/PS08/

Wolfe MD, Rabbi IY, Egesi C et al (2016) Genomewide association and prediction reveals genetic architecture of cassava mosaic disease resistance and prospects for rapid genetic improvement. Plant Genome 9. doi:10.3835/plantgenome2015.11.0118

Wongtiem P, Courtois D, Florin B, Juchaux M, Peltier D, Broun P, Ducos JP (2011) Effects of cytokinins on secondary somatic embryogenesis of selected clone Rayong 9 of *Manihot esculenta* Crantz for ethanol production. Afr J Biotechnol 10:1600–1608

Woodward B, Puonti-Kaerlas J (2001) Somatic embryogenesis from floral tissue of cassava (*Manihot esculenta* Crantz). Euphytica 120:1–6

Wright CE (1965) Field plans for a systematically designed polycross. Rec Agric Res 14:31–41

Wydra K, Zinsou V, Jorge V, Verdier V (2004) Identification of pathotypes of *Xanthomonas axonopodis* pv. *manihotis* in Africa and detection of quantitative trait loci and markers for resistance to bacterial blight of cassava. Phytopathology 94:1084–1093

Xu Y, Crouch JH (2008) Marker-assisted selection in plant breeding: from publications to practice. Crop Sci 48:391–407

Yadav JS, Ogwok E, Wagaba H et al (2011) RNAi-mediated resistance to cassava brown streak Uganda virus in transgenic cassava. Mol Plant Pathol. doi:10.1111/j.1364-3703.2010.00700.x

Zacarias AM, Botha A-M, Labuschagne MT, Benesi IRM (2004) Characterization and genetic distance analysis of cassava (*Manihot esculenta* Crantz) germplasm from Mozambique using RAPD fingerprinting. Euphytica 138:49–53

Zacarias AM, Labuschagne MT (2010) Diallel analysis of cassava brown streak disease, yield and yield related characteristics in Mozambique. Euphytica 176:309–320

Zainuddin IM, Schlegel K, Gruissem W, Vanderschuren H (2012) Robust transformation procedure for the production of transgenic farmer-preferred cassava landraces. Plant Methods 8:24. http://www.plantmethods.com/content/8/1/24

Zehntner L (1919) Estudo sobre algumas variedades de mandiocas Brasileiras. Sociedad Nacionale de Agricultura, Impresa Ingleza-Camerino 61, Rio de Janeiro

Zhang P, Vanderschuren H, Fütterer J, Gruissem W (2005) Resistance to cassava mosaic disease in transgenic cassava expressing antisense RNAs targeting virus replication genes. Plant Biotechnol J 3:385–397

Zhang P, Wang WQ, Zhang GL, Kaminek M, Dobrev P, Xu J, Gruissem W (2010) Senescence-inducible expression of isopentenyl transferase extends leaf life, increases drought stress resistance and alters cytokinin metabolism in cassava. J Integr Plant Biol 52:653–669

Zhao SS, Dufour D, Sánchez T, Ceballos H, Zhang P (2011) Development of waxy cassava with different biological and physico-chemical characteristics of starches for industrial applications. Biotechnol Bioeng. doi:10.1002/bit.23120

Zhu X-G, de Sturler E, Long SP (2007) Optimizing the distribution of resources between enzymes of carbon metabolism can dramatically increase photosynthetic rate: a numerical simulation using an evolutionary algorithm. Plant Physiol 145:513–526

Zhu X-G, Long SP, Ort DR (2010) Improving photosynthetic efficiency for greater yield. Annu Rev Plant Biol 61:235–261

Zidenga T, Leyva-Guerrero E, Moon H, Siritunga D, Sayre R (2012) Extending cassava root shelf life via reduction of reactive oxygen species production1[c][w][OA]. Plant Physiol 4:1396–1407

Chapter 6
Sweetpotato (*Ipomoea batatas* L.)

Robert O.M. Mwanga, Maria I. Andrade, Edward E. Carey, Jan W. Low, G. Craig Yencho, and Wolfgang J. Grüneberg

6.1 History of Cultivation

Cultivated sweetpotato (*Ipomoea batatas* (L.) Lam.) is of neotropical origin. The crop is taxonomically placed in the genus *Ipomoea* of the family Convolvulaceae with over 500 *Ipomoea* species and the *Ipomoea* section Batatas together with 14 wild species, which are nearly all of neotropical origin (Khoury et al. 2015). Sweetpotato is thought to have evolved in Mesoamerica and Northwestern South America, somewhere between the Mexican peninsula Yucatan and the mouth of the Orinoco River (Nishiyama 1971; Austin 1979, 1988a; McDonald and Austin 1990; Zhang et al. 2000). South and Central America certainly has to be recognized as the primary center of genetic

R.O.M. Mwanga
International Potato Center, Kampala, Uganda
e-mail: r.mwanga@cgiar.org

M.I. Andrade
International Potato Center, Maputo, Mozambique
e-mail: m.andrade@cgiar.org

E.E. Carey
International Potato Center, CSIR-CRI, Kumasi, Ghana
e-mail: e.carey@cgiar.org

J.W. Low
International Potato Center, Nairobi, Kenya
e-mail: j.low@cgiar.org

G.C. Yencho
Department of Horticulture, North Carolina State University, Raleigh, NC, USA
e-mail: craig_yencho@ncsu.edu

W.J. Grüneberg (✉)
International Potato Center, Lima, Peru
e-mail: w.gruneberg@cgiar.org

H. Campos, P.D.S. Caligari, *Genetic Improvement of Tropical Crops*,
DOI 10.1007/978-3-319-59819-2_6

diversity (Austin 1978, 1983, 1988b; Austin and Huamán 1996; Huáman and Zhang 1997; Zhang et al. 2000), but cultivated sweetpotatoes in both the northern and southern regions of domestication might represent two prehistoric genepools, with important secondary centers of diversity existing in sub-Saharan Africa, Papua New Guinea, and Indonesia.

The crop has been found in the ruins of the so-far oldest city in the Americas, *Caral* (inhabited from 2600 to 1800 BC), on the Pacific coast of central Peru, and it had already become a crop during the *Tierra Blanca* period (6000–3000 BC) in the coastal areas of the Central Andes, in which irrigation systems for crop cultivation became important for this region of the world. Sweetpotato reached Polynesia, Hawaii, and New Zealand naturally or by early seafarers in pre-Columbian times. The Spanish introduced the crop to the Philippines in the sixteenth century, from whence it spread to other islands and the Asian mainland. By 1594, the crop was recorded in South China, where it was promoted to mitigate the effects of drought on crop production during the Qing Dynasty (ruling from 1644 to 1912), where sweetpotato became in addition to rice one of the most important staple food crops. Portuguese seafarers introduced the crop into Western Mediterranean Europe, Africa, India, and parts of Southeast Asia (Purseglove 1965; O'Brien 1972; Yen 1960, 1976, 1982; Jia 2013). In New Guinea, it developed until the nineteenth century into an important staple food crop. Some evidence indicates that sweetpotato could have reached the New Guinea highlands around 1200 AD (Golson 1976), but the penetration of the crop into Melanesia remains unclear. Secondary centers of diversity evolved on the island of New Guinea (Yen 1974; Austin 1988a) and in East Africa and the West Pacific comprising China, Korea, and Japan (Zhang et al. 2004; Montenegro et al. 2008).

According to FAO, sweetpotato is currently cultivated in 117 countries with 104 million tons of production in 2011. Asia is the world's largest sweetpotato-producing region, with about 80% of annual production, followed by Africa, the Americas, and Oceania with approximately 16%, 3%, and 1% of annual production, respectively (FAOSTAT 2011). However, the cultivated area of the crop in sub-Saharan Africa (SSA) nowadays is nearly the same as compared to the West Pacific (China, Korea, Japan). Trends in area cultivated from 1992 to 2011 by region notably show declines in Asia (from 6.4 to 3.6 million ha) and increases in SSA (from 1.2 to 3.2 million ha). Only 30 countries contribute greater than 99% of annual global sweetpotato production (Grüneberg et al. 2015). Sweetpotato today is used in a variety of ways for food, feed, and processed products, with the principle uses varying by region (Woolfe 1992; Bovel-Benjamin 2007; Padmaja 2009). In developing countries, the crop is mainly grown for homestead food and feed use and for sale to local markets for fresh consumption. The use of both vines and roots for pig feeding is important in China, Vietnam, and Papua New Guinea (Peters 2004). Padmaja (2009) provides details on use of the crop for cattle, poultry, and fish feed. Awareness of the high nutritional value of sweetpotato is driving increasing consumer demand for the crop among health-conscious consumers in the USA and Europe (CSPI 2014). Orange-fleshed sweetpotato (OFSP) with its high provitamin A content can be used effectively to combat vitamin A deficiency

(VAD) (Low et al. 2007; Hotz et al. 2012) among vulnerable populations in the developing world, mainly Africa, South Asia, and Southeast Asia. In all countries in which VAD is a serious public health problem, OFSP is a cost-efficient and sustainable vehicle to alleviate VAD and to improve public health. This holds true even if only small quantities of OFSPs are eaten. OFSP, biofortified with provitamin A, is considered since 2009 by HarvestPlus (Bouis and Islam 2012; Hotz et al. 2012) to be the first biofortified crop ready to go to scale.

6.2 Biology

6.2.1 Flowering

Sweetpotato flowers resemble those of other morning glories, with cultivars differing in size, shape, and color. Flowers occur in axillary inflorescence (compound cyme) of 1–22 buds and open singularly or in pairs or more groups (Jones et al. 1986; Purseglove 1968). Flowers are borne singly or on inflorescences that grow vertically from the leaf axils.

Flowering may be none, sparse, moderate, or profuse. Flower color varies from white to purple. Flower size, shape, limb shape, sepal length, sepal shape, sepal apex, sepal pubescence, sepal color, color of stigma and style, and stigma exertion vary with genotype. Seed capsule set may be none, scarce, sparse, moderate, or profuse (CIP/AVRDC/IBPGR 1991). Flower characters are very important and are not influenced by environmental conditions, but there are big differences among cultivars and their response to environmental conditions. Flowering time and duration, flowering intensity, and seed set of sweetpotato cultivars are strongly influenced by genotype, photoperiod, environmental stress, and trellis-work (wooden poles or metal bars or poles supporting wire or wire mesh on which sweetpotato vines are tied or trained for easy hand pollination and harvesting of capsules (Fig. 6.1) (Millar 1939; Montelaro and Miller 1951; Jones 1980; Eguchi 1996; Fugise et al. 1955; Hsia and Chen 1956). Under normal field conditions, some genotypes do not flower, others produce few flowers, while others flower profusely. The flower is bisexual, with androecium or stamens which are male organs and gynoecium or pistil which forms the female organ (Martin and Ortiz 1967). The base of the pistil has an ovary with a maximum of four ovules. Each capsule can contain a maximum of four seeds, depending on the method of pollination. Polycrosses often have more seeds per flower than hand-crossings. The bisexual flowers open at dawn and close and wilt by early afternoon of the same day (Onwueme 1978). The stigma is receptive only for a few hours in the morning (Jones 1980). Therefore, the chances of failure of pollination are high. The variation in stamen height with respect to the style also affects pollination and seed production.

Fig. 6.1 Sweetpotato controlled pollination in Namulonge, Uganda

6.2.2 Pollen Dispersal and Viability

Insects, mainly bees, are the primary pollinators of sweetpotato (Jones 1980). The stigmas are receptive in the morning. Not much is known about the survival of sweetpotato pollen, but germination can continue for 3–4 h after pollination (Martin and Cabanillas 1966). Systematic studies on pollen viability in sweetpotato would be useful, especially under field conditions (Andersson and de Vicente 2010).

6.2.3 Sexual and Asexual Reproduction and Seed Dispersal

Sweetpotato is mainly outcrossing yet it can also reproduce vegetatively or asexually by stem or vine cuttings in the tropics or adventitious root buds in temperate regions (Anderssson and de Vicente 2010). The storage root also plays a crucial role in asexual reproduction of sweetpotato. Storage roots of some genotypes sprout when left in the soil giving new stems or vines. Despite flowering abundantly, most sweetpotato genotypes set seed poorly due to sterility, numerous self-compatibility, high degrees of self-incompatibility, cross-compatibility and cross-incompatibility challenges, and environmental factors (Stout 1924; Simpson and Ogorzaly 2000; Wang 1964; Anderssson and de Vicente 2010; Jones 1980; Murata and Matsuda 2003). Sexual reproduction is achieved through human intervention via controlled crossings or randomly by bees. Asexual propagation is used by farmers in the production of storage roots. The sexual method is used almost exclusively by plant breeders for the development of improved cultivars. The main dispersal agents of sweetpotato seeds are humans, birds, and water (Bulmer 1966; Andersson and de Vicente 2010; Zhang et al. 2004).

Sweetpotato can show mechanical dormancy (Martin 1946), and the hard testa of its seeds requires scarification prior to germination. Scarification can be achieved through chemical means using concentrated sulfuric acid or by mechanical methods of piercing part of the testa using a sharp object. Under controlled conditions (18 °C, 50% RH), seeds remain viable for over 20 years and can form seed banks in the soil, as evidenced in Brazil (Lebot 2010; Andersson and de Vicente 2010).

6.3 Genetics

Sweetpotato is an autopolyploid highly heterozygotous hybrid propagated by cloning. It is a hexaploid ($2n = 6\times = 90$) (King and Bamford 1937; Jones 1965b, 1974; Magoon et al. 1970). Owing to an even number of chromosome set, a more or less regular meiosis is possible, and true seed set occurs abundantly by cross-pollination (by insects, mainly bees). Generation of new genotypes and selection of best clones from a given population are relatively straightforward. However, changes in populations and sexual reproduction units (breeding nurseries) related to their allele frequencies (genepool) are complex and difficult to describe. Hardy-Weinberg equilibrium distribution of genotypic frequencies is only reached asymptotically and not with the first generation of random mating as in diploids. Meiotic abnormalities are observed which include multivalent formation, translocations, and deletions (Oracion et al. 1990; Magoon et al. 1970), and cytological anomalies result in various degrees of cross- and self-incompatibility (Martin 1965, 1968, 1970). Moreover, a complex sporophytic self- and cross-incompatibility system favors outcrossing, and often cross combinations are impossible to realize. Nevertheless, it appears that breeding populations undergoing intensive recombination and selection change toward more frequent compatibility including self-compatibility (Grüneberg et al. 2015).

The autopolyploid segregation ratios of sweetpotato are usually complex (Shiotani 1988; Jones,1966, 1967b, c). Double reduction is a phenomenon that leads to discrepancies from expected segregation ratios, a problem which does not exist in diploids. In the case of a single dominant allele, the segregation ratios are simple (Poole 1955), and the same is true for self-compatible clones and recessively inherited traits. Poole (1955) studied phenotypic ratios in sweetpotato (rooting vs nonrooting, brown-skinned roots vs cream-skinned roots, ridged vs smooth root surface, orange vs cream flesh, color red vs green stem, flowering vs nonflowering, smooth vs lobed leaf margins). Jones (1967b) published theoretical segregation ratios for qualitative traits and presented four hypotheses (hexasomic, tetradisomic, tetrasomic, disomic) of inheritance. Gallais (2003) describes segregation ratios in the presence of double reduction for hexaploids. Single-locus segregation ratios become more complicated due to dosage effects of dominant alleles (discrete segregation ratios of a single locus become continuous). Kumagai et al. (1990) tested the models of Jones (1967b) and showed that the ß-amylase null trait in storage roots was controlled by one recessive gene and that it was inherited in a hexasomic or tetradisomic manner, but not disomically or tetrasomically. Moreover, resistance to *Sweet potato feathery mottle virus* (SPFMV) and *Sweet potato chlorotic stunt virus* (SPCSV) is thought to be recessively inherited (Mwanga et al. 2002a, b; Grüneberg et al. 2015).

The expression of a recessively inherited attribute is extremely rare in sweetpotato populations, even if the recessive allele has medium frequency (q of 0.3–0.6). Only at high frequencies of the recessive allele ($q > 0.7$) can the desired recessive inherited attribute be observed with elevated frequencies (>10%). This results in the paradox that a recessively inherited attribute is very rarely observed, although the recessive allele is present in the population with medium frequency. Emerging self-compatibility in sweetpotato presents a huge opportunity to increase the number of genotypes for a desired rare and recessively inherited trait (Grüneberg et al. 2015) – a new unique population is formed in which the desired trait is fixed. Regarding the five forces (selection, mutation, gene flow, genetic drift, and nonrandom pairing) changing allele frequencies in populations, it can be generalized that sweetpotato populations exhibit, compared to diploids: lower response to selection for monogenic traits for various types of dominance and purge of recessives that is more difficult (selection); a larger genetic load in the positive and negative sense (mutation); low rates of natural incorporation of new alleles into genepools and extremely difficult incorporation of a recessively inherited attribute into populations (gene flow); pronounced foundation effects with larger potential of very small populations to create new genetic diversity, larger effective population size for the same population size, and lower vulnerability to genetic bottlenecks as well as possibilities to apply greater selection intensity for the same population size (genetic drift); and reduced inbreeding and self-pollination due to cross- and self-incompatibility which additionally enhances heterozygosity (random mating), which is in hexapolyploids already extreme compared to diploids – in a random mating biallelic population, nearly the entire genome is heterozygous in a very wide range of allele frequencies (Gallais 2003 p. 65–66).

For quantitative traits, it is important to obtain estimates of the genetic variance and the phenotypic variance. Among clones, operative broad-sense heritabilities and genotypic and phenotypic correlations determine the response to selection for most traits. Quantitative genetic studies have been conducted to obtain heritability estimates for yield (Bacusmo et al. 1988), root traits (Jones et al. 1969, 1977, 1978, 1986), vine traits (Jones 1969), root quality factors, including dry matter, intercellular space, protein, and baking quality (Jones 1977; Jones et al. 1978, 1987), and resistance to soil insects (Jones et al. 1979) and root-knot nematodes (Jones and Dukes 1980). Martin and Jones (1986) reviewed heritability estimates, while Grüneberg et al. (2015) reviewed variance component estimates for sweetpotato key traits. Narrow-sense heritability, general and specific combining ability, as well as heterosis estimates are of significance in population improvement. Heterosis and transgression are based on additive and nonadditive coaction of alleles and genes. Many parameters are needed to fully explain breeding values since gametes contain three alleles and transfer interactions between alleles to the next generation. Equations to describe the genetic additive variance and the genetic variances due to interactions for higher polyploid species are given by Kempthorne (1957) and Gallais (1981). Narrow-sense heritabilities are not related directly to parent-offspring regressions. Inbreeding progress toward homozygosity is much slower compared to diploids.

Even if plants are self-compatible, it would require seven generations of selfing to reach an inbreeding coefficient of $F = 0.5$ (for the calculations, readers are referred to p. 124 of Gallais 2003). Uncertain degrees of inbreeding of parents make it very difficult to estimate genetic variances and covariances in experiments because inbreeding is not removed by one generation of random mating. The response to selection is determined by the genetic additive variance and the genetic variances due to interactions in ratio to the phenotypic variance in the first generation of selection since the population is not in equilibrium and the response to selection changes after further generations of random mating until the equilibrium is reached. However, when equilibrium is reached, the response to selection is determined by genetic additive variance in ratio to the phenotypic variance as in diploids (Wricke and Weber 1986; Gallais 2003). The highly heterozygous sweetpotato genome nurtures the expectation that heterosis is very important for sweetpotato performance and that systematic exploitation of heterosis can improve the efficiency of population improvement (Grüneberg et al. 2015). However, heterosis increments can only be estimated since development of homozygous inbred lines by selfing to estimate the full amount of heterosis is illusory (other technologies such as those in potato via monohaploids are not available). However, there is greater stability of heterosis between two populations in derived hybrid populations, and breeding schemes in sweetpotato that exploit heterosis might allow enhanced inbreeding within genepools for quality and resistances without sacrificing heterozygosity for quantitative trait performance.

6.4 Germplasm Relations and Utility of Crop Wild Relatives for Breeding

The evolution of cultivated sweetpotato is poorly understood (Nishiyama et al. 1975; Rajapakse et al. 2004; Srisuwan et al. 2006; Roullier et al. 2013). Sweetpotato, a dicot, is a member of the highly diverse morning glory family Convolvulaceae (Austin 1977). It is the only cultivated species in this family, being placed in the genus *Ipomoea*, section Batatas (Ting et al. 1957; Austin 1988a; Bohac et al. 1995). The genus *Ipomoea* is a large genus that includes ca. 600–800 species worldwide (McDonald and Mabry 1992; Austin and Huáman 1996; Austin and Bianchini 1998). Cultivated sweetpotato's closest wild relatives are in the section *Eriospermum* Hallier f., series *Batatas* (Choisy) D. F. Austin. Section Batatas continues to undergo revision, but it contains approximately 14 species most of which are diploid ($2n = 30$) and a few tetraploids ($4n = 60$) (Austin 1991, 1993; Jones 1990). The section Batatas species include wild *I. batatas* (L.) Lam. [including *I. batatas* var. *apiculata* (Martens and Galeotti) McDonald and Austin], *I. cordatotriloba* Dennstedt, *I. cynanchifolia* Meisn., *I. grandiflora* (Dammer) O'Donell, *I. lacunosa* L., *I. leucantha* Jacquin, *I. littoralis* Blume, *I. ramosissima* (Poir.) Choisy, *I. splendor-sylvae* House, *I. tabascana* McDonald and Austin, *I. tenuissima* Choisy, *I. tiliacea* (Willd.) Choisy in D. C., *I. trifida* (H. B. K.) G. Don, and *I. triloba* L. (Martin and Jones 1973; Khoury et al. 2015). Except for *I. littoralis* (Austin 1991), all species are native to the New World, where they extend from the Southern United States throughout Central America and the Caribbean to South America.

The evolutionary status of sweetpotato and its phylogenetic relationship with related species in the genus *Ipomoea* (L.) are not clear. The most widely held hypothesis is that *I. batatas* evolved from interspecific hybridization between *I. trifida* and *I. triloba* (Austin 1988a). The second is that *I. batatas* developed by polyploidization in *I. trifida* (Jone 1967a; Kobayashi 1984; Kobayashi et al. 1994; Bohac et al. 1992; Freyre et al. 1991; Iwanaga et al. 1991). Recent studies based on evaluation of chloroplast haplotypes and nuclear DNA indicate that it may have been domesticated separately in Central and South America through autopolyploidization of distinct populations of *I. trifida* or a close relative (Roullier et al. 2011, 2013). *I. batatas* is claimed to be found with various ploidy levels (mostly 4×, 3×, 6×) – collected in Ecuador, Colombia, Guatemala, and Mexico (Bohac et al. 1993).

Until relatively recently, it was not even known if cultivated sweetpotato (*I. batatas* $2n = 6\times = 90$) was an allopolyploid or autopolyploid. Cytological and genetic studies of sweetpotato are difficult to conduct due to the genetic complexity of sweetpotato. Both allopolyploidy and autopolyploidy hypotheses on the origin of sweetpotato have been proposed (Shiotani and Kawase 1987; Ukoskit and Thompson 1997). Shiotani and Kawase (1989) postulated the genome constitution of sweetpotato as B1B1B2B2B2B2 and suggested additional homology between the B1 and the B2 genomes, based on the occurrence of frequent formation of tetravalents and hexavalents. Molecular marker evidence from several studies indicates that sweetpotato is an autopolyploid with some degree of preferential pairing present. Ukoskit and Thomson (1997) were the first to use molecular markers to suggest that sweetpotato was an autopolyploid. Kriegner et al. (2001) and Cervantes et al. (2008) confirmed this hypothesis in much more detailed studies using amplified fragment length polymorphisms (AFLP) makers. More recently, Zhao et al. (2013) have offered additional support, as has recent genomic sequencing of *I. trifida* and *I. triloba* (Yencho, personal communication). All of these studies support the predominance of autopolyploid inheritance with some degree of preferential pairing.

The outbreeding polyploid nature of sweetpotato has undoubtedly led to the remarkable amount of diversity present in the cultivated sweetpotato genepool (see Traits below). Practically, because all of this diversity is within *I. batatas*, this means that breeders have tremendous access to a wide range of variability to generate and select from and in the near-term breeders and will continue to exploit this diversity for sweetpotato improvement. However, the crop wild relatives (CWR) of cultivated sweetpotato, most notably those in the series *Batatas*, also have the potential to contribute to breeding objectives for this important crop (Kobayashi and Miyazaki 1976; Khoury et al. 2015). However, uncertainty in species boundaries and their phylogenetic relationships, the limited availability of CWS that cross with cultivated sweetpotato, and the difficulty of introgression of genes from CWR into adapted phenotypes have constrained their utilization. Mostly Chinese and Japanese breeders have led the way in using CWR of sweetpotato as a source of genes for batatas improvement.

Khoury et al. (2015) provide an excellent review of the distribution and genetic resource potential of CWR for sweetpotato improvement. Many sweetpotato CWR can be hybridized with the crop through controlled pollinations, somatic cell, and/or ovule culture techniques (Diaz et al. 1996). Crosses involving *I. tabascana*, *I. trifida*, *I. triloba*, *I. littoralis*, *I. grandifolia*, *I. lacunosa*, *I. leucantha*, and wild *I. batatas* in particular have resulted in relatively viable progeny (Jones and Deonier 1965; Nimmakayala et al. 2011). The wild conspecific and *I. trifida* have been documented for their contribution to increases in protein and starch content and nematode and sweetpotato weevil (SPW) resistance (Iwanaga 1988; Shiotani et al. 1994). To date, species that have been examined for use in crop improvement include *I. trifida* and *I. littoralis* for yield and sweetpotato weevil, scab [*Elsinoë batatas* (Saw.) Viegas et Jenkins], and black rot disease (*Ceratocystis fimbriata* Ell. et Halst.) resistance, *I. grandifolia* for sweetpotato stem nematode and sweetpotato virus disease (SPVD) resistance, and *I. triloba* for drought tolerance, root-rot resistance, and foliar fungal disease resistances (Iwanaga 1988; Jarret et al. 1992; Mont et al. 1993; Komaki 2004; Zhang and Liu 2005; Nimmakayala et al. 2011). However, it should be noted that challenges in crossing many of the CWR with cultivated sweetpotato are not insignificant, due to differences in ploidy and numerous interspecific incompatibility barriers. As mentioned by Khoury et al. (2015), in their summary of the potential of CWR for sweetpotato improvement, "Further research combining morphological studies, trait evaluations, and genetic diversity analyses is likewise critically needed for elucidating species boundaries and highlighting accessions of particular value for use in breeding."

6.5 Crossing Approaches

Natural pollination is mostly mediated by insects in the morning hours when many species, mainly Hymenoptera, visit the flowers. These insects transfer pollen, but honeybees and bumblebees are the main pollinators of sweetpotato. Most sweetpotato genotypes flower naturally during the short days in the tropics. However, where flowering is a problem, several techniques have been developed to promote sweetpotato flowering and fruit and seed production. These are short photoperiod (8–11.5 h of intense light), moderate temperature (20–25 °C), relative humidity (over 75%), limited water supply and grafting (sweetpotato scions are grafted onto rootstocks with profuse flowering such as *Ipomoea nil* or *I. setosa*, or a free-flowering sweetpotato clone), trellises (structusres such as tripods, stakes, and wire mesh can serve as trellises (Fig. 6.1), and sweetpotato vines are fixed to the trellises with strings or pins, and excessive vegetative branches are eliminated starting from the ground), growth regulators (gibberellic acid (GA_3 or GA_7) and ethephon (2-chloroethyl phosphonic acid), in combination with a short-day treatment and using scions from mature mother plants), overwintering and vine girdling (making a slanting incision across the stem at about 15 cm from the ground and cutting vines about two-thirds through re-girdling must be continued every 10–15 days when weather conditions favor vegetative growth), pesticide sprays, soil fertilization

(fertilization is not usually required, but careful application of boron, magnesium and iron, medium to low doses of nitrogen and potassium, and high dose of phosphorus could be used to induce flowering), the "bouquet" method (three cuttings, 80 cm long, containing a large number of flower buds, placed in bottles filled with a nutrient solution, 4-12-4 NPK stock solution 10% strong, and 2 ppm of boric acid, pollination is done in the greenhouse), and genetic selection (profuse flowering habit and seed set are selected for). In practice, a combination of these methods and knowledge of the compatible and incompatible groups or parents are used in complement, and details of each technique are given by Zosimo (1999).

In breeding clonally propagated crops, normal clonal propagation is broken by generating true seeds, which results in a new population and genetic variation. Subsequent propagation steps are asexually propagated clonally with selection of superior genotypes (Grüneberg et al. 2009). Superior clones are crossed and used to generate true seeds. The process leads to recurrent cycles of recombination and selection and results in a combination of good traits in genotypes which were in different genotypes before selection. However, in the medium and long term, recurrent selection also results in generation of new genotypes with trait performance outside of the distribution range of previous populations (Grüneberg et al. 2015).

In recurrent selection cycles, high genetic gains across traits can only be achieved by structuring plant breeding into two components: variety development and population improvement. For autopolyploid crops, Gallais (2003) proposed, in addition to new information about their population genetics, a comprehensive breeding scheme comprising variety development and population improvement. Variety development aims at selection of the best or very few best clones (maximum response to selection and complete or nearly complete exploitation of the genetic variation). Population improvement aims at selection of the "best" parents to generate new genetic variation around an improved population mean (in practice the population mean across all traits for which the breeder desires improvement). Variety development and selection for the "best" clone for the current needs of clients are relatively straightforward, and what is the "best" is usually best known locally on the ground. However, population improvement or identifying the best parents to create a new and better population for future selections is a challenge in sweetpotato, as it is for all other clonally propagated crops. Population improvement is indeed complex and should be carried out by an interconnection of breeders for an agro-geographic zone. It often requires more resources and capacities than small- to medium-sized breeding programs can usually afford. More details on crossing for variety development are given by Gruneberg et al. (2015) and Martin and Jones (1986).

6.6 Breeding Methods

Depending on the pollination and propagation biology, various options exist on how to breed a crop (Schnell 1982). For variety development, sweetpotato is treated as a clonally propagated crop and for population improvement as an open-pollinated crop with options to systematically exploit heterosis.

The general principle of breeding clonally propagated crops is to break normal clonal propagation by generating true seeds (Jones 1965a). All subsequent propagation steps are by cloning, and selection is carried out among clones (Grüneberg et al. 2009). Sweetpotato breeding is largely driven through the public sector, supported to a varying extent by policies and to a minor extent by the needs of industry. It is carried out nearly exclusively by the public sector (national agricultural research system/NARS, agricultural research institute/ARI, and universities), and implementation does not require huge investments, as can be seen from the history of sweetpotato breeding (Martin and Jones 1986). All successful sweetpotato breeding programs initiated in the past century such as those at Louisiana State University (LSU), North Carolina State University (NCSU), the Xuzhou Sweetpotato Research Center (XSPRC), and NaCRRI had one characteristic in common – that they intensified recombination (usually by polycross seed nurseries comprising 20–30 parents producing 10,000–100,000 true seed) and conducted gene-pool separation (recombination of parents adapted to local needs). Decentralization of sweetpotato breeding appears to be a key of breeding success due to appropriate consideration of GxE interactions mainly derived by biotic and abiotic stresses, consumer preferences, as well as delegation of decision-making responsibilities with respect to farmer needs and consumer preferences to regions and subregions. Within the Sweetpotato Action for Security and Health in Africa (SASHA) project, supported by the Bill and Melinda Gates Foundation, it was possible for CIP in SSA to unite single breeding programs loosely around breeding platforms (Grüneberg et al. 2015) to exchange information and breeding material, coordinate breeding operations and fundraising as well as joint seed systems (especially basic propagation), and disseminate varieties into the markets across countries.

Traditionally, each surviving seed plant is cloned (occasionally with selection among seed plants, e.g., for SPVD) and raised as A-clones in observation plots (visual screening of general clone performance) or evaluation plots of three to five plants (recording of data on specific traits of each clone). In the past most breeders were using only one location at the early breeding stages due to the restrictions of the propagation coefficient and breeding budget but also because how plant breeding students were taught at universities. A-clones were only selected for highly heritable traits such as general performance (growth type, root size, shape, and color), resistance to pests and diseases, harvest index, and dry matter and nutritional quality. In the next season, B-clones were planted from selected A-clones in larger plots in two to three rows, at least two plot replications, and in one to two locations. The B-clone stage is traditionally the beginning of selection for low heritability traits. At the beginning of the C-clone and D-clone selection stages, the breeding population is reduced to between 30 and 300 clones. While the number of clones in later selection stages is further reduced, those selected clones are tested in more environments and plot replications. The plots for C-clones and D-clones are three to seven row plots. All important agronomic traits are determined, including taste and postharvest characteristics. Furthermore, yield stability parameters are determined such as (1) slope of the regression line and (2) deviations from regression, as well as conducting an additive main effect and multiplicative interaction (AMMI) analysis in those cases where the regression model does not fit (Fox et al. 1997). Finally, two or more clones are selected to enter into officially variety release trials when they are in at least one key trait 10–20% superior relative to check clones (usually outstanding varieties in farming systems).

Two (perhaps three) innovations are in the process to change sweetpotato breeding. The first two (accelerated breeding and heterosis exploiting breeding schemes) are conventional, and the third is molecular breeding (marker-assisted and genomic selection). *Accelerated breeding schemes* (ABS) target early breeding stages, and where the variance component due to genotypes by years is not large and important, respectively, temporal variation of test environments is to be replaced aggressively by spatial variation of test environments – more locations compensate for reduction in test years. In ABS the A-clones are planted directly in small plots (1 m row) in several locations (usually 2–3) without plot replications but with a grid of check clones (usually two check clones). All traits to be considered in early breeding stages are evaluated in these environments/locations sequentially or simultaneously – farmer participation is highly recommended at least in one environment and easily without risk hot spots of stresses can be used to determine the risk to fail among clones. After aggregating information across one to two environments for highly heritable traits and two to four environments for less heritable traits, clones are selected to enter directly again into recombination and into later breeding stages. In case of CIP, even later breeding stages are carried out only across 2 years to enter after 4 years into the official variety test system, which varies from country to country. The ABS has been rapidly adopted among national agricultural research system (NARS) breeders (at least 12 countries by 2014) in SSA with large numbers of variety releases (Grüneberg et al. 2015; Mwanga et al. 2016; Andrade et al. 2016). Heterosis exploiting breeding schemes in sweetpotato have so far only been implemented at CIP in Peru for the so-called global orange-fleshed sweetpotato (OFSP) population for earliness and wide adaptation (impact so far cannot be estimated). However, expectations are high, for example, such breeding schemes could allow breeders in high SPVD pressure zones to apply more inbreeding for SPVD resistance (two partially inbred genepools) without sacrificing heterozygosity (hybrid population) for yield and stability performance. Such a scheme is currently tested at the breeding platform in Uganda – for further genetic details, preliminary results, and discussion, see Grüneberg et al. (2015). Finally, we want briefly to note marker-assisted and genomic selection (see more in the section genomic tools for sweetpotato in this chapter). Molecular markers for SPVD resistance would be extremely helpful to breeders because the attribute exerts extreme unequal pressure across environments; it takes several years to finally decide if a clone is moderate to highly resistant to SPVD, and SPVD results in extreme yield loses rapidly in high SPVD pressure zones and in the long term across all sweetpotato growing environments (AFLP, simple sequence repeat (SSR), and the diversity arrays technology (DArT) markers for SPVD resistance are in the experimental validation process). Genomic selection tools for sweetpotato are still under development at the North Carolina State University and CIP. At CIP in Peru, three populations (phenotyped for 2 years) are maintained as DNA, and in each population, no selection was conducted since the seed stage so that responses to selection by genomic tools can be estimated and different systems experimentally validated. Breeding research in sweetpotato commonly agrees that genomic selection must be considered as an option for the long-term future with tremendous potential for test capacity increase and to predict many traits simultaneously in applied breeding populations, but emphasizes where sweetpotato would have been by putting most efforts into this approach only.

6.7 Traits

Sweetpotato germplasm is highly diverse, with variation for almost every trait considered. Diversity of storage root skin and flesh color, leaf pigmentation and shape, and vine color and growth habit are seen in economically important cultivars (including landraces and bred varieties) and are illustrative of the genetic variation in the crop. Sweetpotato is largely an obligate outcrosser, due to sporophytic self-incompatibility, and its highly heterozygous hexaploid genome is a source of many surprises for the sweetpotato breeder. Sweetpotato also has a relatively high tendency to mutate, accentuated further in many places by propagation using adventitious sprouts from storage roots (Villordon and Labonte 1995), and this contributes further to the diversity of the crop (Shiotan 1988; McGregor and Labonte 2006).

Sweetpotato breeding programs have been active in a number of countries since the mid-twentieth century, producing numerous cultivars for various markets and uses (Grüneberg et al. 2015; Zhang et al. 2009). Linked to formal seed systems, breeding efforts contribute significantly to increasing productivity and economic and nutritional impact of the crop. In a number of places in SSA where the crop is important at commercial or household level, but where breeding and seed system efforts are still immature, "natural selection" has given rise to superior landrace varieties possessing the production, postharvest, and quality attributes required by producers, consumers, marketers, and processors, and these have come to predominate in production systems (Edmond and Ammerman, 1971). Participatory approaches to selection help to ensure that key traits are considered during the selection process (Grüneberg et al. 2009; Gibson et al. 2011a).

Some traits such as flesh color (such as orange) are preferred by women and children; high dry matter is more preferred by men than women or children. In some communities, women like to conserve particular traits such as vigor of genotypes to suppress weeds, earliness, and high yield to alleviate food shortages during food scarcity. In some communities, women and children prefer more sweet genotypes than men. Differences between women and men accessing land for high-yielding sweetpotato varieties for sale in some areas, for example, in Central Uganda where men own land and dominate its use and the distribution of the returns from it. There is no generalization on the gender-specific traits in sweetpotato among the different communities on the globe.

6.7.1 Quality Traits

Flesh color, dry matter content, and cooked taste, texture, and aroma are key quality attributes that distinguish utilization classes. Flesh colors include white, yellow, orange (primarily due to beta-carotene), and purple (due to anthocyanins), with variation in intensity of color related to pigment concentration, both within and between genotypes. Storage root dry matter in sweetpotato ranges from less than 20% to over 40%, with starch being a predominant constituent, with variable levels of sugars and good levels of dietary fiber, minerals, and several vitamins (Woolfe 1992). Sweetness of sweetpotato ranges from very sweet to non-sweet due to variability in sugar levels (principally sucrose, followed by glucose and fructose) before cooking and with maltose produced by hydrolysis of starch by amylases during cooking (Kays et al. 2005; Owusu-Mensah et al. 2016). Beta-amylase activity varies among sweetpotato genotypes, with absence of activity recessively inherited (Kumagai et al. 1990). Cooked texture and properties of processed products are affected by amylase activity during preparation, and by physicochemical properties of the starch, including amylose-amylopectin ratio (Zhang and Oates 1999, Nabuuya et al. 2012). Further, amylase activity and sugar content are influenced by harvest date and storage (Adu-Kwarteng et al. 2014; Morrison et al. 1993). Variation also exists in starch pasting temperature, with quick cooking varieties with 20 °C lower pasting temperature, slower retrogradation. and higher digestibility than ordinary cultivars developed which can save on cooking time and be suitable for starch noodles (Katayama et al. 2015). Katayama et al. (2015) reported the quick cooking trait to be recessively inherited, with reduction in the trait with increasing dosage of the wild-type allele.

The quality of sweetpotato for fresh consumption has different generally recognized quality classes preferred in different parts of the world, with the so-called dessert type (moist-textured, sweet, orange-fleshed types) predominant in the USA and staple type (drier-textured, white- or yellow-fleshed types predominant in much of SSA, Asia, and Oceania). Tumwegamire et al. (2011) reported that orange-fleshed landrace varieties from Eastern Africa showed higher dry matter and less sugars than the dessert types and are more widely acceptable to African consumers than the dessert types. Tomlins et al. (2004) reported that starchiness and stickiness were the most discriminating attributes for consumers in Eastern Africa, with starchiness preferred and stickiness not. In South Africa, Laurie et al. (2012) examined sensory attributes, sugars, and instrumental data in relation to consumer acceptability for eating quality of 12 cultivars including OFSP types, reporting that maltose, dry matter, and sweet flavor were preferred. They recommended two varieties for promotion, which combined both good taste and good agronomic performance. In the USA, Lekrisompong et al. (2012) used trained panelists to develop a sensory lexicon for appearance, aroma, flavor, and texture and then related this to consumer acceptance of sweetpotato cultivars with different flesh colors, including orange, yellow, and purple. Though demographic characteristics of consumers were not evaluated, the consumers in this study clustered into three different groups differentiated by appearance, aroma, flavor, and texture attributes. They recommended promotion of unfamiliar color types (yellow and purple, in this case) on the basis of healthy, nutritional attributes, while avoiding varieties with undesirable taste.

The relevance of understanding and meeting consumer expectations is particularly great for the promotion of the relatively unfamiliar OFSP for helping to alleviate vitamin A deficiency worldwide. Tomlins et al. (2012) examined 12 varieties in Uganda including dry, starchy OFSP types and reported that the logarithm of carotenoid content was negatively correlated with dry matter content and positively correlated with traits such as color, "pumpkin" odor, and taste. Tomlins et al. (2007) had previously reported that on average dry, starchy OFSP types were preferred to pale-fleshed sweetpotato by school children and women with preschool children. Though there is a generally recognized negative genetic correlation between dry matter and beta-carotene content, progress can be made to increase the two simultaneously. Further, positive correlations between zinc, iron, and beta-carotene lead to the expectation that progress may be made with biofortification for each of these (Grüneberg et al. 2015), though more slowly for the latter two.

Additional quality classes of increasing interest include purple-fleshed sweetpotato, sweetpotato for fries, sweetpotato for leafy greens, and as feed. Anthocyanin-rich purple-fleshed types are increasingly recognized for their healthy antioxidant properties, as a source of natural colorant, and for attractive snack products. Purple-fleshed sweetpotatoes contain a number of different anthocyanins with varying antioxidant activity. Hu et al. (2016) identified seven peonidin- and cyanidin-based anthocyanins with potent antioxidant activity among 13 anthocyanins in order to help target breeding efforts. Fried sweetpotato products (French fries and crisps/chips) are increasing in popularity. Fry quality trait will clearly be important and may include selection for low reducing sugars and accumulation of acrylamide (Truong et al. 2014; Walter et al. 1997). Additionally, sweetpotato leafy green vegetable is traditionally popular in a number of countries and increasingly recognized as very healthy food (Islam 2016; Johnson and Pace 2010; Thiyagu et al. 2013). Breeding programs for leafy green types focus on culinary quality required by consumers and on consistent production during repeated harvests and not yet on phytochemical constituents (Lee et al. 2007; Thiyagu et al. 2013). Finally, selection for suitability for animal feed (Shumbusha et al. 2014, 2015) has emphasized dual-purpose types which produce both fresh vines and storage roots.

6.7.2 Diseases and Pests

A number of fungal (33), bacterial (6), and viral diseases (>16) along with plant parasitic nematodes (>13) have been reported to attack sweetpotato (Clark et al. 2015). Occurrence of diseases and the extent to which they are actual constraints vary with agroecology, the degree of resistance present in varieties used, and the degree to which cultural practices exacerbate or control the problem. The potential importance of a disease, such as sweetpotato virus disease (SPVD), may be revealed in breeding programs when susceptible genotypes emerge in segregating populations or when susceptible genotypes are introduced. Resistance to many diseases exists in sweetpotato germplasm, and predominant varieties in a region must typically be addressed during the selection process in order to ensure that susceptible varieties are not released.

In the tropics, fungal and bacterial diseases of sweetpotato do not seem to be a major constraint, perhaps in part due to limited long-term storage. In the East African highlands, *Alternaria* leaf spot and stem blight can be severe, but resistant genotypes are readily selected in breeding programs (Grüneberg et al. 2015). Scab, caused by *Elsinoe batatas*, can be a problem in Southeast Asia and Oceania, but resistant breeding populations have been developed where the disease is a constraint (Lebot 2010). Fusarium wilt was important in the USA, but resistant varieties now predominate, and breeding populations have high frequencies of resistance (Martin and Jones 1986). A number of additional fungal and bacterial diseases of economic importance exhibit variation in resistance levels among varieties (Zhang et al. 2009; Martin and Jones 1986). These include sweetpotato pox or soil rot (*Streptomyces ipomoea*), *Fusarium* root rot (*Fusarium solani*), bacterial stem and root rot (*Ralstonia solanacearum*), and black rot (*Ceratocystis fimbriata*) (Clark et al. 2013). Key nematode species are root-knot (*Meloidogyne incognita*) nematode in the USA and stem nematode (*Ditylenchus destructor*) in China, and resistant cultivars exist in each case. High narrow-sense heritability has been reported for fusarium wilt resistance (Collins 1977) and root-knot nematode resistance (Jones and Dukes 1980). High reported heritabilities and high frequencies of expression in progenies are consistent with dominant inheritance and control of resistance by major genes.

Virus diseases of sweetpotato, particularly the sweetpotato virus disease (SPVD) which is devastating particularly in moist, tropical environments, such as those found around the Lake Victoria basin in Eastern Africa (Gibson and Kreuze 2015). A large number of viruses are reported to infect sweetpotato globally (Clark et al. 2012). Most breeding efforts have concentrated on breeding for resistance to viruses, weevils, nematodes, biomass, drought tolerance, dry matter, and acceptability by farmers and consumers (Grüneberg et al. 2015).

6.8 Field Trialing and Selection Approaches

Because sweetpotato is clonally propagated, the challenge for the breeder is to identify superior new genotypes from among often large populations of genotypes generated from crosses among selected parents. Selected superior genotypes may become new varieties and also parents in the next cycle of selection. Storage root yield is typically the most important trait, but must be accompanied by a host of other traits. Sweetpotato yield is highly influenced by environment, with high genotype by environment interactions; therefore, multi-environment evaluations are essential to the identification of stable genotypes with broad adaptation or for specific environments (Collins et al. 1987). Sweetpotato is also, by nature, quite variable, with high variation typically observed among plants of the same genotype within and between plots, which results in consistently high experimental coefficients of variation for yield. The conventional breeding scheme for sweetpotato (and other clonally propagated crops) anticipates initial evaluation for multiple seasons in a single selection environment followed by multi-environment testing. Variance component analysis and practical experience have shown that rapid identification of superior, stable genotypes can be achieved by multi-locational evaluation at the initial generation of field evaluation (Grüneberg et al. 2009, 2015), resulting in selection and release of varieties in 4 years in contrast to 7 or 8 years required under the conventional system. This method has been termed accelerated breeding scheme (ABS).

The standard methodology used under the ABS involves establishment in unreplicated three-plant plots at more than one environment, preferably with one of the environments providing exposure to key stresses, such as drought, lower soil fertility, or disease pressure. The use of three-plant plots (established from cuttings taken from the sprouted seedlings) allows the breeder to assess interplant variability, selecting for genotypes with uniform performance. Standard check varieties are also used, with care given to ensure that the health status of the check varieties is comparable to that of the seedling genotypes, since virus accumulation in planting material results in degeneration from clonal generation to generation, but sweetpotato viruses are not seed transmitted. Nearest neighbor designs, such as that of Westcott check design which make heavy use of checks, are also being investigated to allow adjustment in yields in relation to field variation and provide greater precision in selection. Use of rapid throughput analytical methods such as near-infrared reflectance spectrometry allows selected genotypes at each trial location to be routinely evaluated for a number of quality traits, including dry matter content, starch, sugars, beta-carotene, iron, and zinc (Fig. 6.2). Standard procedures and standards for trait evaluation agreed among a number of breeding programs associated with CIP's breeding efforts have been published (Grüneberg et al. 2010). The use of index selection, such as that proposed by Pesek and Baker (1969), allows for simultaneous improvement of multiple traits including yield, quality, and disease and pest reaction. Electronic tools which assist with data collection and with rapid analysis of results are used by some breeding programs.

Field selection for some priority traits, such as SPVD resistance, is time-consuming and challenging because it may take some time to separate resistant genotypes from those which are healthy because they have escaped infection. Efforts are underway to identify molecular markers to be used for selection of resistant genotypes. Selection for drought tolerance has been done in Mozambique through selection for vine vigor (necessary for planting material survival under drought conditions), and it has been noted that drought-tolerant genotypes have greater stability of harvest index when exposed to water-deficit and non-stressed environments.

Selected clones advance through unreplicated trials (except across sites), replicated preliminary trials (typically 2 replications over a few key locations), and advanced trials, replicated trials with larger plots. Advanced trial selections are also advanced to on-farm trial evaluations. In addition to field evaluations, other key attributes such as taste or utilization quality attributes, storability (under controlled and more challenging conditions, such as in sand), and production of planting material from root sprouts or vines are also essential to consider.

At all stages of selection, it is essential for the breeder to have a good understanding of the needs of producers and industry. This is effectively done through

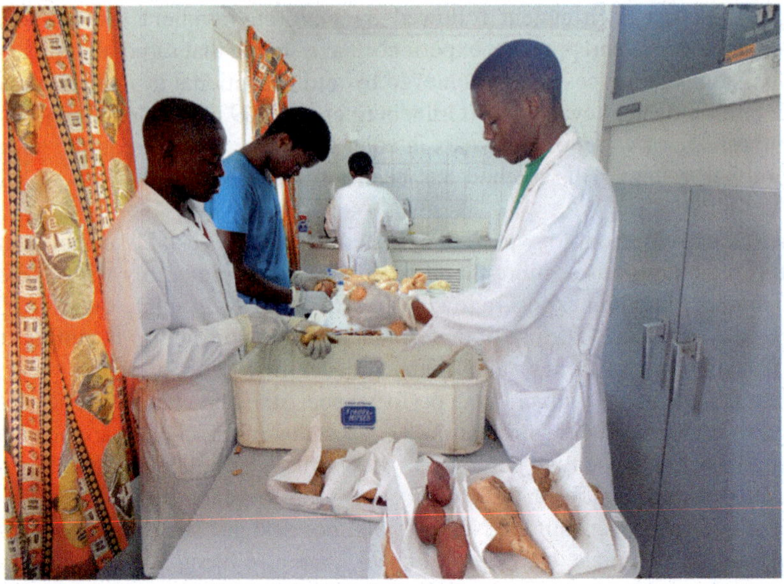

Fig. 6.2 Sample preparation for near-infrared spectroscopy quality analysis at the Agricultural Research Institute of Mozambique (IIAM), Maputo, Mozambique

using a participatory approach, engaging with stakeholders throughout the selection process. Knowledgeable farmers in Uganda have selected an improved variety using participatory plant breeding (Gibson et al. 2011a). However, on-farm evaluation of advanced selections is more typically done, at the advanced trial stage, with engagement of consumers and farmers in assessments. Linkage of breeding efforts and seed dissemination efforts is also recommended, particularly where commercial seed producers can help to gauge suitability of new varieties for market demand.

6.9 Tissue Culture and Clonal Production

Sweetpotato is vegetatively propagated; hence, sweetpotato cultivars are prone to virus infection. Production of virus-free sweetpotato is achieved almost exclusively by meristem culture in vitro. Tissue culture is therefore most useful for producing, propagating, and maintaining virus-free stocks and for the maintenance of germplasm collections. Meristem culture is often coupled with heat treatment of stock plants (e.g., 14 days at 36 °C, or 3–4-week-old in vitro plantlets at 34–37 °C for 1 month) for virus elimination (George 1996; Gaba and Singer 2009).

6.9.1 Meristem Culture

Although other media can be used, sweetpotato meristems can be cultured satisfactorily on Murashige and Skoog (1962) (MS) medium (30 g/l sucrose) or Gamborg et al. (1968) medium. Some workers use MS salts with a wider range of vitamins or employ the same vitamins at a higher concentration, for example, doubling the normal concentration of MS organics. Adjustments can be made to the culture medium to increase survival or increase the growth rate of explants (e.g., pH 5.2 instead of 5.7 MS; 50 instead of 30 g/l sucrose; growth regulator additions such as 1–5 mg/l IAA or NAA + 1–5 mg/l BAP; 0.03 mg/l NAA + 0.3 BAP + 1 mg/l IAA; 0.1 mg/l NAA + 0.1 mg/l kin +1 mg/l GA_3) (George 1996).

Plants are virus cleaned, indexed by grafting sweetpotato cuttings onto indicator plants (*I. setosa* and *I. nil*), and symptoms are evaluated after 4 weeks; positive symptom observation is followed by nitrocellulose membrane enzyme-linked immunosorbent assay (NCM-ELISA) with available antisera. Nucleic acid spot hybridization, PCR, and real-time PCR are optionally used to confirm the presence of some viruses for which antisera are not available. Following initial plant health checks, infected accessions undergo virus elimination and are subsequently rechecked.

6.9.2 Node Cultures

Elongated shoots are produced when nodal segments of sweetpotato are cultured on growth regulator-free MS medium and incubated at 25–28 °C in a 16-h photoperiod (15–60 μmol m^{-2} s^{-1}). Shoots are redivided into nodal sections for subculture. If 10 mg/I ABA is added to MS medium, the growth of axillary shoots is completely inhibited. However, explants remain viable for periods of up to 1 year and are able to produce shoots when transferred to a regulant-free medium (George 1996; Gaba and Singer 2009). There are other sweetpotato micropropagation methods (photoautotrophic micropropagation, somatic embryos for synthetic seed mass production, bioreactors for mass multiplication of sweetpotato nodes in liquid culture, and sweetpotato somatic embryos) described by Gaba and Singer (2009).

Somaclonal variation is a problem in vegetatively propagated plants. Micropropagation by nodal cuttings (from a natural bud) is the safest procedure in terms of controlling the appearance of somaclonal variation. Cloned plants from buds were more genetically uniform than plants propagated from repeated adventitious organogenesis from stored sweetpotato storage roots (Villordon and LaBonte (1996). Somaclonal variation in in vitro regenerated plants from apices during virus cleaning and propagation gave rise to a new cultivar (Moyer and Collins 1983). Genetic stability of cultures maintained in intermediate-term storage conditions for prolonged periods is a concern, but results appear to be satisfactory with current methods (Guo et al. 2001).

The majority of sweetpotato producers in the USA use virus-tested tissue culture seed and supplement their conventional seed produced on farm with certified virus-tested foundation seed. The virus-tested certified seed has had a tremendous positive impact on yield and quality and the sweetpotato industry in the USA and China.

6.10 Seed Production and Marketing

Timely access to quality seed (cutting from a vine) is essential for high yields in any sweetpotato production system. In temperate climates in countries such as China (Zhang et al. 2009) and the USA (Smith et al. 2009), sweetpotato roots are kept and sprouted as the major source of planting material. In tropical climates, seed is usually propagated from cuttings from vines.

Vitality of sweetpotato seed is largely determined by pests and diseases. Vines transmit numerous pests and diseases during vegetative propagation. Most important across all regions are viruses, and among the 20 viruses infecting sweetpotato, SPCSV together with SPFMV results rapidly in devastating yield loses. In regions with pronounced dry seasons, sweetpotato weevils (*Cylas* spp. and *Euscepes postfaciatus*) infest shoots of vine cuttings with eggs and larvae. Moreover, seed health is affected by *Alternaria* in tropical highlands, *Fusarium* in temperate subtropics, and nematodes (mainly *Meloidogyne* spp.) in areas with intensive sweetpotato cultivation. Seed health status can be improved by combined actions of crop rotation, weed control, virus vector-free seed production environments, and specialization including seed coating with insecticides and virus testing.

National research programs have the mandate for breeder seed for new varieties and are responsible for the production of pre-basic seed, which entails assuring high-quality pathogen-free seed stocks. The most common method used for production of pre-basic seed starts with virus cleaning with meristem culture in vitro combined with thermotherapy (Gaba and Singer 2009). In tropical areas, tissue culture plantlets are then hardened and further multiplied in protected screenhouses, with periodic virus indexing to assure that quality is maintained. In the USA, farmers save a portion of their roots for seed the next season. The roots are pre-sprouted in storage by raising the temperature and humidity 2–3 weeks before planting them out in "hotbeds," covered by ventilated plastic mulch. In this system, 23 kg of roots produce approximately 500 sprouts or splits. Six to 8 weeks are required to produce transplants large enough to plant in production fields. Growers supplement this on-farm seed with virus-tested certified cuttings (Dangler 1994; Smith et al. 2009).

In formal or semiformal systems in tropical areas, pre-basic seed is multiplied to basic seed on station or to contracted trained farmers often linked to NGOs, community-based organizations, and farmer organizations. Low multiplication rates of sweetpotato compared to grain crops and the perishability and bulkiness of the vines make long-distance transport costly (Low et al. 2009). Hence, setting up a network of decentralized, trained multipliers to serve their surrounding communities, each at least 10–15 km apart, is a strategy being pursued in several sub-Saharan

African countries (McEwan et al. 2015). These multipliers are encouraged to become commercial root producers, as demand for seed often fluctuates. Several tropical countries are now implementing more formal certification or quality-declared seed classifications at this stage in the multiplication process to assure that farmers know what variety they are receiving and their level of quality (McEwan et al., 2012; Namanda 2012). Experiments have demonstrated that virus-free planting material is higher yielding than non-virus-free (Adikini et al. 2016). In China, the introduction of virus-free seed to 80% of a major growing area by public sector extension led to an average yield increase of 30% (Fuglie et al. 1999).

In many tropical countries, most seed comes from the "informal seed" sector. This presents a challenge for the dissemination of improved varieties, particularly if the public sector extension system is weak. Because it is easy to share planting material, private seed companies have showed limited interest in commercializing sweetpotato. In areas with cultivation throughout the year, actions to improve seed systems target NGOs, road shows for community-based organizations (CBOs) and farmer communities, and field days aligned with introductions of new varieties and recommendations to multiply seeds (Gibson et al. 2011b). Recommendations demonstrated include (1) rogueing out SPVD-diseased plants, (2) taking cuttings only from "clean" field areas where no or only limited rogueing was required (whiteflies spread SPCSV only within short distances), (3) using only relatively young crops to obtain seed, (4) preferring apical parts of vines as planting material (such cuttings carry fewer weevil eggs and are physiologically more vigorous), and (5) providing training in local languages.

In areas with a prolonged dry season, the main problem of production is scarcity of planting material when rain starts (Gibson et al. 2011b). Most farmers are willing to buy seed at least in small amounts and travel to obtain planting material. Farmers conserve seed by (1) growing in wetlands and around waterholes, (2) watering with wastewater, (3) resprouting noncommercial roots missed during harvest and taking sprouts from harvested roots, (4) planting in shade, and (5) using late crops that survive dry seasons.

Current research efforts in the tropics are focused on building sustainable seed systems that assure farmers' timely access to quality planting material. Research is focused on improving pre-basic seed production through (1) reducing the time for virus removal or "cleanup" through new tools, (2) understanding the costs of tissue culture multiplication of plantlets versus screenhouse multiplication of pathogen-tested cuttings, and (3) promoting the establishment of rotation funds so that national programs can maintain core functions. Two technologies have been developed which have strengthened local multiplication efforts. First, the net tunnel is relatively low cost in some sub-Saharan African countries (\$150/6 m^2 curved net) to protect basic stock from virus infection (Shulte-Geldermann et al. 2012). Second, the "Triple S" method (Namanda and Gibson 2015) developed for areas with dry seasons lasting longer than 3 months relies on keeping the small but healthy roots at harvest time as the "seed source" for the next season. The roots are stored in layers of cool sand in an appropriate container and then planted out and fenced 6–8 weeks before the rains start. On average, each root produces 40 cuttings. Detailed descriptions of recommended practices are provided in a manual that is part of a course on "Everything you ever wanted to know about sweetpotato" (Stathers et al. 2013).

6.11 Molecular Genetics and Genomics

Molecular genetic and genomic research in sweetpotato is not as advanced compared to the other major staple crops (e.g., rice, wheat, maize, cassava, and potato). For example, codominant molecular marker resources such as SSR and single-nucleotide polymorphism (SNP) markers are not yet widely available for sweetpotato, a high-quality reference genome for sweetpotato has only recently become available, and the number of scientists conducting genomic research in sweetpotato is limited.

Molecular genetic and genomic research trails behind that of the other major staple crops are largely due to two factors. First, sweetpotatoes have traditionally been viewed as a "poor person" or "orphan" crop. Therefore, until recently, it has received relatively limited attention compared to the major staple crops, and the public (e.g., universities, the CGIAR, and NGOs) and private sectors (e.g., seed companies and agro-based industries) have not invested significant resources toward sweetpotato improvement. Second, cultivated sweetpotato is a highly heterozygous, outcrossing, asexually propagated hexaploid ($2n = 6x = 90$) species, burdened with extremely complex genetics and trait segregation patterns. Its breeding system is also encumbered with numerous self-compatibility, self-incompatibility, cross-compatibility, and cross-incompatibility challenges (Togari and Kawahara 1946; Wedderburn 1967; Williams and Cope 1967). These constraints make sweetpotato breeding very difficult, and it has been hard to efficiently adapt the strategies and molecular genetic tools developed for the diploids to the polyploid crops. However, this situation has begun to change, and significantly more resources are being devoted to the improvement of sweetpotato in both the private and public sectors, and we now have new molecular genetic and genomic resources which are coming on line that promise to mitigate some of the breeding challenges unique to this critical food security crop.

6.11.1 Genetic Engineering of Sweetpotato

For detailed reviews of the advances in genetic engineering of sweetpotato and the use of related biotechnological strategies for sweetpotato improvement, see (Prakash et al. 1991; Prakash 1994; Okada et al. 2002a, b; Yi et al. 2007; Kreuze et al. 2009 ; Mwanga and Ssemakula 2011). Regeneration of sweetpotato from cultured tissues or cells is notoriously difficult, and efficient protocols are improving but generally lacking. Because of this, biotechnological applications of sweetpotato are underdeveloped compared to other major crops. In the 1990s, attempts of sweetpotato transformation experienced slow progress (Prakash and Varadarajan 1992; Gama et al. 1996; Otani et al. 1998; Saito et al. 1998; Kimura et al. 1999). Today, more efficient and reliable transformation protocols are becoming available, but there are still major hurdles to overcome (Song et al. 2004; Lim et al. 2007; Kreuze et al. 2009). Genetic transformation through the use of direct gene transfer and gene editing techniques are promising strategies for introducing novel traits in sweetpotato in cases where no conventional or marker-assisted breeding (MAB) solutions exist. In sweetpotato, the most critical target traits for these strategies include sweetpotato weevil and virus disease resistance, but other opportunities such as the modification of nutritional traits, decreased amylase content, tolerance against oxidative and chilling stress, and starch and the introduction of herbicide resistance are also potential opportunities currently being explored (Qaim 1999; Kreuze et al. 2009). However, no commercial production of GM sweetpotato has yet been reported.

6.11.2 *Use of Molecular Markers in Sweetpotato Breeding*

6.11.2.1 Genetic Diversity Analysis

Genetic diversity analysis of sweetpotato has become fairly common. Diversity analyses are being used to improve our understanding of the genetic relations among germplasm and for parental materials in many breeding programs (Buteler et al. 2002; Hu et al. 2003; Wang et al. 2011; Zhao et al. 2013). To date, sweetpotato genetic diversity studies have been conducted using randomly amplified polymorphic DNAs (RAPDs) (Gichuki et al. 2003, 2005; Jarret and Austin 1994), AFLP (Elameen et al. 2008), and SSR markers (Buteler et al. 1999; Yada et al. 2010, 2015; Tumwegamire et al. 2011). SSR markers have been the most widely used in genetic diversity analysis of sweetpotatoes. Previous studies by Yada et al. (2010) and Rodriguez-Bonilla et al. (2014) showed that SSR markers revealed the highest level of polymorphism due to the codominance nature and high numbers of alleles per locus. Tanaka et al. (2010) reported on the development of cleaved amplified polymorphic sequence (CAPS)-based markers for identification of sweetpotato cultivars. Thirteen primer pairs were designed from the exon sequences of 11 sweetpotato genes to amplify fragments containing an intron. Digestion of the amplified products with restriction enzymes having different recognition sites resulted in 27 polymorphic marker fragments. These were used to distinguish 60 Japanese sweetpotato cultivars. Among the genes used for primer design, the gene encoding the dihydroflavonol 4-reductase (DFR) showed the largest degree of polymorphism, and it is the only report on the development of CAPS-based markers in sweetpotato.

6.11.2.2 Genetic Linkage Mapping and QTL Studies

Genetic linkage mapping in sweetpotato is becoming more frequent. However, to the best of our knowledge, no breeding programs are actively using MAB for sweetpotato improvement. Thompson et al. (1997) were the first to demonstrate the feasibility of linkage mapping in sweetpotato. To do this, they studied two biparental progenies. The first was a cross between the University of Maryland breeding line

MD-708 and the Mississippi State University variety "Vardaman" (MD-708 × "Vardaman"). The second was a cross between "Vardaman" and the USDA US Vegetable Research Laboratory variety "Regal" ("Vardaman" × "Regal"). These populations had 170 and 76 progeny, respectively. Using 100 RAPD primers, they were able to identify 134 polymorphic markers with 74 (60%) segregating 1:1. These were the first linked molecular markers found in sweetpotato, and they demonstrated that the construction of a linkage map in polyploid sweetpotato was feasible.

Zhao et al. (2013) provided a summary of the sweetpotato linkage maps published to date. Table 6.1 provides an update of the linkage mapping studies of sweetpotato that we are aware of. All the linkage maps of sweetpotato have been constructed using an F_1 pseudo-testcross mapping strategy suitable for highly heterozygous parental materials that has been described by Grattapaglia and Sederoff (1994) and the single-dose, simplex fragment approach for the mapping of polyploids originally proposed by Wu et al. (1992), augmented by the use of informative duplex and triplex markers. Using this strategy, single-dose markers are identified via segregation ratio tests and two maps, one from each highly heterozygous parent, and are constructed based on the segregation of single-dose markers (Ukoskit and Thompson 1997; Kriegner et al. 2003; Cervantes-Flores et al. 2008; Li et al. 2010). Marker types used to date include RAPD, AFLP, ISSR, SSR, sequence-related amplified polymorphism (SRAP), and retrotransposon insertion polymorphism (RTISP) markers.

Quantitative trait loci (QTL) studies of sweetpotato are rather limited, and the authors have uncovered only four such studies. Cervantes et al. (2008, 2010) were the first to publish detailed QTL analyses of sweetpotato for resistance to root-knot nematodes (RKN) and for dry matter, starch, and β-carotene content. Single-point analysis of variance and interval mapping revealed seven consistently significant QTL in "Tanzania" and two significant QTL in "Beauregard" associated with resistance to RKN. Based on molecular and phenotypic data, it was hypothesized that RKN resistance in the "Tanzania" × "Beauregard" (TB) cross was conferred by several genes. When all the markers (seven from "Tanzania" and two from "Beauregard") were included in a regression model, these markers explained 40% of the variation in RKN egg masses observed. In the same population, QTL analysis revealed the presence of 13 QTL for storage root dry matter content, 12 QTL for starch content, 8 QTL for ß-carotene content, and 18 QTL for yield. Multiple QTL regression models developed for segregation of alleles in each parent explained 15–24% of the variation in dry matter content, 17–30% of the starch content, 17–35% of ß-carotene content, and 12–30% of the variation in yield (Cervantes et al. 2010).

Li et al. (2010) detected one QTL for starch content in the map of "Zhengshu 20," which explained 7.7% of the variation. More recently, Zhao et al. (2013) reported on QTL analyses of sweetpotato using AFLP and SSR markers and a mapping population consisting of 202 individuals derived from a broad cross between "Xushu 18" and "Xu 781." They identified many QTL for the storage root dry matter content. Using a combination of interval mapping and multiple QTL model, they were able to identify a total of 27 QTL for dry matter content,

Table 6.1 Summary of published sweetpotato linkage map studies to date (December 2016)

Parents	Type of markers[a]	No. of linkage groups	No. of markers	Map length (cM)	Marker density (cM/marker)	Reference
Vardaman × Regal ($n = 76$)						Ukoskit and Thompson (1997)
Vardaman	RAPD	18	102	5024	49.3	
Regal	RAPD	16	94	6560	69.8	
Tanzania × Bikilamaliya ($n = 94$)						Kriegner et al. (2003)
Tanzania	AFLP	90	632	3656	5.8	
Bikilamaliya	AFLP	80	435	3012	6.9	
Tanzania × Beauregard ($n = 240$)						Cervantes-Flores et al. (2008)
Tanzania	AFLP	86	1166	5792	4.5	
Beauregard	AFLP	90	960	5276	4.8	
Nancy Hall × Tainung 27 ($n = 119$)						Chang et al. (2009)
Tainung 27 × Nancy Hall ($n = 112$)						
Nancy Hall	ISSR	9	37	479.8	12.9	
Tainung 27	ISSR	12	47	853.5	17.7	
Luoxushu 8 × Zhengshu 20 ($n = 240$)						Li et al. (2010)
Luoxushu 8	SRAP	81	473	5802	10.2	
Zhengshu 20	SRAP	66	328	3968	12.0	
Xushu 18 × Xu 781 ($n = 202$)						Zhao et al. (2013))
Xushu	AFLP + SSR	90	2077	8185	3.9	
Xu 781	AFLP + SSR	90	1954	8152	4.2	
"Purple Sweet Lord" × 90IDN-47 ($n = 98$)						Monden et al. (2015)
Purple Sweet Lord	RTISP	43	154	931.5	11.6	
90IDN-47	RTISP	47	158	734.3	9.8	

[a] *RAPD* randomly amplified polymorphic DNA, *AFLP* amplified fragment length polymorphisms, *ISSR* inter-simple sequence repeat, *SRAP* sequence-related amplified polymorphism, *RTISP* retrotransposon insertion polymorphism

explaining 9.0–45.1% of the variation. Most (78%) of the QTL had a positive effect on the dry matter variation.

Yield-related traits in sweetpotato are generally thought to be complex quantitative traits influenced by the two QTL studies that have looked at yield and have generally confirmed this observation. Cervantes (2006) observed 18 QTL for storage root yield in the TB mapping population. The QTL for yield due to segregation in "Tanzania" and "Beauregard" were mapped to 11 and 7 different regions, respectively. Both positive and negative yield effects were observed at the QTL, with the complete model for all significant regions explaining approximately 30% and 12% of the total variation in yield, respectively. Chang et al. (2009) also studied yield-related traits in reciprocal "Nancy Hall" × "Tainung 27" mapping population with 119 and 112 progenies, respectively, mapped with 37 (Nancy Hall) and 47 (Tainung 27) inter-simple sequence repeat (ISSR) markers. This study lacked sufficient markers to cover the genome of sweetpotato, but they reported 15 and 9 significant QTL (LOD > 2.5) associated with yield traits in the NH × TN27 and TN27 × NH crosses, respectively.

6.11.3 The Future of Molecular Genetics and Genomics in Sweetpotato

The abovementioned studies represent important steps as the sweetpotato breeding community seeks to integrate modern molecular genetic and genomics tools into MAB of sweetpotato. However, the sweetpotato breeding community still has a long way to go to realize MAB in sweetpotato. Yoon et al. (2015) recently published a review article entitled "Current Status of Sweetpotato Genomics Research" in the Journal of Plant Biotechnology. It is written in Japanese; however, an English translation is available on the Japanese journal's website (http://www.e-sciencecentral.org/articles/SC000013785#ref23-jpb-42-3-161).

Yan et al. (2015) recently published the complete chloroplast genome and a gene expression atlas for sweetpotato using the cultivar Xushu 18. Next-generation sequencing (NGS) methods (Illumina HiSeq 2000) were used to conduct this research, and they reported a circular molecule of 161,303 bp in length with the typical quadripartite structure of large (LSC) and small (SSC) single-copy regions separated by a pair of inverted repeats (IRs). The chloroplast DNA of sweetpotato contained a total of 145 genes, including 94 protein-encoding genes of which there are 72 single-copy and 11 double-copy genes. Phylogenetic analyses were conducted based on 77 protein-coding genes from 33 taxa. This and related research will undoubtedly contribute to a better understanding of the evolution of the genus *Ipomoea* (L.) and the origin of cultivated sweetpotato.

During the last 5 years, two new projects – the China-Japan-Korea Trilateral Research Association of *Sweetpotato* (*TRAS*) Sweetpotato Genome Sequencing Consortium and the Genomic Tools for Sweetpotato (GT4SP) Improvement Project – have also begun to provide valuable new genomic resources for the breeding community, which will provide a solid foundation for MAB in sweetpotato.

The TRAS genome sequencing initiative was established in 2012 with six organizations: The Institute of Sweetpotato Research, Chinese Academy of Agricultural Sciences (China); China Agricultural University (China); the Korea Rural Development Administration; the Korea Research Institute of Bioscience and Biotechnology; the National Agriculture and Food Research Organization (Japan); and the Kazusa DNA Research Institute (Japan). Research on genetic map construction and genome sequencing of two lines of *I. trifida* and the hexaploid sweetpotato

"Xushu 18" has been underway since 2014. Hirakawa et al. (2015) recently published the first de novo genome assemblies of two wild relative diploid species of *I. trifida,* the selfed line Mx23Hm and the highly heterozygous line 0431-1. The genome assembly data, annotations, gene models, and SNPs of *I. trifida* are available at the Sweetpotato GARDEN (http://sweetpotato-garden.kazusa.or.jp). The genome sequence reads obtained by Illumina HiSeq 2000 are available from the DDBJ Sequence Read Archive (DRA) under the accession numbers DRR023905-DRR023907 (Mx23Hm) and DRR023898- DRR023904 (0431-1) (Hirakawa et al. 2015). In addition to the diploid sequencing hexaploid, sequencing has begun using the cultivar Xushu 18. Both of these efforts are using a combination paired end and mate-pair libraries of different insertion sizes and Illumina HiSeq and PacBio sequencing platforms along with a wide variety of assembly routines.

Similar to the TRAS initiative, the GT4SP project has assembled an international team with expertise in applied breeding, crop production, molecular genetics and genomics, and bioinformatics and database management to develop next-generation breeding tools. The GT4SP initiative (https://sweetpotatogenomics.cals.ncsu.edu/) was launched with support from the Bill & Melinda Gates Foundation in 2014. The overarching goals of the project are to "develop modern genomic, genetic, and bioinformatics tools to facilitate crop improvement and improve genetic gains in sweetpotato, an important food security and cash crop with highly recognized potential to alleviate hunger, vitamin A deficiency, and poverty in sub-Saharan Africa (SSA), and predominantly grown in small plot holdings by poor women farmers."

The GT4SP project is led by North Carolina State University (USA) in partnership with the International Potato Center; the Boyce Thomson Institute at Cornell University; Michigan State University; the University of Queensland-Brisbane, Australia; the Uganda National Agricultural Research Organization, National Crops Resources Research Institute; and the Ghana Council for Scientific and Industrial Research, Crops Research Institute. To date, the GT4SP project has developed high-quality reference genomes of the *I. trifida* (NCNSP-0306) and *I. triloba* (NCNSP-0323) lines using Illumina mate pair, PacBio, BioNano optical mapping, and linkage mapping approaches. The genomes were publically released in August 2016 and are available at the Sweetpotato Genomics Resource website (http://sweetpotato.plantbiology.msu.edu/). This resource is continually being updated as new information becomes available.

The GT4SP team is exploring the efficiency of genotyping by sequencing (GBS) and the Diversity Arrays Technology DArTseq SNP development platforms using several cultivated sweetpotato mapping populations. To manage the tremendous amount of phenotypic and genotypic data generated by the GT4SP project, a web-based breeder-friendly database has been established (https://sweetpotatobase.org/). Currently, (August 2017), the SweetpotatoBase (SPbase) database features about 5,900 accession entries and 2,000 trials from several breeding programs. The GT4SP database and capacity development teams are actively training breeders on the use of SPbase and the use of electronic data capture methods in the field such as the use of the Field Book (://wheatgenetics.org/fieldbook) and DataCollector apps for collecting and managing phenotyping and genotyping data generated by the project in the USA, Ghana, Uganda, and Kenya.

Collectively, the achievement of the TRAS and GT4SP projects, and other ongoing efforts yet to be reported will provide valuable resources needed to bring sweetpotato breeding into the genomics era and contribute to the progress of MAB in this critical food security crop and our understanding of the genus *Ipomoea* in general.

6.12 Conclusions and Outlook

Sweetpotato has traditionally been viewed as a "poor person's crop" or "orphan crop," and it has attracted limited attention compared to other staple crops. However, during the last decade, this perception has changed, and it is widely acknowledged that sweetpotato has great potential to contribute to the alleviation of malnutrition and hunger in the developing world. Orange-fleshed sweetpotato, in particular, with its high provitamin A content, has become a prominent example of the effectiveness of biofortified staple crops to combat vitamin A deficiency. Similarly, increasing awareness of the nutritional value of sweetpotato is driving consumer demand among health-conscious consumers globally, and its potential use in a wide range of value-added human and animal products is widely recognized. As the public and private sectors learn more about the benefits and opportunities of sweetpotato, investment in more crop improvement to increase our understanding of the importance and exploiting the great untapped potential of the crop should be amplified.

The sweetpotato breeding community is on the right track of integrating modern molecular genetic and genomics tools into MAB of the crop. However, the sweetpotato breeding community still has a long way to go to realize MAB in sweetpotato. Collectively, all sweetpotato genomics' ongoing efforts and those in the pipeline yet to be reported are important in providing valuable resources needed to bring sweetpotato breeding into the genomics era and contribute to the progress of MAB in this critical food security and nutrition crop and our understanding of the genus *Ipomoea* in general.

Acknowledgment This research was undertaken as part of the CGIAR Research Program on Roots, Tubers and Banana (RTB) and HarvestPlus, part of the CGIAR Research Program on Agriculture for Nutrition and Health (A4NH). It has also received financial support from the Bill and Melinda Gates Foundation and USAID.

References

Adikini S, Mukasa SB, Mwanga ROM, Gibson RW (2016) Effects of sweet potato feathery mottle virus and sweet potato chlorotic stunt virus on the yield of sweet potato in Uganda. J Phytopathol 164:242–254

Adu-Kwarteng E, Sakyi-Dawson EO, Ayernor GS, Truong VD, Shih FF, Daigle K (2014) Variability of sugars in staple-type sweet potato (*Ipomoea batatas*) cultivars: the effects of harvest time and storage. Int J Food Prop 17:410–420. doi:10.1080/10942912.2011.642439

Andersson MS, de Vicente MC (2010) Gene flow between crops and their wild relatives. John Hopkins University Press, Baltimore

Andrade MI, Alvaro A, Menomussanga J, Makunde GS, Ricardo J, Wolfgang J, Grüneberg WJ, Eyzaguirre R, Low J, Ortiz R (2016) 'Alisha', 'Anamaria', 'Bie', 'Bita', 'Caelan', 'Ivone', 'Lawrence', 'Margarete', and 'Victoria' Sweetpotato. Hortscience 51:597–600

Austin DF (1977) Hybrid polyploids in *Ipomoea* section *batatas*. J Hered 68:259–260

Austin DF (1978) The *Ipomoea batatas* complex – I. Taxon. Bull Torrey Bot Club 105:114–129

Austin DF (1979) An infrageneric classification for *Ipomoea* (Convolvulaceae). Taxon 28:359–361

Austin DF (1983) Variability in sweetpotato in America. Proc Am Soc Hortic Sci Trop Reg 7(Part B):15–26

Austin DF (1988a) The taxonomy, evolution and genetic diversity of sweet potatoes and related wild species. In: Gregory P (ed) Exploration, maintenance and utilization of sweet potato genetic resources. International Potato Center, Lima, pp 27–59

Austin DF (1988b) Nomenclature changes in the *Ipomoea batatas* complex (Convolvulaceae). Taxon 37:184–185

Austin DF (1991) *Ipomoea littoralis* (Convolvulaceae) – taxonomy, distribution and ethnobotany. Econ Bot 45:251–256

Austin DF, Bianchini RS (1998) Additions and corrections in American *Ipomoea* (Convolvulaceae). Taxon 47:833–838

Austin DF, Huáman Z (1996) A synopsis of *Ipomoea* (Convolvulaceae) in the Americas. Taxon 45:3–38

Austin DF, de la Puente F, Contreras J (1991) *Ipomoea tabascana*, an endangered tropical species. Econ Bot 45:435

Austin DF, Jarret RL, Tapia C, de la Puente F (1993) Collecting tetraploid *I. batatas* (Linnaeus) Lamarck in Ecuador. FAO/IBPGR Plant Genet Res Newsl 91(92):33–35

Bacusmo JL, Collins WW, Jones A (1988) Effects of fertilization on stability and yield components of sweetpotato clones. J Am Soc Hortic Sci 113:261–264

Bohac JR, Jones A, Austin DF (1992) Unreduced pollen: proposed mechanism of polyploidization of sweet potato (*Ipomoea batatas*). Hortic Sci 27:611

Bohac JR, Austin DF, Jones A (1993) Discovery of wild tetraploid sweetpotatoes. Econ Bot 47:193–201

Bohac JR, Dukes PD, Austin DF (1995) Sweet potato. In: Smartt J, Simmonds NW (eds) Evolution of crop plants, 2nd edn. Longman Scientific and Technical, Harlow, pp 57–62

Bouis H, Islam Y (2012) Delivering nutrients widely through biofortification: building on orange sweet potato – scaling up in agriculture, rural development and nutrition, Focus 19, Brief 11. International Food Policy Research Institute (IFPRI), Washington, DC. http://dspace.africaportal.org/jspui/bitstream/123456789/33024/1/focus19_11.pdf. Accessed 22 Nov 2016

Bovel-Benjamin AC (2007) Sweet potato: a review of its past, present, and future role in human nutrition. Adv Food Nutr Res 52:1–59

Bulmer R (1966) Birds as possible agents in the propagation of the sweet potato. Emu 65:165–182

Buteler MI, Jarret RL, Labonte DR (1999) Sequence characterization of microsatellites in diploid and polyploid Ipomoea. Theor Appl Genet 99:123–132

Buteler M, Labonte D, Jarret R, Macchiavelli, R (2002) Microsatellite-based paternity analysis in polyploid sweetpotato. J Am Soc Hort Sci 127:392–396

Cervantes-Flores J (2006) Development of a genetic linkage map and QTL analysis in sweetpotato. PhD. thesis. North Carolina State University, Raleigh. p 181

Cervantes-Flores JC, Yencho GC, Kriegner A, Pecota KV, Faulk MA, Mwanga ROM, Sosinski S (2008) Development of a genetic linkage map and identification of homologous linkage groups in sweetpotato using multiple-dose AFLP markers. Mol Breed 21:511–532

Cervantes-Flores J, Sosinski B, Pecota K, Mwanga ROM, Catignani G, Truong V, Watkins R, Ulmer M, Yencho G (2010) Identification of quantitative trait loci for dry matter, starch and β-carotene content in sweetpotato. Mol Breed 5:1–16

Chang K, Lo H, Lai Y, Yao P, Lin K, Hwang S (2009) Identification of quantitative trait loci associated with yield-related traits in sweet potato (*Ipomoea batatas*). Bot Stud 50:43–55

CIP/AVRDC/IBPGR, International Potato Center, Asian Vegetable Research and Development Center, and the International Board for Plant Genetic Resources (1991) In: Huaman Z (ed) Descriptors for sweetpotato. International Board for Plant Genetic Resources, Rome, pp 43–84
Clark CA, Davis JA, Abad JA, Cuellar WJ, Fuentes S, Kreuze JF, Gibson RW, Mukasa SB, Tugume AK, Tairo FD, Valkonen JPT (2012) Sweetpotato viruses: 15 years of progress on understanding and managing complex diseases. Plant Dis 96:168–185
Clark CA, Ferrin DM, Smith TP, Holmes GJ (eds) (2013) Compendium of sweetpotato diseases, pests, and disorders, 2nd edn. APS Press, The American Phytopathological Society, St. Paul, p 160
Clark CA, Ferrin DM, Smith TP, Holmes GJ (2015) Diseases of sweetpotato (*Ipomoea batatas* [L.] Lam.) http://www.apsnet.org/publications/commonnames/Pages/Sweetpotato.aspx. Accessed 11/12/2016
Collins WW (1977) Diallel anlaysis of sweet potatooes for resistance to fusarium wilt. J Am Soc Hortic Sci 102:109–111
Collins WW, Wilson LG, Arrendell, Dickey LF (1987) Genotype x environment interactions in sweetpotato yield and quality factors. J Am Soc Hortic Sci 112:579–583
CSPI 2014 reference is CSPI (Center for Science in the Public Interest) (2014) Top 10 foods. http://www.cspinet.org/nah/10foods_bad.html
Dangler JM (1994) Sweetpotato foundation programs provide growers with high-quality pathogen-free seedstock. HortTechnology 4(3):224–227
Diaz J, Schmiediche P, Austin DF (1996) Polygon of crossability between eleven species of *Ipomoea*: section *Batatas*. Euphytica 88:189–200
Edmond JB, Ammerman GR (1971) Sweetpotoes: production, processing, and marketing. AVI Publishing Co, Westport
Eguchi Y (1996) Flowering and seed production in the sweetpotato. In: Ramanatha Rao V (ed) Proceedings of the workshop on the formation of a network for the conservation of sweet potato biodiversity in Asia, Bogor, Indonesia, May 1–5, 1996. International Plant Genetic Resources Institute (IPGRI), Rome, pp 86–92
Elameen A, Fjellheim S, Larsen A, Rognli OA, Sundheim L, Msolla S, Masumba E, Mtunda K, Klemsdal SS (2008) Analysis of genetic diversity in a sweet potato (Ipomoea batatas L.) germplasm collection from Tanzania as revealed by AFLP. Genet Resour Crop Evol 55:97–408
FAOSTAT (2011) FAOSTAT data. Available online: http://faostat.fao.org/
Fox PN, Cross J, Romagosa I (1997) Multi-environment testing and genotype X environment interaction. In: Kempton RA, Fox PN (eds) Statistical methods for plant variety evaluation. Chapman and Hall, London, pp 112–129
Freyre R, Iwanga M, Orjeda G (1991) Use of *Ipomoea trifida* (H.B.K.) G. Don. germplasm for sweetpotato improvement. 2. Fertility of synthetic hexaploids and triploids with 2n gametes of *I. trifida*, and their interspecific crossability with sweet potato. Genome 34:209–214
Fugise K, Yunoue T, Chishiki T (1955) Studies on the habits of flowering and seed setting in the varieties of sweetpotato (in Japanese with English summary). Bull Kyusu Agric Exp Sta 3:109–142
Fuglie KO, Zhang L, Salazar LF, Walker TS (1999) Economic impact of virus-free sweetpotato planting material in Shandong province, China. International Potato Center, Lima
Gaba V, Singer S (2009) Germplasm conservation and conservation by tissue culture. In: Loebenstein G, Thottappilly G (eds) The sweetpotato. Springer Science + Business Media BV, Houten, pp 65–80
Gallais A (1981) Quantitative genetics and breeding methods. Proceedings of the fourth meeting of the Section Biometrics in Plant Breeding, Poitiers, France, 2–4 September 1981
Gallais A (2003) Quantitative genetics and selection theory in autopolyploid plants. INRA, Paris
Gama MICS, Leite RP, Cordeiro AR, Cantliffe DJ (1996) Transgenic sweetpotato plants obtained by *Agrobacterium tumefaciens*-mediated transformation. Plant Cell Tissue Organ Cult 46:237–244
Gamborg O, Milla R, Ojima K (1968) Nutrient requirements of suspension cultures of soyabean root cells. Exp Cell Res 50:151–158

George EF (1996) Plant propagation by tissue culture in practice, part 2, 2nd edn. Exegetics Limited, Edington

Gibson RW, Kreuze JF (2015) Degeneration in sweetpotato due to viruses, virus-cleaned planting material and reversion: a review. Plant Pathol 64:1–15

Gibson RW, Mpembe I, Mwanga ROM (2011a) Benefits of participatory plant breeding (PPB) as exemplified by the first-ever officially released PPB-bred sweet potato cultivar. J Agric Sci 149:625–632

Gibson R, Namanda S, Sindi K (2011b) Sweetpotato seed systems in East Africa. J Sustain Agric 35:870–844

Gichuki ST, Berenyi M, Zhang D, Hermann M, Schmidt J, Glössl J, Burgh K (2003) Genetic diversity in sweetpotato (Ipomoea batatas (L.) Lam.) in relationship to geographic sources as assessed with RAPD markers. Genet Resour Crop Evol 50:429–437

Gichuki S, La Bonte D, Burg K, Kapinga R, Simon JC (2005). Assessment and Genetic Diversity, Farmer Participatory Breeding and Sustainable Conservation of Eastern Africa Sweet Potato. Germ-Plasm. Annual report April 2004 - March 2005

Golson J (1976) Archaeology and agricultural history in the New Guinea Highlands. In: Sieveking G, Longworth IH, Wilson KE (eds) Problems in economic and social archaeology. Duckworth, London, pp 201–220

Grattapaglia D, Sederoff R (1994) Genetic linkage maps of *Eucalyptus grandis* and *E. urophylla* using a pseudo-testcross mapping strategy and RAPD markers. Genetics 137:1121–1137

Grüneberg WJ, Mwanga R, Andrade M, Espinoza J (2009) Selection methods part 5: breeding clonally propagated crops. In: Ceccarelli S, Guimarães EP, Weltzien E (eds) Plant breeding and farmer participation. FAO, Rome, pp 275–322

Grüneberg WJ, Eyzaguirre R, Espinoza J, Mwanga ROM, Andrade M, Dapaah H, Tumwegamire S, Agili S, Ndingo-Chipungu FP, Attaluri S, Kapinga R, Tinh N, Xie K, Tjintokohadi K, Carey T, Low J (2010) Procedures for the evaluation and analysis of sweetpotato trials. http://www.sweetpotatoknowledge.org/files/procedures-for-the-evaluation-and-analysis-of-sweetpotato/. Accessed 27 Jan 2017

Grüneberg WJ, Ma D, Mwanga ROM, Carey EE, Huamani K, Diaz F, Eyzaguirre R, Guaf E, Jusuf M, Karuniawan A, Tjintokohadi K, Song YS, Anil SR, Hossain M, Rahaman E, Attaluri SI, Somé K, Afuape SO, Adofo K, Lukonge E, Karanja L, Ndirigwe J, Ssemakula G, Agili S, Randrianaivoarivony JM, Chiona M, Chipungu F, Laurie SM, Ricardo J, Andrade M, Rausch Fernandes F, Mello AS, Khan MA, Labonte DR, Yencho GC (2015) Advances in sweetpotato breeding from 1993 to 2012. In: Low J, Nyongesa M, Quinn S, Parker M (eds) Potato and sweetpotato in Africa: transforming the value chains for food and nutrition security. CABI, Wallingford, pp 3–68

Guo XD, Zhou MD, Wang Y (2001) In vitro conservation of sweetpotato germplasm. In: Rao VR, Herman M (eds) Conservation and utilization of sweetpotato diversity in Asia. International Plant Genetic Resources Institute, Selangor Darul Ehsan, pp 16–24

Hirakawa H, Okada Y, Tabuchi H, Shirasawa K, Watanabe A, Tsuruoka H, Minami C, Nakayama S, Sakamoto S, Kohara M, Kishida Y, Fujishiro T, Kato M, Nanri K, Komaki A, Yoshinaga M, Takahata Y, Tanaka M, Tabata S, Isobe SN (2015) Survey of genome sequences in a wild sweet potato, Ipomoea trifida (HBK) G Don DNA Res 22:171–179

Hotz C, Loechl C, Lubowa A, Tumwine JK, Ndeezi G, Nandutu Masawi A, Baingana R, Carriquiry A, de Brauw A, Meenakshi JV, Gilligan DO (2012) Introduction of β-carotene-rich orange sweet potato in rural Uganda resulted in increased vitamin a intakes among children and women and improved vitamin A status among children. J Nutr 142:1871–1880

Hsia W, Chen C (1956) Studies on inducing the hard blooming varieties of sweetpotato to bloom. J Agric Assoc China. New Series 14:22–24

Hu J, Nakatani M, Garcia Lalusin A, Kuranouchi T, Fujimura T (2003) Genetic analysis of sweetpotato and wild relatives using inter-simple sequence repeats (ISSRs). Breed Sci 53:297–304

Hu YJ, Deng LQ, Chen JW, Zhou SY, Liu S, Fu YF, Yang CX, Liao ZH, Chen M (2016) An analytical pipeline to compare and characterise the anthocyanin antioxidant activities of purple sweet potato cultivars. Food Chem 194:46–54

Huáman Z, Zhang D (1997) Sweetpotato. In: Fuccillo DA, Sears L, Stapleton P (eds) Biodiversity in trust: conservation and use of plant genetic resources in CGIAR Centres. Cambridge University Press, Cambridge, UK, pp 29–38

Islam S (2016) Some bioactive constituents, antioxidant, and antimutagenic activities in the leaves of *Ipomoea batatas* Lam. genotypes. Am J Food Sci Technol 4:70–80. doi:10.12691/ajfst-4-3-3

Iwanaga M (1988) Use of wild germplasm for sweetpotato breeding. In: Gregory P (ed) Exploration, maintenance, and utilization of sweetpotato genetic resources. International Potato Center, Lima, pp 199–210

Iwanaga M, Freyre R, Orjeda G (1991) Use of *Ipomoea trifida* (H.B.K.) G. Don. germplasm for sweetpotato improvement. 1. Development of synthetic hexaploids of *I. trifida* by ploidy-level manipulations. Genome 34:201–208

Jarret RL, Austin DF (1994) Genetic diversity and systematic relationships in sweetpotato (*Ipomoea batatas* (L.) Lam.) and related species as revealed by RAPD analysis. Genet Resour Crop Evol 41:165–173

Jarret RL, Gawel N, Whittemore A (1992) Phylogenetic relationships of the sweetpotato *Ipomoea batatas* (L.) Lam. J Am Soc Hortic Sci 117:633–637

Jia R (2013) Weather, shocks, sweet potatoes and peasant revolts in historical China. Econ J 124:92–118

Johnson M, Pace RD (2010) Sweet potato leaves: properties and synergistic interactions that promote health and prevent disease. Nutr Rev 68:604–615

Jones A (1965a) A proposed breeding procedure for sweetpotato. Crop Sci 5:191–192

Jones A (1965b) Cytological observations and fertility measurements of sweetpotato (*Ipomoea batatas* (L.) Lam.) Proc Am Soc Hortic Sci 86:527–537

Jones A (1966) Morphological variability in early generations of a randomly intermating population of sweet potatoes (*Ipomoea batatas* (L.) Lam.) Univ Ga Agric Sta Technol Bull NS 56:1–31

Jones A (1967a) Should Nishiyama's K123 (*Ipomoea trifida*) be designated I. batatas? Econ Bot 21:163–166

Jones A (1967b) Theoretical segregation ratios of qualitatively inherited characters for hexaploid sweetpotato (*Ipomoea batatas*). Technical Bulletin No. 1368. USDA

Jones A (1967c) Theoretical segregation ratios of qualitatively inherited characters for hexaploid sweetpotato (*Ipomoea batatas*). Technical Bulletin No. 1368. USDA

Jones A (1969) Quantitative inheritance of ten vine traits in sweetpotatoes. J Am Soc Hortic Sci 94:408–411

Jones A (1974) Chromosome numbers in genus *Ipomoea*. J Hered 55:216–219

Jones A (1977) Heritability estimates of seven root traits in sweetpotatoes. J Am Soc Hortic Sci 102:440–442

Jones A (1980) Sweet potato. In: Fehr WR, Hadley HH (eds) Hybridization of crop plants. American Society of Agronomy, Madison, pp 645–655

Jones A (1986) Sweetpotato heritability estimates and their use in breeding. Hortic Sci 21:14–17

Jones A, Dukes PD, Cuthbert FP Jr (1977) Pesticides increase true seed production of sweet potato. HortScience 12:154–167

Jones A, Schalk JM, Dukes PD (1987) Control of soil insect injury by resistance in sweet potato. J Am Soc Hortic Sci 112:195–197

Jones A (1990) Unreduced pollen in a wild tetraploid relative of sweet potato. J Am Soc Hortic Sci 115:512–516

Jones A, Deonier MT (1965) Interspecific crosses among *Ipomoea lacunosa, I. ramoni, I. trichocarpa* and *I. triloba*. Bot Gaz 126:226–232

Jones A, Dukes PD (1980) Heritability of sweetpotato resistances to root knot caused by *Meloidogyne incognita* and *M. javanica*. J Am Soc Hortic Sci 105:154–156

Jones A, Steinbaur CE, Pope DT (1969) Quantitative inheritance of ten root traits in sweetpotatoes. J Am Soc Hortic Sci 94:271–175

Jones A, Dukes PD, Schalk JM (1978) Heritability estimates for resistance in sweetpotato soil insects. J Am Soc Hortic Sci 104:424–426

Jones A, Dukes PD, Schalk JM (1979) Heritability estimates for resistance in sweetpotato soil insects. J Am Soc Hortic Sci 104:424–426

Jones A, Dukes PD, Schalk JM (1986) Sweetpotato breeding. In: Bassett MJ (ed) Breeding vegetable crops. AVI Pub. Co, Westport, pp 1–35

Katayama K, Tamiya S, Sakai T, Kai Y, Ohara-Takada A, Kuranouchi T, Yoshinaga M (2015) Inheritance of low pasting temperature in sweetpotato starch and the dosage effect of wild-type alleles. Breed Sci 65:352–356. doi:10.1270/jsbbs.65.352

Kays SJ, Wang Y, McLaurin WJ (2005) Chemical and geographical assessment of the cultivated sweetpotato clones of the world. J Am Soc Hortic Sci 130:591–597

Kempthorne O (1957) An introduction to genetic statistics. Wiley, New York

Khoury CK, Heider B, Castañeda-Álvarez NP, Achicanoy HA, Sosa CC, Miller RE, Scotland RW, Wood JR, Rossel G, Eserman LA, Jarret RL, Yencho GC, Bernau V, Juarez H, Sotelo S, de Haan S, Struik PC (2015) Distributions, ex situ conservation priorities, and genetic resource potential of crop wild relatives of sweetpotato [*Ipomoea batatas* (L.) Lam., I. series Batatas]. Front Plant Sci. doi:10.3389/fpls.2015.00251

Kimura T, Otani M, Noda T, Ideta O, Shimada T, Saito A (1999) Decrease of amylase content in transgenic sweetpotato. Breed Res 1:142

King JR, Bamford R (1937) The chromosome number in *Ipomoea* and related genera. J Hered 28:279–282

Kobayashi M (1984) The *Ipomoea trifida* complex closely related to sweet potato. In: Shideler SF, Rincon H (eds) Proceedings of the 6th symposium of the International Society of Tropical Root Crops. CIP, Lima, pp 561–568

Kobayashi RS, Miyazaki T (1976) Sweet potato breeding using wild related species. Proc 4th Symp Int Soc Trop Root Crop 1:53–57

Kobayashi RS, Bouwkamp JC, Sinden SL (1994) Interspecific hybrids from cross incompatible relatives of sweetpotato. Euphytica 80:159–164

Komaki K (2004) Breeding value of wild species closely related to sweetpotato. Proceedings of international workshop on production, utilization and development of sweetpotato, Korea, pp 164–172

Kreuze JF, Valkonen JPT, Ghislain M (2009) Genetic engineering. In: Loebenstein G, Thottappilly G (eds) The sweetpotato. Springer, Berlin

Kriegner A, Cervantes JC, Burg K, Mwanga ROM, Zhang DP (2001) A genetic linkage map of sweetpotato [Ipomoea batatas (L.) Lam.] based on AFLP markers. In: Partners in research for the 21st century. Int. Potato Center Progr. Rpt. 1999–2000. p 303–314

Kriegner A, Cerventes JC, Burg K, Mwanga ROM, Zhang D (2003) A genetic linkage map of sweetpotato [Ipomoea batatas (L.) Lam.] based on AFLP markers. Mol Breed 11:169–185

Kumagai T, Umemura Y, Baha T, Iwanaga M (1990) The inheritance of ß-amylase null in storage roots of sweetpotato, *Ipomoea batatas* (L.) Lam. Theor Appl Genet 79:369–376

Laurie SM, Faber M, Calitz FJ, Moelich EI, Mullerd N, Labuschagne MT (2012) The use of sensory attributes, sugar content, instrumental data and consumer acceptability in selection of sweet potato varieties. J Sci Food Agric. (wileyonlinelibrary.com). doi:10.1002/jsfa.5932

Lebot V (2010) Sweet potato. In: Bradshaw JE (ed) Root and tuber crops, handbook of plant breeding, vol 7. Springer Science+Business Media, LLC, New York, pp 97–125

Lee JS, Ahn YS, Chung MN, Kim HS, Jeong BC, Bang JK (2007) Varieties for sweetpotato tips production as a vegetable. Kor J Breed Sci 39:224–231

Leksrisompong P, Whitson ME, Truong VD, Drake MA (2012) Sensory attributes and consumer acceptance of sweet potato cultivars with varying flesh colors. J Sens Stud 27:59–69

Li AX, Liu QC, Wang QM, Zhai H, Yan WZ, Zhang HY, Li M (2010) Mapping QTLs for starch content in sweetpotato. Mol Plant Breed 8:516–520. (in Chinese)

Lim S, Kim JH, Kim SH, Kwon SY, Lee HS, Kim JS, Cho KY, Paek KY, Kwak SS (2007) Enhanced tolerance of transgenic sweetpotato plants that express both CuZnSOD and APX in chloroplasts to methyl viologen-mediated oxidative stress and chilling. Mol Breed 19:227–239

Low JW, Arimond M, Osman N, Cunguara B, Zano F, Tschirley D (2007) A food-based approach introducing orange-fleshed sweet potatoes increased vitamin A intake and serum retinol concentrations in young children in rural Mozambique. J Nutr 137:1320–1327

Low JW, Lynam J, Lemaga B, Crissman C, Barker I, Thiele G, Namanda S, Wheatley C, Andrade M (2009) Sweetpotato in Sub-Saharan Africa. In: Loebenstein G, Thottappilly G (eds) The Sweetpotato. Springer Science+Business Media B.V, Dordrecht, pp 359–390

Magoon ML, Krishnan R, Baii KV (1970) Cytological evidence on the origin of sweetpotato. Theor Appl Genet 40:360–366

Martin JA Jr (1946) Germination of sweet potato seed as affected by different methods of scarification. Proc Am Soc Hortic Sci 47:387–390

Martin FW (1965) Incompatibility in the sweet potato. A review. Econ Bot 19:406–415

Martin FW (1968) The system of self-incompatibility in *Ipomoea*. J Hered 59:263–267

Martin FW (1970) Self and interspecific incompatibility in the Convolvulaceae. Bot Gaz 131:139–144

Martin FW, Cabanillas E (1966) Post-pollen-germination barriers to seed set in sweet potato. Euphytica 15:404–411

Martin FW (1988) Genetic and physiological basis for breeding and improving the sweet potato. In: Institut Scientifique de Recherche Agronomique (INRA) (ed) Proceeding 7th symposium of the international society for tropical root crops, Gosier, Guadeloupe, July 1–6, 1985. INRA, Paris, pp 749–761

Martin FW, Jones A (1973) The species of *Ipomoea* closely related to the sweetpotato. Econ Bot 26:201–215

Martin FM, Jones A (1986) Breeding sweet potatoes. Plant Breed Rev 4:313–345

Martin FW, Ortiz S (1967) Anatomy of the stigma and style of sweet potato. New Phytol 66:109–113

McDonald JA, Austin DF (1990) Changes and additions in *Ipomoea* section *Batatas* (Convolvulaceae). Brittonia 42:116–120

McDonald JA, Mabry TJ (1992) Phylogenetic systematics of New World *Ipomoea* (Convolvulaceae) based on chloroplast DNA restriction site variation. Plant Syst Evol 180:243–259

McEwan M, Namanda S, Lusheshanija D (2012) Whose standards matter? Piloting quality declared planting material inspection guidelines for sweetpotato in Lake Zone, Tanzania. 16th triennial symposium of the international society for tropical root crops. Abeokuta, Nigeria

McEwan M, Almekinders C, Abidin PE, Andrade M, Carey EE, Gibson RW, Naico A, Namanda S, Schulz S (2015) Can small still be beautiful? Moving local sweetpotato seed systems to scale in sub-Saharan Africa. In: Low J, Nyongesa M, Quinn S, Parker M (eds) Potato and sweetpotato in Africa: transforming the value chains for food and nutrition security. CABI, Wallingford, pp 289–310

McGregor CE, Labonte DR (2006) Differential expression of genes between storage roots of sweetpotato cultivars jewel and white jewel. J Am Soc Hortic Sci 131:798–805

Millar JC (1939) Further studies and technique used in sweetpotato breeding in Louisiana. J Hered 30:485–492

Mont J, Iwanaga M, Orjeda G, Watanabe K (1993) Abortion and determination of stages for embryo rescue in crosses between sweetpotato, *Ipomoea batatas* Lam. (2n=6x=90) and its wild relative, *I. trifida* (H.B.K.) G. Don. (2n=2x=30). Sex Plant Reprod 6:176–182

Montelaro J, Miller JC (1951) A study of some factors affecting seewd setting in the sweet potato. Proc Am Soc Hortic Sci 57:329–334

Montenegro Á, Avis C, Weaver A (2008) Modeling the prehistoric arrival of the sweet potato in Polynesia. J Archaeol Sci 35:355–367

Monden Y, Yamamoto A, Shindo A, Tahara M (2014) Efficient DNA fingerprinting based on the targeted sequencing of active retrotransposon insertion sites using a bench-top high-throughput sequencing platform. DNA Res 21:491–498

Morrison T, Pressey R, Kays SJ (1993) Changes in alpha- and beta-amylase during storage of sweet potato lines with varying starch hydrolysis potential. J Am Soc Hortic Sci 118:236–242

Moyer JW, Collins WW (1983) "Scarlet" sweet potato. Hortscience 18:111–112

Murashige T, Skoog F (1962) A revised medium for rapid growth and bio-assay with tobacco tissue cultures. Physiol Plant 15:473–479

Murata T, Matsuda Y (2003) Histological studies on the relationship between the process from fertilization to embryogenesis and the low seed set of sweet potato, *Ipomoea batatas* (L.) Lam. Breed Sci 53:41–50

Mwanga ROM, Yencho GC, Moyer JW (2002a) Diallel analysis of sweetpotatoes for resistance to sweetpotato virus disease. Euphytica 128:237–248

Mwanga ROM, Moyer J, Zhang D, Carey EE, Yencho GC (2002b) Nature of resistance to sweetpotato virus diseases. Acta Hortic 583:113–119

Mwanga ROM, Ssemakula G (2011) Orange-fleshed sweet potatoes for food, health and wealth in Uganda. Intern J Agric Sust 9:42–49

Mwanga ROM, Kyalo G, Ssemakula GN, Niringiye C, Yada B, Otema MA, Namakula J, Alajo A, Kigozi B, Makumbi RNM, Ball A, Grüneberg WJ, Low JW, Yencho GC (2016) 'NASPOT 12 O' and 'NASPOT 13 O' sweetpotato. Hortscience 51:291–295

Nabubuya A, Namutebi A, Byaruhanga Y, Narvhus J, Wicklund T (2012) Potential use of selected sweetpotato (Ipomea batatas Lam) varieties as defined by chemical and flour pasting characteristics. Food Nutr Sci 3:889–896. http://dx.doi.org/10.4236/fns.2012.37118

Namanda S (2012) Current and potential systems for maintaining sweetpotato planting material in areas with prolonged dry seasons: a biological, social and economic framework. PhD Thesis, University of Greenwich, London, UK

Namanda S, Gibson RW (2015) Shortage of sweetpotato planting material caused by prolonged dry seasons in Africa: strategies to increase its availability in Uganda. In: Low J, Nyongesa M, Quinn S, Parker M (eds) Potato and sweetpotato in Africa: transforming the value chains for food and nutrition security. CABI, Wallingford, pp 322–329

Nimmakayala P, Vajja G, Reddy UK (2011) In: Kole C (ed) "Ipomoea," in wild crop relatives: genomic and breeding resources. Springer, Berlin

Nishiyama I (1971) Evolution and domestication of the sweet potato. Bot Mag Tokyo 84:377–387

Nishiyama I, Niyazaki T, Sakamoto S (1975) Evolutionary autoploidy in the sweet potato (*Ipomoea batatas* (L.) Lam.) and its progenitors. Euphytica 24:197–208

O'Brien P (1972) The sweet potato: its origin and dispersal. Am Anthropol 74:342–365

Okada Y, Nishiguchi M, Saito A, Kimura T, Mori M, Hanada K, Sakai J, Matsuda Y, Murata T (2002a) Inheritance and stability of the virus-resistant gene in the progeny of transgenic sweet potato. Plant Breed 121:249–253

Okada Y, Saito A, Nishiguchi M, Kimura T, Mori M, Matsuda Y, Murata T (2002b) Microprojectile bombardment-mediated transformation of sweet potato (*Ipomoea batatas* (L.) Lam). SABRAO J Breed Genet 34:1–8

Onwueme IC (1978) The tropical tuber crops: yams, cassava, sweet potato and cocoyams. Wiley, New York. 234 p

Oración MZ, Niwa K, Shiotani I (1990) Cytological analysis of tetraploid hybrids between sweet potato and diploid *Ipomoea trifida* (H.B.K.) Don. Theor Appl Genet 80:617–624

Otani M, Shimada T, Kimura T, Saito A (1998) Transgenic plants production from embryogenic callus of sweetpotato (*Ipomoea batatas* (L.) Lam.) using *Agrobacterium tumefaciens*. Plant Biotechnol 15:11–16

Owusu-Mensah E, Oduro I, Ellis WO, Carey EE (2016) Cooking treatment effects on sugar profile and sweetness of eleven-released sweet potato varieties. J Food Process Technol 7:4. http://dx.doi.org/10.4172/2157-7110.1000580

Padmaja G (2009) Uses and nutritional data of sweetpotato. In: Loebenstein G, Thottappilly G (eds) The sweetpotato. Springer Science & Business Media BV, The Netherlands, pp 189–234

Pesek J, Baker RJ (1969) Desired improvement in relation to selection indices. Can J Plant Sci 49:803–804

Peters D (2004) Use of sweet potato in pig production in Asia: agricultural and socio-economic aspects, animalscience.com Reviews, No. 4. Pig News Inf 25(1):25N–34N

Poole CF (1955) Sweet potato genetic studies. Tech Bull 27 Hawaii Agr Exp Sta, Honolulu

Prakash CS (1994) Sweet potato biotechnology: progress and potential. Biotechnol Dev Monit 18:18–22

Prakash CS, Varadarajan U (1992) Genetic transformation of sweet potato by particle bombardment. Plant Cell Rep 11:53–57

Prakash CS, Varadarajan U, Kumar AS (1991) Foreign gene transfer to sweet potato (*Ipomoea batatas*). Hortic Sci 26:492

Purseglove J (1965) The spread of tropical crops. In: Baker HK, Stebbins GL (eds) The genetics of colonizing species. Academic, New York, pp 375–386

Purseglove JW (1968) Tropical crops, dicotyledons, vol 1. Longman Scientific and Technical, Harlow. 719 p

Qaim M (1999) The economic effects of genetically modified orphan commodities: projections for sweetpotato in Kenya. ISAAA and Center for Development Research [ZEF], Bonn. 47 p

Rajapakse S, Nilmalgoda SD, Molnara M, Ballarda RE, Austin DF, Bohac JR (2004) Phylogenetic relationships of the sweetpotato in *Ipomoea* series *Batatas* (Convolvulaceae) based on nuclear-amylase gene sequences. Mol Phylogenet Evol 30:623–632

Rodriguez-Bonilla L, Cuevas HE, Montero-Rojas M, Bird-Pico F, Luciano-Rosario D, Siritunga D (2014) Assessment of genetic diversity of sweet potato in Puerto Rico. PLoS One 9:e116184

Roullier C, Rossel G, Tay D, McKey D, Lebot V (2011) Combining chloroplast and nuclear microsatellites to investigate origin and dispersal of New World sweet potato landraces. Mol Ecol 20:3963–3977

Roullier C, Kambouo R, Paofa J, McKey D, Lebot V (2013) On the origin of sweet potato (*Ipomoea batatas* (L.) Lam.) genetic diversity in New Guinea, a secondary centre of diversity. Heredity 110:594–604

Saito A, Kimura T, Ideta O, Mori M, Nishiguchi M (1998) Transgenic sweet potato (*Ipomoea batatas* L. (Lam)) exhibiting resistance to sweet potato feather mottle potyvirus. Sweet Potato Res 7:4

Schnell FW (1982) A synoptic study of the methods and categories of plant breeding. Z Pflanzenzüchtung 89:1–18

Shiotani I (1988) Genomic structure and the gene flow in sweet potato and related species. In: Exploration and maintenance and utilization of sweet potato genetic resources. First Planning Conference, Lima, Peru. International Potato Centre (CIP), Lima, pp 61–73

Shiotani I, Kawase T (1987) Synthetic hexaploids derived from wild species related to sweet potato. Jpn J Breed 37:367–376

Shiotani I, Kawase T (1989) Genomic structure of the sweet potato and hexaploids in *Ipomoea trifida* (H.B.K) Don. Jpn J Breed 39:57–66

Shiotani I, Huang ZZ, Sakamoto S, Miyazaki T (1994) The role of the wild *Ipomoea trifida* germplasm in sweet potato breeding. Acta Hortic (ISHS) 380:388–398

Shulte-Geldermann E, Agili S, Ndolo P, Low JW (2012) Net tunnels to protect sweetpotato planting material from disease a guide to construct and maintain tunnels: a guide to construct and maintain tunnels. International Potato Center, Nairobi

Shumbusha D, Ndirigwe J, Kankundiye L, Musabyemungu A, Gahakwa G, Ndayemeye PS, Mwanga ROM (2014) 'RW11-17', 'RW11-1860', 'RW11-2419', 'RW11-2560', 'RW11-2910', and 'RW11-4923' sweetpotato. Hortscience 49:1349–1352

Shumbusha D, Ndirigwe J, Kankundiye L, Musabyemungu A, Mwanga ROM (2015) Development of dualpurpose sweetpotato varieties through participatory breeding in Rwanda. In: Low J, Nyongesa MS, Quinn S, Parker M (eds) Potato and sweetpotato in Africa: transforming the value chains for food and nutrition security. CABI, Wallingford, pp 79–87

Simpson BB, Ogorzaly MC (2000) Economic botany: plants in our world, 3rd edn. McGraw-Hill, New York. 544 p

Smith TP, Stoddard S, Shankle M, Schultheis J (2009) Sweetpotato production in the United States. In: Loebenstein G, Thottappilly G (eds) The sweetpotato. Springer Science + Business Media BV, Houten, pp 287–323

Song GQ, Honda H, Yamaguchi KI (2004) Efficient *Agrobacterium tumefaciens*–mediated transformation of sweet potato (*Ipomoea batatas* (L.) Lam.) from stem explants using a two- step kanamycin-hygromycin selection method. In Vitro Cell Dev Pl 40:359–365

Srisuwan S, Sihachakra D, Siljak-Yakovlev S (2006) The origin and evolution of sweet potato (*Ipomoea batatas* lam.) and its wild relatives through the cytogenetic approaches. Plant Sci 171:424–433

Stathers T, McEwan M, Gibson R, Mwanga R, Carey E, Namanda S, Abidin E, Low L, Malinga J, Agili S, Andrade M, Mkumbira J (2013) Everything you ever wanted to know about sweetpotato: reaching agents of change ToT manual, sweetpotato seed systems. Int Potato Center Nairobi Kenya 3:93–142

Stout AB (1924) The flowers and seed of sweetpotatoes. J N Y Bot Gard 25:153–168

Tanaka Y, Brugliera F, Kalc G, Senior M, Dyson B, Nakamura N, Katsumoto Y, Chandler S (2010) Flower color modification by engineering of the flavonoid biosynthetic pathway: practical perspectives. Biosci Biotechnol Biochem 74(9):1760–1769

Thiyagu D, Rafii MY, Mahmud TMM, Latif MA, Malek MA, Sentoor G (2013) Genotype by environment assessment in sweetpotato as leafy vegetable using AMMI model. Pak J Bot 45(3):843–852

Ting YC, Kehr AE, Miller JC (1957) A cytological study of the sweetpotato plant *Ipomoea batatas* (L.) Lam. and its related species. Am Nat 91:197–203

Thompson PG, Hong LL, Ukoskit K, Zhu S (1997) Genetic linkage of randomly amplified polymorphic DNA (RAPD) markers in sweetpotato. J Am Soc Hortic Sci 122:79–82

Togari Y, Kawahara U (1946) Studies on the self- and cross-incompatibility in sweetpotato (a preliminary report). I. On the different grades of compatibility among the compatible matings. Imp Agr Exp Sta Tokyo Bull 52:1–19. [in Japanese]

Tomlins K, Rwiza E, Nyango A, Amour R, Ngendello T, Kapinga R, Rees D, Jolliffe F (2004) The use of sensory evaluation and consumer preference for the selection of sweetpotato cultivars in East Africa. J Sci Food Agric 84:791–799

Tomlins K, Ndunguru G, Stambul K, Joshua N, Ngendello T, Rwiza E, Amour R, Ramadhani B, Kapande A, Westby A (2007) Sensory evaluation and consumer acceptability of pale-fleshed and orange-fleshed sweetpotato by school children and mothers with preschool children. J Sci Food Agric 87:2436–2446

Tomlins K, Owori C, Bechoff A, Menya G, Westby A (2012) Relationship among the carotenoid content, dry matter content and sensory attributes of sweet potato. Food Chem 131:14–21

Truong VD, Pascua YT, Reynolds R, Thompson RL, Palazoğlu TK, Mogol BA, Gökmen V (2014) Processing treatments for mitigating acrylamide formation in sweetpotato french fries. J Agric Food Chem 2014(62):310–316. http://dx.doi.org/10.1021/jf404290v

Tumwegamire S, Kapinga R, Rubaihayo PR, LaBonte DR, Grüneberg WJ, Burgos G, zum Felde T, Carpio R, Pawelzik E, Mwanga ROM (2011) Starch, sucrose, β-carotene, iron, zinc, calcium, and magnesium in East African sweetpotato [*Ipomoea batatas* (L.) Lam.] germplasm. Hortscience 46:348–357

Ukoskit K, Thompson PG (1997) Autopolyploidy versus allopolyoloidy and low-density randomly amplified polymorphic DNA linkage maps of sweetpotato. J Am Soc Hortic Sci 122:822–828

Villordon AQ, Labonte DR (1995) Variation in randomly amplified DNA markers and storage root yield in 'Jewel' sweetpotato clones. J Am Soc Hortic Sci 120:734–740

Villordon AQ, Labonte DR (1996) Genetic variation among sweetpotatoes propagated through nodal and adventitious sprouts. J Am Soc Hortic Sci 121:170–174

Walter WM Jr, Collins WW, Troung VD, Fine TI (1997) Physical compositional and sensory properties of French fry-type products from five sweet potato selections. J Agric Food Chem 45:383–388

Wang H (1964) A study of the self- and cross-incompatibilities in the sweetpotato in Taiwan (Formosa). Proc Am Soc Hortic Sci 84:424–430

Wang YW, Samuels TD, Wu YQ (2011) Development of 1,030 genomic SSR markers in switchgrass. Theor Appl Genet. 2011 122(4):677–686

Wedderburn MM (1967) A study of hybridisation involving the sweet potato and related species. J Plant Growth Regul 16:69–75

Williams DB, Cope FW (1967) Notes of self-incompatibility in the genus *Ipomoea* L. Proc Int Symp Trop Root Crop 1:16–30

Woolfe JA (1992) Sweet potato: an untapped food resource. Cambridge University Press, Cambridge, UK
Wricke G, Weber WE (1986) Quantitative genetics and selection in plant breeding. Walter de Gruyter, Berlin
Wu KK, Burnquist W, Sorrells ME, Tew TL, Moore PH, Tanksley SD (1992) The detection and estimation of linkage in polyploids using single-dose restriction fragments. Theor Appl Genet 83:294–300
Yada B, Tukamuhabwa P, Wanjala B, Dong-Jin K, Skilton RA, Alajo A, Mwanga ROM (2010) Characterization of Ugandan sweet potato germplasm using fluorescent labelled simple sequence repeat markers. Hortscience 45(2):225–230
Yada B, Brown-Guedira G, Alajo A, Ssemakula GN, Mwanga ROM, Yencho GC (2015) Simple sequence repeat marker analysis of genetic diversity among progeny of biparental mapping population of sweetpotato. Hortscience 50(8):1143–1147
Yan L, Lai X, Li X, Wei C, Tan X, Zhang Y (2015) Analyses of the complete genome and gene expression of chloroplast of sweet potato [Ipomoea batata]. PLoS ONE 10: e0124083
Yen D (1960) The sweet potato in the Pacific: the propagation of the plant in relation to its distribution. J Polynesian Soc 69:368–375
Yen DE (1974) The sweet potato and Oceania: an essay in ethnobotany, Bernice Pauahi Bishop Museum Bulletin (USA) No. 236. Bishop Museum Press, Honolulu. 389 p
Yen DE (1976) Sweet potato *Ipomoea batatas* (Convolvulaceae). In: Simmonds NW (ed) Evolution of crop plants. Longman Scientific and Technical, London, pp 42–45
Yen DE (1960) Sweetpotato in historical perspective. In: Villareal RL, Griggs TD (eds) Sweetpotato. Proceedings of the first international symposium, AVRDC, Publ. No. 82–172. pp 17–30
Yi G, Shin YM, Choe G, Shin B, Kim YS, Kim KM (2007) Production of herbicide-resistant sweet potato plants transformed with the bar gene. Biotechnol Lett 29:669–675
Yoon UH, Jeong JC, Kwak SS, Yang JW, Kim TH, Lee HU, Nam SS, Hahn JH (2015) Current status of sweetpotato genomics research. J Plant Biotechnol 42:161–167
Zhang SS, Liu LF (2005) Utilization of the wild relatives in sweetpotato breeding. In: Liu QC (ed) Sweetpotato breeding and industrialization in China. China Agricultural University Press, Beijing, pp 29–32
Zhang T, Oates CG (1999) Relationship between α-amylase degradation and physico-chemical properties of sweet potato starches. Food Chem 65:157–163
Zhang D, Cervantes J, Huamán Z, Carey E, Ghislain M (2000) Assessing genetic diversity of sweet potato (*Ipomoea batatas* (L.) Lam.) cultivars from tropical America using AFLP. Genet Resour Crop Evol 47:659–665
Zhang D, Rossel G, Kriegner A, Hijmans R (2004) AFLP assessment of diversity in sweetpotato from Latin America and the Pacific Region: its implications on the dispersal of the crop. Genet Resour Crop Evol 51:115–120
Zhang L, Wanga Q, Wanga Q (2009) Sweetpotato in China. In: Loebenstein G, Thottappilly G. (eds) 2009The sweetpotato. Springer Science+Business Media B.V., Dordrecht, pp 325–358
Zhao N, Yu X, Jie Q, Li H, Li H, Hu J, Zhai H, He S, Liu Q (2013) A genetic linkage map based on AFLP and SSR markers and mapping of QTL for dry-matter content in sweetpotato. Mol Breed 32:807–820
Zósimo H (ed) (1999) Sweetpotato germplasm management (*Ipomoea batatas*) training manual. International Potato Center, Lima

Chapter 7
Bananas and Plantains (*Musa* spp.)

Allan Brown, Robooni Tumuhimbise, Delphine Amah, Brigitte Uwimana, Moses Nyine, Hassan Mduma, David Talengera, Deborah Karamura, Jerome Kuriba, and Rony Swennen

7.1 Introduction

With a production of 145 million metric tons worldwide (worth 26.5 billion Euro), banana (*Musa* spp.) is one of the world's most important staple food crops and arguably the world's most popular fruit in terms of international trade (FAO 2014). Banana and plantains (*Musa* spp.), collectively referred to here as bananas, are grown in more than 135 countries and found in most tropical and subtropical regions around the world. While industrialized nations view banana primarily as a dessert

A. Brown (✉) • H. Mduma • R. Swennen
International Institute of Tropical Agriculture, Arusha, Tanzania
e-mail: a.brown@cgiar.org; h.mduma@cgiar.org; r.swennen@cgiar.org

R. Tumuhimbise • J. Kuriba
National Agricultural Research Organization, Kampala, Uganda
e-mail: rtumuhimbise@hotmail.com; jkubiriba2012@gmail.com

D. Amah
International Institute of Tropical Agriculture, Ibadan, Nigeria
e-mail: D.Amah@cgiar.org

B. Uwimana • M. Nyine
International Institute of Tropical Agriculture, Kampala, Uganda
e-mail: B.Uwimana@cgiar.org; m.nyine@cgiar.org

D. Talengera
National Agricultural Research Laboratories, Kampala, Uganda
e-mail: dtalengera@yahoo.com

D. Karamura
Bioversity International, Kampala, Uganda
e-mail: d.karamura@cgiar.org

H. Campos, P.D.S. Caligari, *Genetic Improvement of Tropical Crops*,
DOI 10.1007/978-3-319-59819-2_7

item, many regions of the developing world consider cooking bananas and plantains as essential staples that contribute significantly to the caloric intake of low-income subsistence farmers. Although sensitivity to photoperiod has been noted in certain cultivars (Fortescue et al. 2011), banana is an almost nonseasonal crop that reliably provides a carbohydrate source year-round which makes it vitally important to both nutrition and food security. Propagation by farmers is commonly through suckers or side shoots originating from lateral buds at the base of the main plant. Multiple fungal and bacterial pathogens present serious constraints to production of bananas, as does the occurrence of insects and nematodes (Jones 1999). Viral diseases caused by banana streak virus (BSV), cucumber mosaic virus (CMV), banana bract mosaic virus (BBMV), and the emerging banana bunchy top virus (BBTV) are also receiving increased attention (Kumar et al. 2015). The predominance of these biotic agents differs from region to region, but most are found throughout the banana production regions in Asia, Africa, and the Americas and represent common targets for plant improvement worldwide. As with all crops, abiotic factors associated with climate change such as drought and heat stress also present considerable challenges to production (van Asten et al. 2011; Wairegi et al. 2010), but arguably the single greatest constraint to genetic improvement is the narrow genetic basis of most cultivated bananas (Hippolyte et al. 2012) and the physiological and reproductive barriers of the plant itself (Ssesuliba et al. 2008; Fortescue and Turner 2004; Dumpe and Ortiz 1996). Reproductive barriers limit sexual recombination in banana and hinder plant improvement. While all of the seed-bearing progenitors of modern banana cultivars are diploid in nature, those that have been cultivated for consumption are primarily seedless triploids. Female fertility of triploids has been described, but seed set is generally extremely limited which complicates breeding efforts and intensives resources and time required to develop superior varieties with enhanced resistance to multiple biotic agents and abiotic agents. Banana improvement is further complicated by parthenocarpy, reduced male fertility in some cultivars, low seed viability, irregular meiotic behavior, long generation times, and diverse genomic configurations (Ortiz 2013, 2015; Ortiz and Swennen 2014). To date, the limited progress that has been achieved in banana breeding has occurred through crossbreeding approaches that involve hybridization followed by phenotypic selection among half sibs and/or full sib progenies.

7.2 Banana Classification

Banana is a monocotyledon herbaceous plant represented by three genera (*Musa, Ensete, and Musella*) within the family Musaceae of the order Zingiberales (De Langhe et al. 2009). The genus *Ensete* consists of monocarpic, unbranched herbaceous plants that rarely produce suckers and are used for food, fiber, and ornamental purposes. They resemble banana, but their oversized, edible corms and wide-spreading and immensely long, paddle-shaped leaves with crimson midribs make them very distinctive. Their fruits are similar in appearance to those of banana, but are dry, seedy, and inedible (Deckers et al. 2001). The most recognizable member of this genus is perhaps the false or Abyssinian banana (*E. ventricosum*) that plays a significant role in Ethiopian agriculture and food security (Tsegaye and Struik 2002).

The genus *Musa* consists of cultivated triploid cultivars and clones propagated through vegetative methods with limited genetic variation beyond what could be expected through somaclonal variation (and perhaps epigenetics) and the diploid wild progenitors of these cultivars that are capable of sexual recombination. More than 60 species within four recognized sections of the genus *Musa* have been described, but the taxonomy of *Musa* and the relationship between wild and cultivated bananas are far from settled (De Langhe et al. 2009; Janssens et al. 2016). Almost all diploid species are native to Southeast Asia, from India and Thailand to New Guinea and Queensland, Australia (Simmonds 1987). Edible bananas, with the exception of the Fe'i group of the Australimusa section, are derived almost exclusively from two species, *Musa acuminata* and *Musa balbisiana* of the section *Musa* (Dodds 1945). *M. acuminata* and *M. balbisiana* are diploid ($2n= 2x= 22$) in their base genomic complements and designated as AA and BB, respectively (Simmonds and Shepherd 1955). In addition to monospecific cultivars (AA, BB), interspecific diploid clones (AB) are also recognized. Higher-order combinations of the AA and BB base genomes arose through chromosome restitution at meiosis, to produce distinct groupings at the triploid (AAA, AAB, ABB groups) and occasionally tetraploid levels (Simmonds 1962). Triploids, due to their optimal vigor and seedless characteristic, are the preferred configuration for most consumers throughout the world (Simmonds 1987). The rare tetraploid cultivars tend to be physically larger but have relatively small bunches, while most diploid cultivars tend to be weaker plants with smaller bunches (De Langhe 1986). Edible parthenocarpic diploids are however cultivated in certain regions such as Tanzania (Simmonds 1962) where the Mchare (or Mshale) diploids are preferred for the unique texture characteristics of the fruit.

Generally, modern classification systems of banana tend to follow Simmonds and others (Simmonds and Shepherd 1955; Stover and Simmonds 1987) and are based on ploidy status and the relative contribution of the two genomes. Simmonds (1962, 1966) suggested that the formal Latin nomenclature should be replaced by a ploidy-based nomenclature in which the cultivar is referred to by the genus and a genomic grouping (e.g., *Musa*, AAA Group, "Gros Michel"). Cultivars are placed in higher-level groupings based on the number of chromosomes and the species that contribute to their genetic makeup (AA, BB, AAA, AAB, and ABB) (Karamura et al. 2012). Simmonds and Shepherd (1955) utilized 15 taxonomic characters specific to *M. balbisiana* and *M. acuminata* to assign cultivars to groups, and this classification scheme has been periodically updated (Stover and Simmonds 1987).

Below the level of group, cultivars are assigned to clusters of subgroups that are characterized by a representative member. For example, "Cavendish" and "Gros Michel" are considered separate subgroups under the AAA grouping along with several mutants and variants derived from these economically important cultivars. The grouping (AAA) also includes all of the economically important East African Highland cooking bananas. While this classification system may be convenient, it appears to lack hierarchical, biological, or economic significance, for example, Mysore, Pome, and Plantain are all recognized subgroup clusters within the AAB group, but they are utilized for different purposes, and subsequent results of molecular and morphological diversity studies suggest that they are genetically distinct and likely have arisen from dissimilar parentage (De Langhe et al. 2010; Christelová et al. 2017).

Likely, cultivars within the AAA, AAB, and ABB groupings arose from multiple hybridization events followed by subsequent backcrossing to various AA, BB, and AB progenitors which results in an unequal chromosomal allocation at meiosis (De Langhe et al. 2010). It has been suggested that this phenomenon could explain the unequal and nonadditive chromosomal complementation which has been observed among interspecific hybrids (d'Hont et al. 2000). If indeed cultivars within groupings arose from multiple hybridization events, it suggests that classification may be more dependent on specific diploid progenitors than on traditional groupings based on ploidy. Morphological characteristics and nuclear and cytoplasmic molecular markers have been used to differentiate the progenitor *M. acuminata* diploids into several subspecies that correspond to specific geographic ranges from mainland Asia to the archipelagoes of Indonesia, New Guinea, and the Philippines (Hippolyte et al. 2012; Carreel et al. 2002; Perrier et al. 2011). Currently, there are eight recognized diploid AA subspecies that include roughly from West to East: *M. acuminata* ssp. *burmannica*, *M. acuminata* ssp. *siamea*, *M. acuminata* ssp. *truncata*, *M. acuminata* ssp. *malaccensis*, *M. acuminata* ssp. *zebrina*, *M. acuminata* ssp. *microcarpa*, *M. acuminata* ssp. *errans*, and *Musa acuminata* ssp. *banksii*. The subspecies have contributed important diploid parents to modern breeding programs, but much work remains toward evaluating and preserving germplasm that has not been readily accessed due to logistic or political reasons in past collecting expeditions.

Considerable efforts have been made over the past few decades to preserve, characterize, and provide access to genetic resources of *Musa*. Banana germplasm for use in breeding is distributed through the Biodiversity International Musa Germplasm Transit Centre which oversees more than 1500 accessions. The center secures available banana germplasm for long-term conservation and holds the collection in trust for the benefit of future generations under the auspices of the Food and Agriculture Organization of the United Nations. The conserved germplasm is placed in the Multilateral System of Access and Benefit Sharing of the International Treaty on Plant Genetic Resources for Food and Agriculture. All accessions have been indexed, conserved in vitro (Van den houwe et al. 1995), and most stored under cryopreservation (Panis et al. 2005). Characterization of germplasm occurs in both field trials and at the molecular level. Passport and characterization data is freely available through the Musa Germplasm Information System (MGIS) (https://www.crop-diversity.org/mgis/).

7.3 Banana Breeding Objectives

The primary objective of most banana breeding programs is the uniform production of large bunches that meet the regional qualitative and quantitative demands of growers. These demands include superior fruit quality, high suckering ability, short stature, and enhanced root systems that provide effective soil anchorage and efficient uptake of water and minerals. Other agronomic traits such as photosynthetic efficiency and rapid cycling are also important breeding objectives for increased yield. The relative importance of these objectives varies across geographic regions, among subgroups of banana, and with the intended final use of the product. In recent years, the anticipated and realized threats of pests and diseases have resulted in increased emphasis placed on identifying and utilizing improved sources of host-plant resistance to pests and diseases, particularly in regard to the Sigatoka complex, multiple races of Fusarium wilt, bacterial wilt, bunchy top, nematodes, and weevils.

In Uganda, banana breeding focuses largely on the improvement of East African Highland (cooking) bananas (EAHBs) (AAA). The expected yield and plantation life of these bananas has significantly declined, in no small part due to pests (such as banana weevils and nematodes) and diseases (including black Sigatoka and bacterial wilt). Some of the key breeding objectives by the National Banana Research Program of the Uganda National Agricultural Research Organization (NARO), in partnership with the International Institute of Tropical Agriculture (IITA), have been to identify and integrate host-plant resistance to the Sigatoka complex, weevils, and nematodes from wild diploid progenitors into elite EAHB backgrounds. A generalized criterion for selection of EAHB based on agronomic traits is presented in Table 7.1.

Table 7.1 Characteristics of the ideotype of East African Highland cooking bananas

Trait	Description
Yield potential	>25 t/ha/year
Bunch weight	>15 kg
Plant height	<3 m
Time of flowering	210–270 days
Time of bunch maturity	90–120 days
Number of hands	8–12/bunch
Number of fingers	100–190/bunch
Fruit finger circumference	10–15 cm
Fruit finger length	13–20 cm
Suckering ability	75% follower sucker growth at harvest
Root system	Vigorous (fast growing, deep, and branched)
Bunch orientation	Pendent
Reaction to prevalent diseases	Resistant to the black Sigatoka complex and bacterial wilt
Reaction to prevalent pests	Resistant to weevils and nematodes
Reaction to drought stress	Resistant/tolerant

7.4 Constraints to Banana Breeding

As previously discussed, the greatest constraint to banana genetic improvement is the limited production of viable seeds due to polyploidy, female sterility, and other factors affecting seed production in triploid and diploid banana. Female sterility has been intensified as a consequence of human selection for parthenocarpy in banana. Simmonds (1962) first suggested that continuous clonal propagation of diploids has led to an accumulation of structural chromosomal changes (translocations, inversions, and other events) that restrict normal meiosis and pollen fertility and reduce expected recombination. Specific abnormalities such as translocations have been noted in the diploids "Pisang Lilin" and "Pisang Jari Buaya," but the extent to which this phenomenon occurs throughout *Musa* is still not well understood. Adeleke et al. (2004) observed that in general, a higher incidence of univalent formation was related to low pollen fertility in both diploids and triploids. Sterility in plantains and EAHB triploid bananas has been associated with meiotic irregularities and uneven number of chromosomes, as well as to environmental factors and to influences of individual genotypes (Swennen and Vuylsteke 1993; Ssebuliba et al. 2000). Seed yield is influenced by time of pollination, environmental conditions, genetic variation in female fertility, differences observed among pollinations made between the basal and distal hand, and variation associated with the relative contributions of the *acuminata* and *balbisiana* genomes (Simmonds 1962). Sathiamoorthy and Rao (1980) observed increased seed set with proportional contributions from the *balbisiana* genome and speculated that the factors for seed sterility have accumulated preferentially in the *M. acuminata* genome (Simmonds 1962). The use of embryo rescue has significantly improved seed germination with observations of up to 30% increases in viable embryos (Swennen and Vuylsteke 1993).

A further complication associated with improvement of cultivated bananas is that the highly heterozygous state of the parents results in extremely variable progeny that makes predictions of progeny performance on the basis of parental phenotypes unreliable (Ortiz 2000). The progeny from crosses can include mixtures of ploidy levels and sometimes aneuploids. Oselebe et al. (2006) reported that while progeny of 2×–2× crosses were almost exclusively diploid (99.7%), those of mixed ploidy crosses tended to include individuals that varied in their chromosomal compliment. The same study observed that the direction of the cross impacted results. When the diploid is used as the maternal parent in a 2×–4× cross, over 96% of progeny are diploids, but when the tetraploid is used as the maternal parent (4×–2×), the observed progeny is predominately triploid (94%) with varying degrees of other ploidy levels between the diploid and the pentaploid observed. While these mixed ploidy progenies provide a mechanism for enhancing genetic diversity and recombination, they also necessitate the use of early screening of ploidy levels.

7.5 Breeding Strategies

Plant breeding provides one of the highest returns of investment in agricultural research, and while banana production has benefited from these investments, few banana and plantain cultivars acceptable to farmers and consumers were produced prior to the 1980s (Roux 2001). Triploid bananas are preferred by growers as they commonly display the most advantageous combination of fruit and vegetative characters (De Langhe 1986; Stover and Simmonds 1987), but it was generally assumed that these triploid cultivars (such as "Cavendish") were effectively sterile (Stover and Buddenhagen 1986). Persistent efforts, however, from multiple breeding programs including IITA and the Honduran Foundation for Agricultural Research (FHIA) have demonstrated that viable embryos could be produced from what were previously considered recalcitrant triploids by making hybridizations with selected pollen of diploid banana (Aguilar-Moran 2013). FHIA successfully produced 40 viable tetraploid embryos utilizing "Cavendish" as a female, although this effort required an almost Herculean task that included the pollination of over 20,000 bunches. By the end of the twentieth century, efforts to improve banana focused primarily on the use of improved diploid and synthesized tetraploid gene pools to develop secondary triploids of bananas and plantains (Fig. 7.1).

Under this breeding scheme, the identification of improved diploids that provide donor traits of interest and cultivated triploids with superior quality characteristics takes on a vital role in synthesizing superior secondary triploids. Conventionally, it was assumed that when crosses were made between male diploids and female triploids to obtain tetraploid progeny, all three sets of maternal chromosomes were trans-

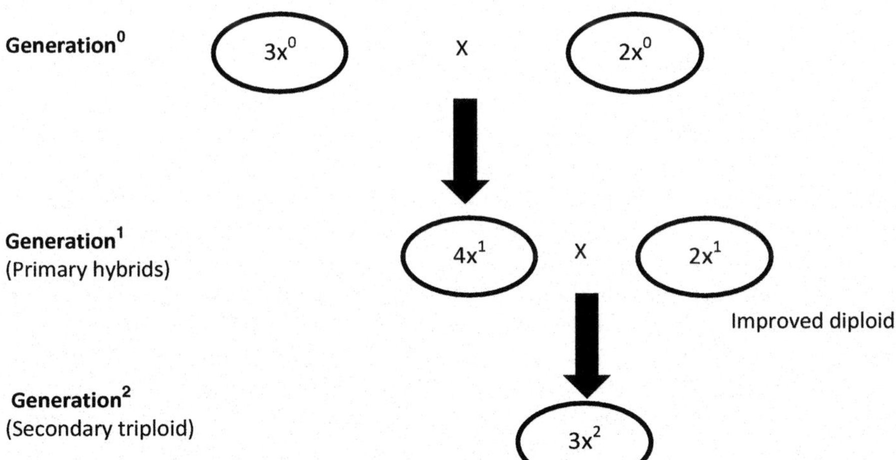

Fig. 7.1 The scheme of the banana breeding process whereby initial crosses are carried out between triploid landrace ($2n = 3x$) and diploid ($2n = 2x$)

ferred intact to the tetraploid offspring with recombination only occurring as a result of the contribution of diploid male parent (Dodds 1943). Vuylsteke et al. (1993), however, noted that tetraploid progeny from such crosses displayed variation in disease resistance, morphological traits, and growth and yield parameters that were inconsistent with this hypothesis. It was further suggested that segregation and recombination during modified megasporogenesis leading to the formation of 2n eggs in the triploid parent perhaps better explained the observed results. With the advent of new genomic resources and tools, this phenomenon needs to be further investigated in order to better understand the extent of sexual recombination in banana breeding.

This breeding strategy has been adopted by multiple programs including FHIA in Honduras which has produced a generation of acceptable tetraploids that are still currently being used in breeding (Rowe and Rosales 1993). An example includes FHIA 21, a tetraploid derived from AVP-67 French plantain that is still being utilized in plantain improvement. IITA has also utilized this approach successfully to introgress alleles for resistance/tolerance to key pest and diseases in high-yielding hybrids derived from preferred plantain cultivars (Ortiz et al. 1995; Tenkouano and Swennen 2004; Vuylsteke et al. 1993). Plantain hybrid releases include PITA 14 and PITA 17 (primary tetraploids) and more recently PITA 21, PITA 23, and PITA 24 (secondary triploids) all derived from three seed-fertile triploid French plantains Obino L'ewai, Bobby Tannap, and Mbi Egome. Their common attributes include BLS resistance/tolerance and good bunch characteristics (Tenkouano et al. 2011) and early suckering (Vuylsteke et al. 1993). These IITA hybrids have currently been distributed to ten countries in Africa and three countries in Central and South America for evaluation and adoption.

At NARO, tetraploids (AAAA) were synthesized from EAHBs (AAA) by crossing to the wild-seeded, fertile male parent Calcutta 4 (AA) that is used by many programs as a source of resistance to multiple pests and diseases. A number of these tetraploids developed such as 365K-1, 1201K-1, 917K-2, 660K-1, 1438K-1, and 222K-1 25 were fundamental in the development of 27 NARITA triploid banana hybrids by NARO and IITA (Tushemereirwe et al. 2015). These banana hybrids were selected from early evaluation trials based largely on resistance to black Sigatoka and bunch size and subsequently advanced to the preliminary yield trials in Uganda. NARITAs are currently under evaluation for agronomic, sensory, pest, and disease resistance traits in multi-environment trials in Uganda and Tanzania.

An alternative breeding scheme has been suggested by Vakili (1968) and involves the polyploidization of diploid hybrids or cultivars through the use of colchicine or oryzalin to obtain tetraploids for crossing with 2× lines to generate triploids. This approach is currently being pursued by multiple breeding programs and shows considerable promise (Bakry et al. 2009). According to Tenkouano et al. (2011), the two schemes conform to differences in breeding philosophies. The former can be viewed as evolutionary breeding as it attempts to mimic the developmental pathway of *Musa* by crossing female triploid landraces to diploid accessions of *M. acuminata* or *M. balbisiana*, while the latter can be considered reconstitutive breeding as it utilizes the most likely diploid ancestors or relatives of triploid landraces for chromosome doubling to create improved triploids.

7.5.1 Development of Improved Diploids

Regardless of the breeding scheme, the narrow genetic variability and limited fertility among cultivated triploid bananas make diploid bananas vital to genetic improvement. A number of fertile improved diploids with varying degrees of disease resistance have been released by IITA and FHIA (Tenkouano et al. 2003; Rowe and Rosales 1993). Diploid improvement has almost exclusively been through the use of *M. acuminata* cultivars such as "Calcutta 4" (*M. acuminata*), a source of resistance to the Sigatoka complex, yellow Sigatoka, fusarium wilt, banana weevil, and burrowing nematodes (Ortiz 2015). Decades of breeding utilizing this material have resulted in the production of improved diploid lines which combine disease/pest resistance, short stature, and interesting bunch characteristics (Tenkouano et al. 2003; Krishnamoorthy and Kumar 2005). Developing further improved diploids that possess multiple sources of resistance, while preserving the quality characteristics of preferred triploid would greatly increase the efficiency of breeding efforts that are hindered by the constraints previously described.

M. balbisiana diploid cultivars have made limited contributions to breeding due to the presence of endogenous banana streak virus (eBSV) sequences originating from the B genome that are activated under the appropriate conditions (Iskra-Caruana et al. 2014). Progenies of interspecific *acuminata/balbisiana* hybridizations have often been associated with the occurrence of banana streak disease, and this has resulted in the underutilization of the B genome which could be an important source of drought tolerance and resistance alleles not found in *M. acuminata* (Bakry et al. 2009). Recently, Noumbissie et al. (2016) reported segregation of the eBSV sequences in the progeny of crosses between the tetraploid hybrid CRBP 39 (+eBSV) and the AA male parent Pahang (−eBSV), from which resulted triploid eBSV-free offspring. Umber et al. (2016) also documented the successful creation of diploids free of eBSV alleles from *M. balbisiana* diploids suggesting that recombination between *M. acuminata* and *M. balbisiana* can be accomplished for improvement of both cultivated banana and plantain without concern of introducing banana leaf streak.

7.5.2 Breeding Methodology and Evaluation of Hybrids

Production of viable seed through hybridization is critical for the success of breeding and is dependent on residual fertility of triploid cultivars. Practical aspects of artificial hybridization have been described by Tenkouano et al. (2011). Hybridizations are made through manual pollinations in the early hours of the morning when pollen availability is not a limiting factor. It can take up to several months to obtain seeds from a desired cross, and production of seeds is generally poor and has been reported in the range of 0.3–21.7 seeds per bunch (Swennen and Vuylsteke 1993) (Fig. 7.2). Seeds obtained from crosses also germinate poorly, and it is standard practice by most programs to recover hybrids through in vitro culture (Bakry 2008) (Fig. 7.3).

Fig. 7.2 Male banana flowers open in the evening and are ready for pollination early in the morning

Fig. 7.3 Bract is pulled back and pollen applied directly to the receptive female flowers

Bunches are harvested prior to physiological maturity, generally when the first signs of yellow color are observed in the distal fingers. Bunches are left to ripen in protected sheds, and seeds are extracted and surface-sterilized for embryo culture to avoid seed/embryo desiccation. Embryos are extracted from dissected seeds under aseptic conditions and cultured on artificial culture media (Bakry et al. 2009; Uma et al. 2011). Longitudinal excisions are made on the seed to expose the embryo beneath the micropyle. The embryo is placed on sterile culture medium and incubated in the dark for germination. Embryos typically germinate between 5 and 20 days after which time they are transferred to an environment with appropriate light and dark cycles for shoot and root development. Well-developed seedlings can then be cloned to replicate one to four rooted plantlets or transferred to a screenhouse for conditioning prior to field evaluations. More research on pollen production, pollen tube growth, and embryo viability is required to better understand issues associated with poor seed production and to optimize conditions that will lead to better seed yield (Uma et al. 2011). In particular, detailed knowledge of floral biology and seed development is crucial for recovery of progeny from crosses (Fortescue and Turner 2011).

In the field, new hybrids are subjected to early evaluation trials (EETs) with limited replication (one to five plants). EETs are observed generally for two cycles during which a few simply inherited traits such as bunch size and orientation, Sigatoka resistance, seed production, and ploidy level are evaluated. Plants that show promise in EETs are further evaluated in preliminary yield trials (PYTs) where replicated clones of selected hybrids are evaluated over two cycles. PYTs involve more detailed evaluation for additional, complex traits such as yield and disease resistance. Finally, superior-performing plants from PYTs are cloned in significant numbers to allow for multi-locational evaluation trials (MET) that often include direct input from farmers (Tenkouano et al. 2011). In theory, this process can take a minimum of 7 years to produce a superior banana hybrid, although in practice this time frame is often exceeded.

7.6 Applied Biotechnology

7.6.1 Molecular-Assisted Breeding

Due to the breeding constraints previously discussed, the use of molecular markers holds considerable promise in improving the efficiency of banana breeding but is currently not routinely used in most breeding programs. Efforts toward the development and use of molecular markers have been greatly facilitated by the recent release and refinement of the draft genomic sequence of the double haploid *M. acuminata* cultivar "Pahang" (A genome) (D'Hont et al. 2012; Martin et al. 2016) and a draft sequence of *M. balbisiana* "Pisang Klutuk Wulung" (B genome, Davey et al. 2013). Multiple transcriptome data sets have also been published (Li et al. 2013; Wang et al. 2012), and of these publications, over 45,000 expressed sequence tags, and 34,000 annotated genes associated with *Musa* are currently available through NCBI-EST database. These genomic resources have contributed to the availability of multiple classes of markers summarized in Tables 7.1 and 7.2. Molecular markers provide genetic "landmarks" for tagging important traits in plant breeding, conducting linkage analysis and estimates of genetic diversity, facilitating gene introgression through marker-assisted studies, providing validation of taxonomy and cultivar identification, and estimating evolutionary and speciation events.

Table 7.2 A summary of molecular markers utilized in banana research and their research applications

Marker application	Marker type	Reference
Molecular systematics	Isozymes, SSR, DArTs, RFLP and ITS	Perrier et al. (2011), Christelová et al. (2017), and Simmonds (1966)
Genetic diversity studies	RAPD, SSR, AFLP and MSAP	Karamura et al. (2016), Kitavi et al. (2016), Nyine and Pillay (2011), Opara et al. (2010) Wang et al. (2007), Noyer et al. (2005), Creste et al. (2004), Ude et al. (2003), Ude et al. (2002), Pillay et al. (2001), Crouch et al. (2000), and Crouch et al. (1999)
Genome characterization	RAPD, RFLP, ITS, dCAP and IRAP	
Cultivar identification and pedigree tracking	Isozymes, RFLP, SSR, RAPD, EST-SSR and ISSR	Mbanjo et al. (2012), Hippolyte et al. (2012), and Raboin et al. (2005)
Linkage analysis	Isozyme, RAPD, RFLP, AFLP, SSR AS-PCR and DArTs	Mbanjo et al. (2012), Hippolyte et al. (2010) and Fauré et al. (1993)
Genome-wide association studies and marker-assisted selection	Isozymes, dCAP and SNP	Sardos et al. (2016), Umber et al. (2016), and Noumbissié et al. (2016)

7.6.1.1 Association of Molecular Marker with Important Genes

The tagging of significant genes contributing to traits of interest with genic or linked markers allows for screening of plant germplasm at the earliest stages of development. The association of these markers with important traits can come through classical linkage or association studies or through candidate gene approaches that leverage recently available genomic resources. An example of the candidate gene approach is provided by Emediato et al. (2009) who amplified homologues of black leaf streak disease resistance genes in *Musa* through the use of degenerate primers based on genes from other crops. The study successfully amplified sequence differences between the diploid *M. acuminata* cultivars "Calcutta 4" (resistant) and "Pisang Berlin" (susceptible). This work followed the earlier identification of 50 distinct tagged nucleotide binding site-leucine-rich repeat (NBS-LRR) resistance gene analogs in cultivar "Calcutta 4" by Miller et al. (2008). Wang et al. (2012) used pooled DNA from *Fusarium oxysporum* f.sp. *cubensis* (Foc TR4)-resistant and susceptible cultivars to identify randomly amplified polymorphic DNA (RAPD) markers that could distinguish between resistant and susceptible cultivars. Two RAPD markers were converted to sequence characterized amplified region (SCAR) markers which could be amplified in Foc TR4-resistant banana genotypes ("Williams 8818-1" and Goldfinger), but not in five tested susceptible banana cultivars. Work on this continues at the national banana program in Brazil (EMBRAPA) and shows great promise in providing an early screen for resistance to Foc TR4 (Silva et al. 2016).

As previously discussed, endogenous banana streak virus (eBSV) limits the extensive use of the B genome in banana breeding, but the tagging of this sequence has opened possibilities of greater utilization in the future. Lheureux et al. (2003) mapped the eBSV sequence using amplified fragment length polymorphism (AFLP) markers, and Noumbissié et al. (2016) used simple sequence repeat (SSR) markers and eBSV-specific PCR markers to identify hybrids lacking the eBSV sequence. Umber et al. (2016) successfully identified infectious and noninfectious BSV alleles using derived cleaved amplified polymorphic sequences (dCAPS). These studies suggest that these markers can be used early in the breeding process as diagnostic markers for eBSV-free B genome hybrids that will greatly enhance breeding efforts. While the progress shown by these early efforts is promising, markers associated with traits of economic importance need to be validated in broader germplasm pools over multiple years to ensure that they will prove to be reliable and stable and that genotypic predictions at the early stages of screening will be highly correlated with plant phenotypes at full maturity under field conditions.

7.6.1.2 Linkage, Association Mapping, and Genomic Selection

Genetic linkage maps provide opportunities for gene identification and a mechanism for understanding the inheritance pattern of both qualitative and quantitative traits. Mapping requires appropriate plant populations of known structure derived from parents that differ significantly in traits of interest, a set of markers segregating in the given population that provides substantial coverage of all chromosomes, and the careful collection of phenotypic data from multiple years and preferably locations. Linkage mapping has not gained significant practical application in banana breeding. This could be attributed in part to limitations inherent in marker technologies and analysis (Foolad 2007; Pillay et al. 2012), to previously described chromosomal abnormalities in banana that inhibit recombination and to contribute ambiguous assignment of marker location. To date, principally F1 and F2 diploid populations have been utilized due to difficulties associated with developing double haploid or recombinant inbred lines in banana.

The first genetic mapping population in banana was reported in 1993 (Fauré et al. 1993) and consisted of 92 F_2 progeny (AA) derived from an F_1 hybrid (SFB5) of the cross "SF265" (CIRAD-IRFA II.04.20.004.020) × "banksii" (CIRAD-IRFA II.04.01.004.001). Seventy-seven loci consisting of RAPDs, isozymes, and restriction fragment length polymorphisms (RFLPs) were mapped onto 15 linkage groups spanning 606 cM. Segregation distortion was associated with 36% of the mapped loci and was biased toward the "banksii" parent. Hippolyte et al. (2010) published a more saturated map using an F_1 diploid "AA" population created from a cross between "Borneo" and "Pisang Lilin." The map was constructed using 426 markers (SSR and DArT). Separate maps were constructed for markers that segregated from each of the heterozygous parents, and a synthetic map was constructed that spanned 11 linkage groups and represented 1197 cM. Three regions of this synthetic map were inconsistent between the two parents and were attributed by the authors to structural rearrangements. Subsequent mapping projects have also noted such incongruities, and while these suggestions are supported by cytogenetic evidence such as multivalent pairing (Shepherd 1999), much work needs to be done to verify this hypothesis and determine the extent that such phenomenon occurs across *Musa* spp. Mbanjo et al. (2012) produced the most recent map utilizing an F_1 population consisting of crosses between 6142-1×8075-7 and 6142-1-S×8075-7. Two maternal (6142-1 and 6142-1-S) and one paternal (8075-7) maps were generated using diversity array technology (DArT), SSR, and allele-specific PCR (AS-PCR) markers. As with other maps, considerable (41%) segregation distortion was observed at marker loci.

Association mapping has been proposed as an alternative to conventional linkage mapping. In this strategy, a panel of genotypes from unrelated population (or a population with known genetic substructure) is utilized to identify associations between molecular markers that are in linkage disequilibrium with genetic loci affecting phenotypes. Molecular markers that are distributed throughout the genome such as single nucleotide polymorphisms (SNP) are preferred for genome-wide association studies (GWAS). Sardos et al. (2016) demonstrated the technique by assembling a GWAS panel of 104 AA accessions using 5544 SNP markers derived from genotyping by sequencing (GBS) and publicly available phenotypic data on parthenocarpy. The study identified 13 genomic regions associated with parthenocarpy, and multiple candidate genes in these regions corresponded with putative growth regulators and genes associated with gametophyte development and female sterility in other plant species.

Genomic selection (GS) is a form of marker-assisted selection that utilizes high-density molecular markers such as SNPs to provide coverage of the whole genome, ensuring that all quantitative trait loci (QTL) are in linkage disequilibrium with at least one marker (Hayes and Goddard 2010). GS estimates the genomic breeding value of individual genotypes in a large segregating population utilizing one of several GS models (Meuwissen et al. 2001). GS is less concerned with the identification of individual QTL as it is with developing appropriate models to enhance selection efficiency. As the cost of generating marker data becomes increasingly more affordable, GS has become an attractive alternative to many breeding programs (Lorenz et al. 2011; Crossa et al. 2010). Currently, efforts are underway to evaluate GS as a strategy to improve banana by generating appropriate breeding models for the improvement of EAHB (Nyine et al. 2016).

7.6.1.3 Estimating Genetic Diversity and Evolutionary Events

Estimates of genetic diversity determine to a large extent the potential of plant improvement that can be anticipated and can also provide guidance to breeders as to the appropriate parents to use in breeding schemes. Estimates based on phenotypic or morphological characters have long been used in banana (Karamura 1998), but often these estimates can be biased by environmental influences as well as the sometimes complimentary and polygenic nature of underlying genetic factors. Molecular markers avoid these issues as they are highly heritable and have supplemented or replaced the usage of such measurements in most plant species where they are available. In banana, several classes of molecular markers have been used to estimate diversity among populations of varying size representing regional collections and breeding program germplasm. These include RFLPs (Jarret et al. 1993; Bhat et al. 1995), RAPDs (Bhat et al. 1995; Crouch et al. 2000; Pillay et al. 2001; Ude et al. 2003; Nyine and Pillay 2011), AFLPs (Ude et al. 2002, 2003; Noyer et al. 2005; Wang et al. 2007; Opara et al. 2010), SSRs (Kaemmer et al. 1997; Crouch et al. 1999; Tenkouano et al. 1999; Noyer et al. 2005; Creste et al. 2004; Hippolyte et al. 2012; Kitavi et al. 2016; Karamura et al. 2016), sequence-related amplified polymorphisms (SRAPs) (Wei et al. 2011; Valdez-Ojeda et al. 2014), DArT (Risterucci et al. 2009), and methylation-sensitive amplification polymorphism (MASP) (Noyer et al. 2005).

Estimates of genetic diversity generated from these studies vary with the class and number of markers used and with the genotypes selected for inclusion in any given study, but a few generalized observations can be made: (1) Diversity estimates based on phenotypic measurements are often poorly correlated with molecular estimates (Crouch et al. 2000); (2) despite considerable phenotypic or morphological variation among regional *Musa* landraces, they tend to have limited genetic variation when assayed with molecular markers. For example, East African

Highland bananas (EAHBs) have been classified into five clades (clone sets) based on their end use and morphological distinctiveness (Karamura 1998). Studies focusing on EAHB using both RAPDs and SSR markers, however, found limited evidence to support the significant variation either within or between these clades (Pillay et al. 2001; Kitavi et al. 2016; Karamura et al. 2016). This led to the suggestion that EAHB arose from a single hybridization event that has subsequently been acted on by a series of somatic mutations and influenced by natural and directed selection leading to many distinct cultivars. Presumably, the finite numbers of markers used are unable to distinguish among the clades. Utilizing different classes of molecular markers can sometimes reveal variation in populations where it has not been previously noted. In plantain landraces of West Africa, RAPD, SSR, and amplified fragment length polymorphism (AFLP) markers displayed few polymorphisms (Crouch et al. 2000). However, when HpaII and MspI methylation-sensitive amplified markers were used, polymorphism (MSAP) profiles revealed three clusters (Noyer et al. 2005) and a genetically distinct subset of plantains from Cameroon (Ude et al. 2003).

Molecular markers have played important roles in determining the evolutionary history of cultivated banana and establishing links to diploid progenitors. Whether the breeder utilizes an evolutionary or reconstitutive approach to banana improvement (discussed in a previous section) plays a vital role in effectively combining novel resistance traits with quality characteristics desired by growers. Perrier et al. (2011) detail the available molecular, archaeology, genetic, and linguistic evidence for this important aspect of breeding. Of particular interest to the dessert banana industry has been the observation that East African diploid bananas appear to have played an important evolutionary role in the development of "Cavendish" and "Gros Michel," the most widely used cultivars that have dominated the banana export industry over the past century (Raboin et al. 2005; Risterucci et al. 2009).

7.7 Conclusions

While progress has been made toward genetic improvement since the first formal programs were established almost a hundred years ago, in some aspects the breeding of banana is still in its infancy when compared to the improvement of other important staple crops. In no small part, this is due to the physical and reproductive constraints of the plant itself, but there is room for optimism as these constraints appear to be neither absolute nor prohibitive. Molecular markers, the availability of additional genomic resources, and ongoing studies elucidating the floral and reproductive biology of banana hold great promise for the next hundred years of banana improvement.

Genetic engineering has not been discussed in this chapter, but the early work in this arena also suggests that it also has the potential to make an important contribution to *Musa* improvement through the introduction of genetic factors not found within cultivated or wild *Musa* germplasm (Tripathi et al. 2012).

Further work is needed on predicting the performance and combining ability of male and female parents in *Musa* improvement. Tenkouano et al. (2012) reported the significance of additive genetic effects on expression of bunch weight, fruit filling time, fruit length, plant height, and number of leaves and nonadditive effects for suckering behavior and fruit circumference in 3× hybrids obtained from plantain derived 4×–2× crosses. They further suggested that maternal general combining ability (GCA) accounted for the additive genetic variation for plant height and number of leaves, while paternal GCA effects accounted for fruit filling time, bunch weight, and fruit length. On the other hand, specific combining ability (SCA) effects were observed for all traits, except fruit filling time, suggesting that additional genetic gain could be achieved through recombinative heterosis for such traits. They concluded that increased bunch weight and faster cycling are inherited from the 2× male parent, while plant height, number of leaves, and suckering behavior are inherited from the 4× female parent which should guide parental selection for 4×–2× crossbreeding. More of this information is needed to efficiently guide the decision-making efforts of breeders and allow them to allocate limited resources.

Finally, in the popular press, there has been considerable alarm in recent years as to the future of banana in the face of an evolving pathogen (Foc TR4) that threatens the production of much of the world's dessert banana production. In some ways, the economic damage that this pathogen will likely cause can be viewed as self-inflicted in nature. The export industry has demonstrated an overdependence on monoculture and complacency in regard to breeding that has significantly contributed to creating an environment conducive to the selection and spread of novel pathogenic races. This is a lesson that should have been learned more than a half century before when a similar threat was encountered by a different race of the same pathogen. Banana breeding efforts to improve the industry standard "Gros Michel" were curtailed or sidelined when a suitable replacement ("Cavendish") was selected from existing stock. Hopefully, the current crises will provide an impetus and serve as a reminder to all that proactive breeding programs are the most efficient and cost-effective frontline defense against current and evolving threats to production.

Acknowledgments This research was undertaken with the support of the Belgium Government, IITA, the Bill and Melinda Gates Foundation, the CGIAR Research Program on Roots, Tubers and Banana (RTB), USAID and HarvestPlus, part of the CGIAR Research Program on Agriculture for Nutrition and Health (A4NH).

References

Adeleke MT, Pillay M, Okoli BE (2004) The relationships between meiotic irregularities and fertility in diploid and triploid *Musa* L. Cytologia 69:387–393

Aguilar Morán JF (2013) Improvement of Cavendish banana cultivars through conventional breeding. Acta Hortic 986:205–208

van Asten PJA, Fermont AM, Taulya G (2011) Drought is a major yield loss factor for rainfed East African highland banana. Agric Water Manag 98:541–552

Bakry F (2008) Zygotic embryo rescue in bananas. Fruits 63:111–115
Bakry F, Carreel F, Jenny C, Horry JP (2009) Genetic improvement of banana. In: Jain SM, Priyadarshan PM (eds) Breeding plantation tree crops: tropical species. Springer, New York, pp 3–50
Bhat KV, Jarret RL, Rana RS (1995) DNA profiling of banana and plantain cultivars using random amplified polymorphic DNA (RAPD) and restriction fragment length polymorphism (RFLP) markers. Electrophoresis 16:1736–1745
Carreel F, De Leon DG, Lagoda P et al (2002) Ascertaining maternal and paternal lineage within *Musa* by chloroplast and mitochondrial DNA RFLP analyses. Genome 45:679–692
Christelová P, De Langhe E, Hřibová E et al (2017) Molecular and cytological characterization of the global Musa germplasm collection provides insights into the treasure of banana diversity. Biodivers Conserv 26:801–824
Creste S, Neto AT, Vencovsky R et al (2004) Genetic diversity of Musa diploid and triploid accessions from the Brazilian banana breeding program estimated by microsatellite markers. Genet Resour Crop Evol 51:723–733
Crossa J, de los Campos G, Pérez P et al (2010) Prediction of genetic values of quantitative traits in plant breeding using pedigree and molecular markers. Genetics 186:713–724
Crouch JH, Crouch HK, Tenkouano A et al (1999) VNTR-based diversity analysis of 2x and 4x full-sib Musa hybrids. Electron J Biotechnol 2:130–139
Crouch HK, Crouch JH, Madsen S et al (2000) Comparative analysis of phenotypic and genotypic diversity among plantain landraces (*Musa* spp., AAB group). Theor Appl Genet 101:1056–1065
D'Hont A, Paget-Goy A, Escoute J et al (2000) The interspecific genome structure of cultivated banana, *Musa* spp. revealed by genomic DNA in situ hybridization. Theor Appl Genet 100:177–183
D'Hont A, Denoeud F, Aury JM et al (2012) The banana (*Musa acuminata*) genome and the evolution of monocotyledonous plants. Nature 488:213–217
Davey MW, Gudimella R, Harikrishna JA et al (2013) A draft *Musa balbisiana* genome sequence for molecular genetics in polyploid, inter-and intra-specific *Musa* hybrids. BMC Genomics 14:683
De Langhe E (1986) Towards an international strategy for genetic improvement in the genus Musa. In: Persley GJ, De Langhe EA (eds) Banana and plantain breeding strategies. Proceedings of an International Workshop, Cairns, Australia. 1–17 October, 1986. INIBAP, Montpellier, p 19–23
De Langhe E, Vrydaghs L, De Maret P et al (2009) Why bananas matter: an introduction to the history of banana domestication. Ethnobot Res Appl 7:165–177
De Langhe E, Hribova E, Carpentier S et al (2010) Did backcrossing contribute to the origin of hybrid edible bananas? Ann Bot 106:849–857
Deckers J, Tessera M, Alemu K, Abate T, Swennen R (2001) Ensete. In: Raemaekers RH (ed) Crop production in tropical Africa. DGIC, Brussels, pp 587–591
Dodds KS (1943) The genetic system of banana varieties in relation to banana breeding. Emp J Exp Agric 11:89–98
Dodds KS (1945) Genetical and cytological studies of Musa. VII. Certain aspects of polyploidy. J Genet 46:161–179
Dumpe BB, Ortiz R (1996) Apparent male fertility in *Musa germplasm*. HortSci 31:1019–1022
Emediato FL, Nunes FA, Teixeira CC, Passos MA, Bertioli DJ, Pappas GJ, Miller RN (2009) Characterization of resistance gene analogs in *Musa acuminata* cultivars contrasting in resistance to biotic stresses. In: Shu QY (ed) Induced plant mutations in the genomics era. FAO, Rome, pp 443–445
FAO (2014) Banana market review and banana statistics 2012–2013, Rep. I3627E/1/01.14. FAO, Rome. http://www.fao.org/docrep/019/i3627e/i3627e.pdf
Fauré S, Noyer JL, Horry JP et al (1993) A molecular marker-based linkage map of diploid bananas (*Musa acuminata*). Theor Appl Genet 87:517–526
Foolad MR (2007) Genome mapping and molecular breeding of tomato. Int J Plant Genomics 2007:64358

Fortescue JA, Turner DW (2004) Pollen fertility in *Musa*: viability in cultivars grown in southern Australia. Aust J Agric Res 55:1085–1091

Fortescue JA, Turner DW (2011) Reproductive biology. In: Pillay M, Tenkouano A (eds) Banana breeding: constraints and progress. CRC Press, Boca Raton, pp 305–331

Fortescue JA, Turner DW, Romero R (2011) Romero evidence that banana (*Musa* spp.), a tropical monocotyledon, has a facultative long-day response to photoperiod. Funct Plant Biol 38:867–878

Hayes B, Goddard M (2010) Genome-wide association and genomic selection in animal breeding. Genome 53:876–883

Hippolyte I, Bakry F, Seguin M et al (2010) A saturated SSR/DArT linkage map of *Musa acuminata* addressing genome rearrangements among bananas. BMC Plant Biol 10:65

Hippolyte I, Jenny C, Gardes L et al (2012) Foundation characteristics of edible *Musa* triploids revealed from allelic distribution of SSR markers. Ann Bot 109:937–951

Iskra-Caruana M, Chabannes M, Duroy PO et al (2014) A possible scenario for the evolution of banana streak virus in banana. Virus Res 186:155–162

Janssens SB, Vandelook F, De Langhe E et al (2016) Evolutionary dynamics and biogeography of *Musaceae* reveal a correlation between the diversification of the banana family and the geological and climatic history of Southeast Asia. New Phytol 210:1453–1465

Jarret RL, Vuylsteke DR, Gawel NJ et al (1993) Detecting genetic diversity in diploid bananas using PCR and primers from a highly repetitive DNA sequence. Euphytica 68:69–76

Jones DR (ed) (1999) Diseases of banana, abaca' and enset. CABI, Wallingford

Kaemmer D, Fischer D, Jarret RL et al (1997) Molecular breeding in the genus *Musa*: a strong case for STMS marker technology. Euphytica 96:49–63

Karamura DA (1998) Numerical taxonomic studies of the East African highland (*Musa* AAA East Africa) in Uganda. Dissertation, University of Reading

Karamura DA, Karamura EB, Tinzaara W (2012) In: Karamura DA, Karamura EB, Tinzaara W (eds) The current classification and naming of the East African highland bananas (Musa AAA) based on Morphological Characteristics in book: Banana cultivar Names, Synonyms and their Usage. Bioversity International, East Africa, pp 6–23

Karamura D, Kitavi M, Nyine M et al (2016) Genotyping the local banana landrace groups of East Africa. Acta Hortic 1114:67–74

Kitavi M, Downing T, Lorenzen J et al (2016) The triploid East African Highland Banana (EAHB) genepool is genetically uniform arising from a single ancestral clone that underwent population expansion by vegetative propagation. Theor Appl Genet 129:547–561

Krishnamoorthy V, Kumar N (2005) Preliminary evaluation of diploid banana hybrids for yield potential, male fertility and reaction to *Radopholus similis*. Plant Gen Res Newsl 141:39–43

Kumar LP, Selvarajan R, Iskra-Caruana M, Chabannes M, Hanna R (2015) Biology, etiology, and control of virus diseases of banana and plantain. In: Loebenstein G, Katis NI (eds) Advances in virus research, vol 91. Academic, Burlington, pp 229–269

Lheureux F, Carreel F, Jenny C et al (2003) Identification of genetic markers linked to banana streak disease expression in inter-specific *Musa* hybrids. Theor Appl Genet 106:594–598

Li C, Shao J, Wang Y et al (2013) Analysis of banana transcriptome and global gene expression profiles in banana roots in response to infection by race 1 and tropical race 4 of *Fusarium oxysporum* f. sp. Cubense. BMC Genomics 14:851

Lorenz AJ, Chao S, Franco G et al (2011) Chap. 2: Genomic selection in plant breeding: knowledge and prospects. Adv Agron 110:77–123

Martin G, Baurens FC, Droc G et al (2016) Improvement of the banana "*Musa acuminata*" reference sequence using NGS data and semi-automated bioinformatics methods. BMC Genomics 17:1–12

Mbanjo EGN, Tchoumbougnang F, Mouelle AS et al (2012) Development of expressed sequence tags-simple sequence repeats (EST-SSRs) for Musa and their applicability in authentication of a *Musa* breeding population. Afric J Biotechnol 11:13546–13559

Meuwissen TH, Hayes BJ, Goddard ME (2001) Prediction of total genetic value using genome-wide dense marker maps. Genetics 157:1819–1829

Miller RNG, Bertioli DJ, Baurens FC et al (2008) Analysis of non-TIR NBS-LRR resistance gene analogs in *Musa acuminata* Colla: isolation, RFLP marker development, and physical mapping. BMC Plant Biol 8:15

Noumbissié GB, Chabannes M, Bakry F et al (2016) Chromosome segregation in an allotetraploid banana hybrid (AAAB) suggests a translocation between the A and B genomes and results in eBSV-free offsprings. Mol Breed 36:1–14

Noyer JL, Causse S, Tomekpe K et al (2005) A new image of plantain diversity assessed by SSR, AFLP and MSAP markers. Genetica 124:61–69

Nyine M, Pillay M (2011) The effect of banana breeding on the diversity of East African Highland banana (Musa, AAA). Acta Hortic 897:225–229

Nyine M, Uwimana B, Swennen R, Batte M, Brown A, Hřibová E, Doležel J (2016) Genomic breeding approaches for East African Bananas. In: Abstracts of the plant and animal genome conference XXIV January 08–13, San Diego, CA

Opara UL, Jacobson D, Al-Saady NA (2010) Analysis of genetic diversity in banana cultivars (*Musa* cvs.) from the South of Oman using AFLP markers and classification by phylogenetic, hierarchical clustering and principal component analyses. J Zhejiang Univ Sci B 11:332–341

Ortiz R (2000) Understanding the *Musa* genome: an update. Acta Hortic 54:157–168

Ortiz R (2013) Conventional banana and plantain breeding. Acta Hortic 986:77–194

Ortiz R (2015) Plant breeding in the omics era. Springer, New York

Ortiz R, Swennen R (2014) From crossbreeding to biotechnology-facilitated improvement of banana and plantain. Biotechnol Adv 32:158–169

Ortiz R, Ferris RSB, Vuylsteke DR (1995) Banana and plantain breeding. In: Gowen S (ed) Bananas and plantains. Springer, New York, pp 110–146

Oselebe HO, Tenkuoano A, Pillay M et al (2006) Ploidy and genome segregation in *Musa* breeding populations assessed by flow cytometry and randomly amplified polymorphic DNA markers. J Am Soc Hortic 131:780–786

Panis B, Piette B, Swennen R (2005) Droplet vitrification of apical meristems: a cryopreservation protocol applicable to all *Musaceae*. Plant Sci 168:45–55

Perrier X, De Langhe E, Donohue M, Lentfer C, Vrydaghs L, Bakry F, Carreel F, Hippolyte I, Horry J-P, Jenny C, Lebot V, Risterucci A-M, Tomekpe K, Doutrelepont H, Ball T, Manwaring J, de Maret P, Denham T (2011) Multidisciplinary perspectives on banana (*Musa* spp.) domestication. PNAS 108:11311–11318

Pillay M, Ogundiwin E, Nwakanma DC et al (2001) Analysis of genetic diversity and relationships in East African banana germplasm. Theor Appl Genet 102:965–970

Pillay M, Ude G, Kole C (eds) (2012) Genetics, genomics and breeding of bananas. CRC, Boca Raton

Raboin LM, Carreel F, Noyer JL et al (2005) Diploid ancestors of triploid export banana cultivars: molecular identification of 2n restitution gamete donors and n gamete donors. Mol Breed 16:333–341

Risterucci AM, Hippolyte I, Perrier X et al (2009) Development and assessment of diversity arrays technology for high-throughput DNA analyses in *Musa*. Theor Appl Genet 119:1093–1103

Roux NS (2001) Mutation induction in *Musa*. In: Jain SM, Swennen R (eds) Banana improvement: cellular, molecular biology, and induced mutations. Science, Enfield

Rowe P, Rosale F (1993) Diploid breeding at FHIA and the development of Goldfinger (FHIA-01). InfoMusa 2:9–11

Sardos J, Rouard M, Hueber Y et al (2016) A genome-wide association study on the seedless phenotype in banana (*Musa* spp.) reveals the potential of a selected panel to detect candidate genes in a vegetatively propagated crop. PLoS One 11:5

Sathiamoorthy S, Rao VNM (1980) Pollen production in relation to genome and ploidy in banana clones. Proc Nat Semi Banana Prod. Tech., TNAU, Coimbatore, pp 65–66

Shepherd K (1999) Cytogenetics of the genus *Musa*. International Network for the Improvement of Banana and Plantain, Montpellier

Silva PRO, de Jesus ON, Bragança CAD et al (2016) Development of a thematic collection of *Musa* spp accessions using SCAR markers for preventive breeding against Fusarium oxysporum f. sp cubense tropical race 4. Genet Mol Res 15:5017765

Simmonds NW (1962) The evolution of the bananas. Longmans, London

Simmonds NW (1966) Bananas, 2nd edn. Longmans, London

Simmonds NW (1987) Classification and breeding of bananas. In: Persley G, De Langhe E (eds) Banana and plantain breeding strategies. Proceedings of an International Workshop held at Cairns Australia, 13–17 October 1986. Austrailian Centre for International Agricultural Research, Canberra

Simmonds NW, Shepherd K (1955) The taxonomy and origins of the cultivated bananas. J Linn Soc Lond Bot 55:302–312

Ssebuliba R, Vuylsteke D, Hartman J et al (2000) Towards improving highland bananas. Uganda J Agric Sci 5:36–38

Ssesuliba RN, Tenkouano A, Pillay M (2008) Male fertility and occurrence of 2n gametes in East African highland bananas (*Musa* spp.) Euphytica 164:153–162

Stover RH, Buddenhagen IW (1986) Banana breeding: polyploidy, disease resistance and productivity. Fruits 41:175–191

Stover RH, Simmonds NW (1987) Bananas, Tropical agricultural series, 3rd edn. Longmans, London

Swennen R, Vuylsteke D (1993) Breeding black Sigatoka resistant plantain with a wild banana. Trop Agric 70:74–77

Tenkouano A, Swennen R (2004) Progress in breeding and delivering improved plantain and banana to African farmers. Chron Hortic 44:9–15

Tenkouano A, Crouch JH, Crouch HK et al (1999) Comparison of DNA marker and pedigree-based methods of genetic analysis of plantain and banana (*Musa* spp.) clones. I. Estimation of genetic relationships. Theor Appl Genet 98:62–68

Tenkouano A, Vuylsteke D, Okoro J et al (2003) Registration of TMB2x5105-1 and TMB2x9128-3 diploid banana hybrids with good combining ability, partial resistance to black Sigatoka and resistance to nematodes. Hortscience 38:468–472

Tenkouano A, Pillay M, Ortiz R (2011) Breeding techniques. In: Pillay M, Tenkouano A (eds) Banana breeding: constraints and progress. CRC Press, Boca Raton, pp 181–202

Tenkouano A, Ortiz R, Vuylsteke D (2012) Estimating genetic effects in maternal and paternal half-sibs from tetraploid-diploid crosses in *Musa* spp. Euphytica 185:295–301

Tripathi JN, Muwonge A, Tripathi L (2012) Efficient regeneration and transformation of plantain cv. 'Gonja manjaya' (*Musa* spp. AAB) using embryogenic cell suspensions. In Vitro Cell Dev Biol 48:216–224

Tsegaye A, Struik PC (2002) Analysis of enset (*Ensete ventricosum*) indigenous production methods and farm-based biodiversity in major enset-growing regions of southern Ethiopia. Explor Agric 38:291–315

Tushemereirwe W, Batte M, Nyine M, Tumuhimbise R, Barekye A, Tendo S, Talengera D, Kubiriba J, Lorenzen J, Swennen R, Uwimana B (2015) Performance of NARITA banana hybrids in the preliminary yield trial for three cycles in Uganda. Banana Technical Report, 35p. www.musalit.org/seeMore.php?id=15482

Ude G, Pillay M, Nwakanma D et al (2002) Genetic diversity in *Musa acuminata* Colla and *Musa balbisiana* Colla and some of their natural hybrids using AFLP markers. Theor Appl Genet 104:1246–1252

Ude G, Pillay M, Ogundiwin E et al (2003) Genetic diversity in an African plantain core collection using AFLP and RAPD markers. Theor Appl Genet 107:248–255

Uma S, Lakshmi S, Saraswathi MS et al (2011) Embryo rescue and plant regeneration in banana (*Musa* spp.) plant cell. Tissue Organ Cult 105:105–111

Umber M, Pichaut JP, Farinas B et al (2016) Marker-assisted breeding of *Musa balbisiana* genitors devoid of infectious endogenous banana streak virus sequences. Mol Breed 36:1–11

Vakili NG (1968) Responses of *Musa* acuminata species and edible cultivars to infection by *Mycosphaerella musicola*. Trop Agric 45:13–22

Valdez-Ojeda R, James-Kay A, Ku-Cauich JR et al (2014) Genetic relationships among a collection of *Musa* germplasm by fluorescent-labeled SRAP. Tree Genet Genomic 10:465–476
Van den Houwe I, De Smet K, Tezenas de Montcel H et al (1995) Variability in storage potential of banana shoot cultures under medium term storage conditions. Plant Cell Tissue Organ Cult 42:269–274
Vuylsteke D, Swennen R, Ortiz R (1993) Development and performance of Black Sigatoka-resistant tetraploid hybrids of plantain (*Musa* spp., AAB group). Euphytica 65:33–42
Wairegi LWI, van Asten PJA, Tenywa MM et al (2010) Abiotic constraints override biotic constraints in East African highland banana systems. Field Crop Res 117:146–153
Wang XL, Chiang TY, Roux N et al (2007) Genetic diversity of wild banana (*Musa balbisiana* Colla) in China as revealed by AFLP markers. Genet Resour Crop Evol 54:11251132
Wang W, Hu Y, Sun D et al (2012) Identification and evaluation of two diagnostic markers linked to Fusarium wilt resistance (race 4) in banana (*Musa* spp.) Mol Biol Rep 39:451–459
Wei JY, Liu DB, Wei SX et al (2011) Analysis of genetic diversity in banana cultivars (*Musa* spp.) using sequence-related amplified polymorphism markers. Acta Hortic 897:263–265

Chapter 8
Oil Palm (*Elaeis guineensis*)

Brian P. Forster, Baihaqui Sitepu, Umi Setiawati, Eddy S. Kelanaputra, Fazrin Nur, Heru Rusfiandi, Safrina Rahmah, Jennifer Ciomas, Yassier Anwar, Syamsul Bahri, and Peter D.S. Caligari

8.1 History of Oil Palm Cultivation

Oil palm, *Elaeis guineensis*, Jacq's name, comes from the Greek word *elaion* meaning oil. Oil palm has its origins in the African continent. Our knowledge of the early history of oil palm (*Elaeis guineensis* Jacq.) cultivation is poor and mainly based on records of journeys and explorations in Africa (Rees 1965; Zeven 1965). Portuguese exploration and trade on the West African coast began in 1434, followed by the Dutch and English about 150 years later. The first record of what may be the oil palm is that of Cadamosto (1435–1460; cited in Crone 1937), who wrote: 'There is to be found in this country a species of tree bearing red nuts with black eyes in great quantity, but they are small'. With respect to an oil used for food, he wrote '[It] has three properties, the scent of violets, the taste of our olive oil and a colour which tinges the food like saffron, but is more attractive'. Duarte Pacheco Pereira mentions palm groves near Liberia during his voyage of 1506–1508 and trade in palm oil (*azeite de palma*) near Forcados in Nigeria. Later records also mention palm wine. Opsomer (1956) may be credited as providing the first official description of oil palm and giving the origin as West Africa (Fig. 8.1).

In Southeast Asia, mainly in Indonesia, Malaysia, Thailand and Papua New Guinea, the history of the oil palm cultivation can be traced to just four Dura palms planted at the Bogor Botanical Garden, Java, Indonesia (Hartley 1988; Taniputra et al. 1987). The African origins of these palms are not clear or how they reached

B.P. Forster (✉) • B. Sitepu • U. Setiawati • E.S. Kelanaputra • F. Nur • H. Rusfiandi
S. Rahmah • J. Ciomas • Y. Anwar • S. Bahri • P.D.S. Caligari
Verdant Bioscience, Timbang Deli Estate, Galang, North Sumatra, Indonesia
e-mail: brian.forster@verdantbioscience.com; baihaqi.sitepu@verdantbioscience.com; umi.setiawati@verdantbioscience.com; eddy.setiawan@verdantbioscience.com; fazrin.nur@verdantbioscience.com; heru.rusfiandi@verdantbioscience.com; safrina.rahmah@verdantbioscience.com; jennifer.ciomas@verdantbioscience.com; yassier.anwar@verdantbioscience.com; syamsul.bahri@verdantbioscience.com; peter.caligari@verdantbioscience.com

H. Campos, P.D.S. Caligari, *Genetic Improvement of Tropical Crops*,
DOI 10.1007/978-3-319-59819-2_8

Fig. 8.1 Monument of 150 years of oil palm, Bogor Botanical Garden, Java, Indonesia

Indonesia. Rutgers et al. (1922) suspected that seedlings from Amsterdam were dispatched at the same time as the ones from Mauritius (Bourbon) in February 1848 by D T Pryce, but they may all have come from Amsterdam in March 1848. The descendants from these four palms have been used in various breeding programmes, starting with AVROS (Algemene Vereniging van Rubberplanters ter Oostkust van Sumatra), Banting (Malaysia) and Dami (Papua New Guinea). The first commercial oil palm planting in SE Asia was at Sungai Liput Aceh, Sumatra, Indonesia, in 1911. The last of the four-foundation palm trees at Bogor died in 1992 (after reaching 30–35 m and being struck by lightning), and all that remains of them are botanical specimens (leaves) housed at the Herbarium Bogoriense in Bogor, West Java, Indonesia. Pamin (1988) reported that offspring of the original Bogor palms were planted in Medan in 1975.

In 1884, oil palm was planted in Deli, Sumatra, Indonesia, as ornamental plants on tobacco plantations. The first commercial plantings began in 1911 (Taniputra et al. 1987) with plantations set up by A Hallet in Sungai Liput (Aceh, Sumatra) and Pulau Raja (Asahan, Sumatra) and by K. Schadt in Tanah Itam Ulu, North Sumatra. The key to production success was the introduction of the pollinating weevil in 1981, *Elaeidobius kamerunicus*, which provided good fruit set (see Sect. 8.3).

The development of oil palm plantations on a large scale began in 1920s, both in Africa (DR Congo) and SE Asia (Indonesia and Malaysia) (Corley and Tinker 2003). This development was supported by activities in agronomy, crop protection, plant selection and plant breeding. In DR Congo, the work was pioneered by the Institut National pour l'Etude Agronomique du Congo Belge (INEAC) and in Indonesia by the Algemene Vereniging van Rubberplanters ter Oostkust van Sumatra (AVROS). In Malaysia oil palm plantations were developed by the Department of Agriculture (Corley and Tinker 2003).

Serious breeding activities were initiated by AVROS in the 1920s in Sumatra, Indonesia. These focused on Deli × Deli Dura crosses and crossing Deli Dura with imported materials from Africa (Corley and Tinker 2003). Experimental trials were planted in Sungei Pantjur and Polonia, Sumatra. Dura materials planted at Deli, Sumatra, are well known as Deli Dura and have been used widely around the world as female parents in breeding and commercial production as they have good fruit traits, e.g. thick oil-bearing mesocarp (60% of the fruit) and hence high oil content (Corley and Tinker 2003).

Other Deli Dura lines from Malaysia are characterised by short plant stature, termed 'dumpy' (Sparnaaij et al. 1963); these have a thick trunk diameter and slow growth (Rosenquist 1985). SOCFINDO (a major oil palm company based in Indonesia) also conducted selections of Deli Dura at Mopoli and Bangun Bandar in Sumatra (Pamin et al. 1990). Another significant Dura population was developed by London Sumatra (LONSUM) and planted in Gunung Melayu in 1919, also in Sumatra (Rosenquist 1985; Corley and Tinker 2003). LONSUM also introduced Dura and significantly Pisifera selected breeding material from Dami, Papua New Guinea. The Dami Dura breeding material was selected in a magnesium-deficient environment where there was selection for high leaf (frond 17) magnesium content, low crown disease incidence, high fruit/bunch ratio and high yield.

A famous Pisifera genotype in oil palm breeding is Sungai Pantjur 540 (SP 540) which provides good combining ability with Deli Dura in producing high yielding commercial thin-shelled Tenera. SP 540 was part of a seed transfer from the Eala Botanical Garden, DR Congo, to Yangambi, DR Congo, in 1922 (Pamin et al. 1990); the seed giving rise to SP 540 was derived from a Django Tenera palm, which played an important role in the breeding programmes at Yangambi, DR Congo. Other significant sources of Pisifera are from populations from Nigeria: Calabar and La Me material.

The selection of oil palm in Nigeria began with a survey of 800 palms consisting of various forms and types, grown at Calabar, Nigeria, from 1912 to 1916 (Hartley 1988). The virescent fruit-type trait (see Sect. 8.8) was found in these selection activities (first reported by Smith 1929, in Corley and Tinker 2003). The *virescens* gene controls fruit colour which changes from green to yellow as opposed to the wild-type *nigrescens* in which fruits change from black to red during ripening. Exploitation for the *virescens* gene is currently being carried out by several oil palm research centres as the green-yellow colour change allows better selection for ripe bunches than black-red.

La Me populations were obtained from selections made by M. Houard between 1924 and 1930 in Bingerville, Ivory Coast (Hartley 1988; Cochard et al. 2000; Corley and Tinker 2003). The Institut pour Recherche sur les Huiles et Oleagineux (IRHO; replaced by CIRAD-CP, Centre de Cooperation International en Recherche

Agronomique pour le Dévéloppement – Culture perennes) conducted germplasm exchanges between various stations, IRHO, Ivory Coast, Benin and DR Congo (Rosenquist 1985; Corley and Tinker 2003).

Shell thickness is a major trait in oil palm breeding and is a monogenic trait controlled by the *Sh* gene (Beirnaert and Vanderweyen 1941, see Sect. 8.8). This gene is of huge economic significance in the change from Dura to high yielding Tenera fruit types as the preferred commercial material. The Tenera fruit type has an oil yield increase of 30% over Dura (Hardon et al. 1987; Corley and Lee 1992).

8.1.1 Basic Crop Facts

Oil palm is the most important oil crop in the world with annual yields from eight to ten times greater per hectare than its nearest rivals, oilseed rape/canola and soybean. Oil palm is a crop of the humid tropics and rings the globe 20° from the equator; plantations are among the most profitable land uses in the humid tropics (Sayer et al. 2012). The crop produces two oils from its fruit: crude palm oil (CPO) from the fleshy mesocarp and kernel oil from the seed. Oil palm accounts for over 35% of global edible oil production (Basiron 2007; Corley 1998; Corley and Tinker 2003). In 2005 expanding oil palm plantations accounted for over 8.5 million hectares; by 2010 palm oil production was estimated at 30.4 million tonnes and is predicted to reach 200 million tonnes by 2050 (Barcelos et al. 2015). Population growth along with increasing demand for low-cost, high-quality oil has driven up oil palm production, notably in Malaysia and Indonesia. Palm oil is a tropical crop and is harvested continuously, thus supporting a reliable and stable global market. However, yield per land area, although high relative to other oil crops, has remained largely static over the past decade at about 3–4 tonnes crude oil and kernel oil per hectare per year (FAO statistics) with increased planting areas accounting for increased production. The loss of rainforest to oil palm plantations is a serious concern to environmentalists and is an additional factor driving greater efficiency in the oil palm industry (Sheil et al. 2009; Pye and Bhattacharya 2013). Current yields of palm oil range from 2 to 10 tonnes/hectare/year; however, early physiological studies (Corley 1983; Corley and Lee 1992) estimate that 18 t/ha/year may be achieved.

Cloning of superior palms has been one approach suggested to increase the yield of oil palm. The basic concept is to identify high-performing individuals and to clone these for commercial production. The oil palm plant does not branch and has only one meristem (see Sect. 8.3); thus, vegetative propagation via conventional propagation methods (cuttings/budding) is not possible. However, oil palm can be clonally propagated using tissue culture methods (see Sect. 8.11), and in vitro methods have been developed to produce commercial clonal planting materials. Yields of 11–12 t/ha/year have been reported from clones (Soh et al. 2006, 2009). There are however serious concerns in clonal propagation of oil palm, reviewed by Soh (1986) and Nelson et al. 2009 (see Sect. 8.11). Chief among these is the unreliability in selecting the ortet (with the genetic potential for high yield) for cloning, as this is influenced by the ortet's environment. Also somaclonal variation induced by tissue culture can lead to physiological defects (see Sect. 8.11), and there are major challenges in scaling up production/distribution for the oil palm industry which is often operating in remote locations. Therefore, breeding and seed production remain the main tools for oil palm improvement.

The main products of oil palm are crude palm oil (CPO) and palm kernel oil (PKO). CPO is obtained from mesocarp extraction of oil palm fruit. Generally, the mesocarp contains up to 65% oil (Hartley 1988; Corley and Tinker 2003). With respect to quality, palm oil is generally composed of about 43% palmitic acid, 39% oleic acid, 5% stearic acid and other essential fatty acids such as myristic acid (Siew 2002). Oil palm is the richest plant source for pro-vitamin A and vitamin E (Barcelos et al. 2015). The kernel is the endosperm part of the palm seed which is protected by a shell (endocarp). The kernel generally contains 47–52% oil (Hartley 1988; Corley 1998) of high quality. Palm kernel oil contains 44–50% lauric acid, 15% myristic acid and other essential fatty acids (Corley and Tinker 2003). Together with coconut oil, PKO is the world's largest source of lauric acid (Sambanthamurthi et al. 2000).

Recent data (2015) indicates that India is the largest consumer of palm oil (absorbing 28% world CPO products), followed by Europe (around 22%) and China (22%). The world's largest CPO exporter is Indonesia with a market share of 53%, followed by Malaysia at 38%, with Papua New Guinea, Benin, Guatemala and other countries collectively accounting for 9% (Foreign Agriculture Service/USDA 2015).

As stated earlier, oil palm is a tropical crop grown between latitudes 20°N and 20°S. Within this tropical belt, it can be grown in a wide range of environments, topographies and soil types, but high yields require a minimum and well-distributed rainfall of 1800–2000 mm per year and sunshine of up to 2000 h per year (Hartley 1988). Areas with high rainfall up to 5000 mm per year may still be suitable but require freely draining soils and/or drainage systems. Basic agronomy of oil palm involves palm spacing, normally in a 9 × 9 × 9 m triangular design which results in about 143 palms/ha (Alvarado et al. 2007). Palm spacing may vary depending on topography and frond length. Optimal spacing is designed to maximise land use, reduce the incidence of overlapping fronds and maximise light interception and soil nutrient uptake (Hartley 1988). However, if compact palms are planted, a density of 200 palm/ha may be reached (Alvarado et al. 2007). Plantations have a generation life span of 20–35 years when palms reach 13 m in height, but increasingly the stand per hectare, usually as a result of disease losses, often dictates the replanting date in certain environments.

Fertiliser use is a major part of oil palm agronomy. Intensive cultivation depletes soil nutrients which need to be replenished as fertiliser, the major components of which are nitrogen, phosphorus, potassium and magnesium. Nitrogen and phosphorus are closely correlated to oil palm production; the correct combination of both can increase production up to 20% (Soon and Hong 2001). Nitrogen can also affect the absorption of phosphorus (Tampubolon et al. 1989), while potassium can increase the number and weight of bunches. Magnesium has a major role in the metabolism of phosphate and is a major constituent of chlorophyll and thus photosynthetic efficiency (Rankine and Fairhurst 1999). Approaches to determine fertiliser use efficiency are described by Goh (2011). There is no doubt that agronomy has had, and continues to play, a major role in oil palm production. However, new cultivars are needed to break through yield barriers. Breeding is expected to produce new cultivars with greater yield potential, but to realise this potential tailored agronomic practices will be needed.

Oil palm plantations are often large and grown on an industrial scale; however, oil palm is also an important smallholder crop and this also needs to be considered for breeding. It is a highly profitable crop and its cultivation has alleviated poverty and improved livelihoods of millions of people (Sayer et al. 2012).

8.2 Biology of Oil Palm

Oil palm, *Elaeis guineensis* Jacq., is classified under the family Palmae in the order Palmales, subfamily Arecoideae, in the tribe Cocoseae and the subtribe Elaeidinae. The Arecoideae form the largest subfamily of the Arecaceae and have been classified into 112 genera (Dransfield et al. 2005).

Oil palm is a perennial plant and is relatively long-lived and can reach an age of more than 100 years. Under favourable climatic conditions, the single apical meristem is active continuously, producing a new leaf primordium approximately every 2 weeks in mature palms and every 9 days at the juvenile stage (Corley and Gray 1976). It is a C3 plant and stores photosynthates (carbohydrates) in the trunk. Oil palm is monoecious, i.e. produces separate male and female inflorescences on the same palm in an alternating cycle of variable duration depending on genetic factors, age and particularly environmental conditions (Dransfield and Uhl 1998; Adam et al. 2005, 2011). The alternating cycle is a mechanism which places restrictions on self-fertilisation, thus promoting cross-fertilisation, maintaining heterozygosity and avoiding inbreeding depression (the expression of deleterious homozygous alleles).

Inflorescences (flower bunches) are formed throughout the year in an acropetal sequence in the leaf axils and are produced continually once the plant has passed the juvenile stage (Dransfield and Uhl 1998; Adam et al. 2005). Each inflorescence is a compound spike carried on a stout peduncle (stalk) of about 30–45 cm in length. The spikelets are arranged spirally on a central rachis (stem) in a similar manner to the arrangements of the leaves. Male inflorescences are usually the first to be produced and visible typically 32–36 months after seed germination.

The ovule (female gametophyte) consists of a nucellus and two integuments that are attached to it by the funiculus. The nucellus represents the megasporangium, in which a meiocyte undergoes meiosis forming four megaspores. Only one of the four develops into an embryo sac (the megagametophyte). The mature embryo sac contains eight nuclei, organised into seven cells. These include one egg cell, associated with two synergid cells (together they form the egg apparatus), a large fused central cell with two nuclei and three antipodal cells at the opposite end to the egg apparatus. The embryo sac is classed as a polygonum type, resulting from one megaspore (Kajale and Ranade 1953), although adoxa types have also been reported in oil palm (de Poerck 1950). The integuments around the embryo sac form a micropyle, a narrow canal through which a pollen tube grows into the nucellus and enters the embryo sac via a degenerate synergid (Endress 2011). One of the two sperm cells conveyed by the pollen tube fertilises the egg cell resulting in the zygote, and the other fuses with the central cell, forming the endosperm, thus affecting double fertilisation.

Microspore (male gametophyte) development in oil palm can be divided into seven stages: pollen mother cell (PMC), tetrad, empty, starch-filled, late to uninucleate, binucleate and mature pollen. Meiosis of a PMC results in a tetrad of four uninucleate microspores conjugated to each other. The nuclei of these uninucleate microspores then undergo asymmetrical mitosis resulting in binucleate microspores comprising a vegetative nucleus and a generative nucleus. Oil palm pollen matures and is shed at the binucleate stage. In binucleate species, both the vegetative and generative nuclei of a mature pollen grain travel together along the pollen tube at pollination (pollen germination on the stigma of female flowers). During this migration, the generative nucleus undergoes a second mitotic division to produce two sperm cells one of which is normally associated with the vegetative nucleus and, along with the other sperm cell, form the male germ unit inside the pollen tube (Nasution et al. 2009).

Inefficient pollination was a problem in the early days of oil palm cultivation and resulted in poor fruit set, bunch failure and yield loss. Assisted pollination (hand pollination) was practiced to overcome this problem, before it was realised that oil palm was primarily insect pollinated and not wind pollinated. The introduction of the pollinating weevil from Africa, *Elaeidobius kamerunicus*, in 1981 succeeded in eliminating the problem of low fruit set (Syed et al. 1982). *Elaeidobius kamerunicus* is dependent on the male oil palm inflorescence as a breeding site. Adult weevils lay eggs in anthesing male inflorescences and consume oil palm pollen. The larvae live on and consume the decomposing male inflorescence and pupate within the spikelets of the flowers, emerging as adults about 10 days later (Tuo et al. 2011).

The oil palm fruit is a drupe with a spherical, ovoid or elongated shape. The pericarp is composed of the exocarp (external layer, skin), mesocarp (outer pulp that contains oil) and endocarp (a hard shell that encloses the kernel). The kernel consists of the embryo and endosperm which is packed with kernel oil and other food reserves for the embryo. According to the shell thickness, the fruit type can be divided into three forms: Dura (thick-shelled), Pisifera (shell-less) and Tenera (thin-shelled with a fibre ring) (Fig. 8.2). At maturity the fruit is shed.

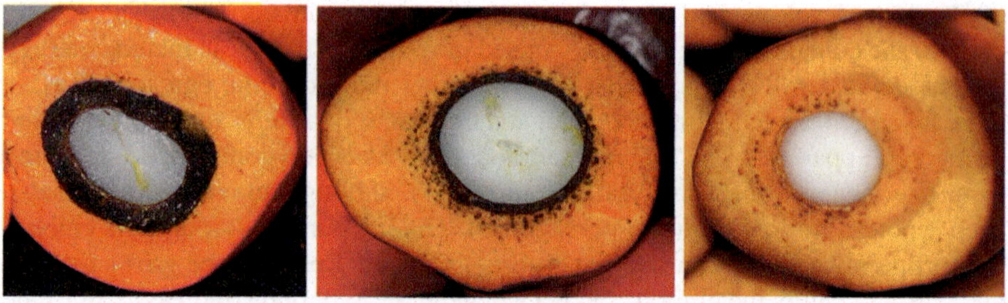

Fig. 8.2 Fruit type of oil palm (**a**) Dura, thick-shelled; (**b**) Tenera, thin-shelled with a fibre ring; (**c**) Pisifera no shell with traces of a fibre ring around the kernel

8.3 Genetics of Oil Palm

Oil palm is a diploid species with a genome consisting of 16 pairs of chromosomes ($2n = 2\times = 32$). Oil palm chromosomes are relatively small, and the genome contains around 3.7 pg DNA, which in sequencing terms equates to about 3.4×10^9 base pairs (Rival et al. 1997; Srisawat et al. 2005). DNA sequencing of the oil palm genome has revealed extensive genetic duplication indicating that oil palm evolved from a tetraploid species (Singh et al. 2013b). Odd ploidy constitutions such as haploids ($n = 16$ chromosomes), triploids ($3n = 48$) and tetraploids ($4n = 64$) have been reported, but these are rare and suffer from sterility. Aneuploids having odd chromosome numbers (more or less chromosomes than the normal euploid complement of 32 chromosomes) have also been reported (Nasution et al. 2013). Although ploidy and aneuploid genetic stocks have been used widely in genetic studies and breeding of other species, e.g. wheat, these genetic stocks are an underdeveloped resource in oil palm breeding, with the exception of haploids (see Sect. 8.10). Most breeding efforts exploit normal diploid germplasm with some inclusion of material from diverse geographic sources (Wening et al. 2012a).

Oil palm is an outbreeder and therefore highly heterozygous. Artificial self-pollination is possible, but repeated rounds of selfing (to produce homozygous lines) may result in inbreeding depression (weak plants) (Hardon 1970; Dumortier et al. 1992). However, in certain more selected breeding materials, inbreeding depression has not been expressed, and this has been explained because deleterious recessive genes have already been largely screened out. Probably the main reason inbred lines have not been developed and tested is the long generation time (minimum of 5 years, although many breeding programmes will have a generation time of at least 10 years). Therefore a programme of eight self-generations would take a minimum of 40 years.

The first high-density genetic map of oil palm was developed by Billotte et al. (2005) using simple sequence repeat (SSR, also known as microsatellites) DNA markers. Genetic maps continue to be developed in oil palm and now contain several types of markers (see Sect. 8.9.4).

The oil palm genome was sequenced and published by Singh et al. (2013a); this is a key step forward in oil palm genetics, the development of functional genomics (assigning genes to functions) and discovering new sequence-based DNA markers. The first sequence-based markers are now available for shell thickness, virescent fruit and mantled fruit (https://www.orionbiosains.com/about-orion; see Sect. 8.9.4).

8.4 Germplasm

Commercial oil palm planting material worldwide is mainly produced from a restricted genetic base, limited mainly to the genetic variation present in the Deli Dura and AVROS Pisifera parental lines. This contrasts with the recommendation frequently made in the literature to increase the genetic diversity of planting materials. Diversity of material is often unique to an oil palm seed supplier and exploited for specific purposes, e.g. ASD's main goal has been to increase oil palm productivity by increasing oil and kernel content, while at the same time reducing vegetative growth (Sterling and Alvarado 2002).

In recent years, in addition to increasing yield, breeding objectives aim to overcome problems in production in oil palm plantations, major targets include:

1. High palm product (crude palm oil and kernel oil) yielding material per unit area
2. Low vegetative vigour, e.g. high harvest index
3. Pest and disease (*Fusarium oxysporum* wilt in Africa and *Ganoderma* in SE Asia) tolerance
4. Good performance over a wide range of environments or selection for specific environments
5. Easy harvesting (see section on novel traits) and the developed for future mechanisation, e.g. high yield per harvester
6. Specific bunch and oil characters for vertically integrated plantation companies which will lever greater value from refining

The genetic base needs to be broadened to realise these goals. As a result, many breeding companies and research stations around the world have taken part in germplasm collections from the centre of diversity of the species in Africa (Nigeria, Cameroon, Angola, Ghana and Tanzania).

Breeding in oil palm started in Nigeria with a very limited genetic base, using Dura (D) material to make D×D crosses for commercial Dura seed production and planting. Later the genetics of shell thickness was understood, and Pisifera pollen parents were used to produce Tenera genotypes from Dura × Pisifera crosses, as Tenera thin-shelled fruits give higher oil yields than the Dura thick-shelled fruit. There is therefore interest in exploring and exploiting genetic variation in Dura, Pisifera and Tenera material.

8.4.1 Dura

8.4.1.1 Deli Dura

Although it is believed that all Deli Dura descended from only four progenitor palms planted in Bogor, Indonesia, the population of Dura expanded to several millions before systematic breeding programmes started in the 1850s. Several countries carried out independent breeding programmes, for example: Indonesia at Gunung Bayu, Pabatu, Dolok Sinumbah, Marihat Baris, Mopoli/Bangun Bandar and Gunung Melayu; Malaysia at Serdang Avenue, Elmina, Ulu Remis and Johor Labis; and the Ivory Coast at Dabou (Rosenquist 1985). Descendants of the four Dura progenitors were planted in the Deli district of Sumatra in 1915 and later introduced into different breeding programmes in Indonesia, Malaysia and Papua New Guinea (Sterling and Alvarado 2002).

8.4.1.2 Other Dura Populations

Other Dura populations have been developed from the following germplasm sources:

Angola: The main advantage offered by the Angola population when used as a female progenitor is reduced stem growth rate. However, a disadvantage is that bunch production is less than in the Deli Dura materials. Angola germplasm combines well when used as female parents with Ekona and Mardi pollen.

Bamenda: Although not extensively tested, the Bamenda population exhibits great potential in relation to bunch production. Annual bunch yields for Bamenda × AVROS progenies are about 200 kg/plant/year with reduced stem growth and acceptable bunch composition. This translates to a high commercial potential of over 1 tonne of oil per hectare per year over conventional standards.

Kigoma: DxP progenies obtained from Kigoma as female with several male genotypes showed a high bunch and oil yield potential, very similar to Tenera materials derived from Deli Dura. This is particularly pronounced when the male source is from Mardi or AVROS origins. In addition, vegetative growth is lower in Tanzanian derivatives than in those of Deli origin.

Other uncommon sources of Dura: An experiment planted in 1990 in Santo Domingo de los Colorados, Ecuador, showed that the Deli and Deli × Angola Dura female lines and the Ekona male lines were very precocious (early flowering); the best specific combinations were the Deli × Yangambi and Kigoma × Ekona.

8.4.2 Pisifera

Given that Pisifera palms are normally female sterile, it is impossible to evaluate their yield production performance from phenotypic individual palm data. The only direct measurements that can be obtained from Pisifera trees are morphological and leaf mineral content. The most commonly recorded traits are leaf area index, stem growth rate, leaf length, leaf emission rate and magnesium content (Breure 1982). The strategy has therefore often been to select the progenies, i.e. progeny test the Pisifera, and on the basis of the performance of their Tenera sibs. Pisifera can be selected from progeny of Tenera × Tenera crosses or Tenera × Pisifera crosses. However, there are fertile Pisifera and therefore Pisifera × Pisifera crosses are possible if a fertile Pisifera is used as the female (Chin 1982, 1988, 1995). However, fertile Pisifera have not been widely used in oil palm breeding programmes, and this may be because there was a perception that fertile Pisifera transmitted thicker shells to their Tenera progenies (Menendez and Blaak 1964); another downside is the difficulty in germinating Pisifera seed using conventional methods.

Thus, selection is usually based on progeny testing to gain estimates of general and specific combining ability. This is currently the only reliable criterion to decide the genetic potential of individual Pisifera, with the aim of selecting them as male parents for commercial oil palm seed production. However, the commitment to testing candidate Pisifera for commercial seed production is justified as a single Pisifera has the annual potential to produce sufficient pollen to produce several million seed per year (Dumortier and Konimor 1999). ASD has been phenotyping progeny from its original introduced D'jongo-derived sources of male parents since 1969 (Sterling and Alvarado 2002). Although phenotyping is important, it is expected that progeny testing will include genotypic selection in the near future as markers for specific traits, including shell thickness becoming established.

There are no reports of fertile Pisifera being planted on a commercial scale, and it would seem that this could only be achieved currently by the production of fertile Pisifera clones because of the difficulty of working with material with no shell in the nursery. Working with shell-less material would require careful handling to prevent high emergence losses caused by fungal infection.

8.4.3 Tenera

Since Pisifera palms are often sterile, the development of Tenera populations is the best way to produce new Pisifera germplasm. Selfing of Teneras will produce Dura (*Sh/Sh*), Tenera (*Sh/sh*) and Pisifera (*sh/sh*) genotypes in a 1:2:1 ratio, respectively.

The most famous TxT progeny was that which produced SP 540 at Sungai Pantjur (originating from D'jongo, Eala, DR Congo). This palm was selfed to produce the progeny Pol 820 in 1931. In 1973 RISPA repeated the selfing of SP 540 and produce another generation of three progenies, all of which were planted at Aek Pantjur, Indonesia, in 1973. The Pisifera descendants from this are now used as males in commercial seed production by RISPA, Lonsum (through Dami material) in Indonesia, by Banting in Malaysia and by Dami in Papua New Guinea (Rosenquist 1985).

8.4.4 Inbred Lines

Inbreeding is the mating of individuals that are related, the most extreme form being selfing. Oil palm is a naturally outcrossing species, but controlled selfing is possible by storing pollen and using this to pollinate subsequent female bunches from the same palm. However, oil palm suffers from inbreeding depression, which can be evident in early generations. Selfed Tenera in Southern Zaire showed degeneration at the seedling stage, giving poor germination and irregular or deformed seedlings, while at Yangambi two selfed progenies gave bunch yields of 33 percent less than normal (outcrossed materials). In Nigeria, comparisons between selfed and crossed material showed reduce cumulative yields of the selfed materials by 13–14%. Data from the Ivory Coast also showed strong inbreeding yield depression (Gascon et al. in Hartley 1988).

In general, it has been found that inbreeding reduces yield but has a greater effect on bunch weight than on bunch number. However, there have been many instances of individual selfings showing little or no inbreeding depression, while in other cases inbreeding has tended to concentrate on either good characters (e.g. dumpiness) or bad characters (e.g. orange spotting and crown disease). The fact that there are large yield differences between selfed progenies suggest that some specific genes or genomic regions play a role in inbreeding depression (Rosenquist 1984).

Despite the difficulties of inbreeding, the development of inbred lines has been pursued in oil palm as these lines represent more homozygous genotypes that can be exploited in genetic studies and the production of F_1 hybrids (see Sect. 8.10).

8.4.5 Germplasm Acquisition and Collections

The need for extensive germplasm collection to broaden the genetic base of oil palm has become an imperative for oil palm breeding. One of the first organisations to start collecting oil palm from its natural centre of diversity was the West African Institute for Oil Palm Research (WAIFOR) based in Nigeria. Progress was interrupted by World War II, but collection was resumed in 1952. Collections were made from the Calabar, Aba and Ufuma regions of Angola starting in 1911, and the material was introduced to breeding programmes. In 1926 WAIFOR received Deli material from SOCFIN-Tanjung Genting and from AVROS, Malaysia. Later, Deli material was also received from Serdang Centre Experimental Station, Malaysia, and IRHO (Richardson and Alvarado 2003). More recently collections were carried out in a wider range of environments, e.g. the highland areas of Afikpo in Ebonyi State, Nigeria. Afikpo lies between latitude 5053′N and longitude 7056′E (Cochard et al. 2009; Okwuagwu 2011).

In 1973, Malaysia under the work of PORIM (currently known as MPOB) started collecting wild material from Africa and initiated surveys to collect oil palm genetic material from the entire centre of origin in Africa. Wild oil palm can be found in

Nigeria, Cameroon, DR Congo, Tanzania, Madagascar, Angola, Senegal, Gambia, Sierra Leone, Guinea and Ghana; all of these countries were explored (Rajanaidu and Jalani 1994).

Major oil palm collections were established at the Ghana Oil Palm Research Institute, as this was a research station of WAIFOR. Breeding programmes started in Ghana in 1967 using these materials. Recent collections were made in the northern Ghanaian regions of Vogoni Korri Nadowli, Vogoni Korri Forest Nadowli, Bugri Corner Koka Bawku, Saaka Bawku and Damango Canteen Bredi Farm (Oykere-Boateng et al. 2008; Sapey et al. 2012). In 2010 oil palm companies in Indonesia formed a consortium to collect materials from Cameroon.

Oil palm collections are typically maintained as living palms. Although oil palm is long-lived, trees will die eventually and living collections take up a lot of space. Conservation programmes can now exploit genetic fingerprinting techniques to reduce the amount of redundancy (duplication of genes) in collections. Oil palm may also be conserved as pollen, which may be stored for 10–20 years in a vacuum at −20 °C. Cryopreservation is also available in oil palm (see Sect. 8.11).

8.5 Crossing Approaches

8.5.1 Pollen Collection and Storage

Pollen is collected from isolated male flower bunches (Fig. 8.3). Inflorescence isolation usually takes place when one third of the spathe has opened. Frond spines should be removed (cleaned) and fronds bent downwards for easier manipulations.

Fig. 8.3 Male inflorescence selection and isolation

The spathes are cut open with a knife and sprayed with formalin (2 ml/L water) to kill any insects present.

Double terylene bags are used to isolate the male inflorescence. Ideally, the inner bag should be new and modified with the insertion of a pipe leading to a small plastic bag in one corner for pollen collection. The second bag is normally a previously used terylene bag. Insecticide granules are placed inside a cotton wool wad and placed round the inflorescence stalk. Old/used terylene bags are applied as an outer protective layer and tied with rubber rope. More than 8 days are then required for complete anther dehiscence (pollen shedding). If the male inflorescence undergoes anthesis in less than 9 days, the male inflorescence must be rejected as the legitimacy of the pollen cannot be guaranteed. Pollen remains viable for at least 6 days after release from the anther (Hardon and Turner 1967). The isolated male inflorescence is normally harvested when one third of the flowers of the inflorescence are open/anthesing. Flowers start to open from the base of the spikelet and usually all flowers will open within 2 days.

The anthesed male inflorescence is stored in a hot room (34–39 °C) for about 18 h to allow all flowers to open and shed pollen into the bottom corner of the inner bag fitted with a collecting bag. After storage in the hot room, the male inflorescence is hand beaten to allow all the pollen to fall downwards for collection. The pollen collected is sieved in a sterile box to separate the pollen from debris; the purified pollen sample is then put in glass ampoules, freeze-dried and vacuum-packed. The vacuumed pollen samples are stored in a freezer below 0 °C.

8.5.2 Measuring Pollen Viability

Pollen samples are normally taken at two stages for viability testing: before processing and after processing/storage. The first step is the preparation of the pollen germination medium (10 g sucrose are dissolved in 100 ml distilled water). The solution is cooled and 15 drops of 5% boric acid added.

Pollen samples are collected using the tip of a needle and stirred into a drop of the liquid culture medium on a microscope slide. A coverslip is then added. The sample is incubated at 34–37 °C for about 3 h and then examined under a microscope. The percentage viability (germinated pollen) is then calculated.

8.5.3 Isolation of the Female Inflorescence

Female flower bunch development is monitored prior to isolation (Fig. 8.4). The outer and inner spathes are removed, and the female inflorescence is sprayed with 2% formalin to kill any pollen and insects present. A wad of cotton wool containing insecticide is then tied to the stalk.

Fig. 8.4 Female inflorescence selection and isolation

After cleaning (removal of the spathes from the inflorescence), the female inflorescence is isolated using double terylene bags. The inner bag is new, and the outer bag is normally a recycled old one. At least 8 days are needed until the female inflorescence is receptive, fully opened flowers. If the female inflorescence becomes receptive in less than 9 days, the female inflorescence must be rejected as uncontrolled pollination may have occurred.

8.5.4 Pollination

Pollination is normally carried out in the morning when the flower's nectar is visible. All tools used for pollination are sterilised using 96% alcohol. Prior to pollination the area around the isolated female inflorescence is sprayed with insecticide. Pollen for pollination is mixed with talcum powder usually in a ratio of 0.03–0.2 g pollen (for pollen with >80% viability) with 1 g talcum powder). Pollination is affected by blowing the pollen/talc preparation into the bag by piercing a hole in the bag window, and the pollination may be carried out on two consecutive mornings to achieve maximum fruit set. The isolation bag is then shaken to help distribute the pollen over the whole inflorescence. The bag is resealed and a label recording the dates and parents is attached. At 21–25 days after pollination, the terylene bag is removed. The bunch is inspected for success (fruit set) 60–67 days after pollination.

8.5.5 Commercial Tenera (Dura × Pisifera) Production

Dura palms are used as seed/female palms, and the Pisifera are used as male/pollen palms. Pollen from selected commercial father palms is collected on a routine basis and stored ready for use. The crossing process is the same as that described above, but may include more rigorous procedures to ensure quality of the commercially produced seed, e.g. high pollen viability and legitimacy testing.

The seed production programme requires selected Dura and Pisifera. The Dura selection is based on progeny performance by DxP progeny trials (see Section on Trials). The Pisifera selection is based on individual Pisifera through Pisifera testing trials (see Sect. 8.9).

Most oil palm seed producers in the world use Deli Dura as mother palms, since they guarantee offspring with superior bunch and fruit characteristics. There are some research stations in Africa that do not use Deli Dura lines as female parents, because major objectives are to obtain *Fusarium oxysporum* wilt resistance and drought tolerance. Methods in seed production are described in Sect. 8.12.

8.5.6 *Elaeis oleifera*

Elaeis oleifera (E.o.) is a palm species originating from South America (Corley and Tinker 2003). Oil produced by *E. oleifera* contains greater oleic acid and linoleic than oil from *Elaeis guineensis* (E.g.) (Mohd Din et al. 2000). Oil of *E. oleifera* is often considered healthier because of its higher percentage of unsaturated fatty acids. Hybrids from E.g. × E.o. crosses have been planted for their resistance to fatal yellowing (Turner 1981) and their higher unsaturated oil properties.

Progeny from interspecific crosses between E.o. × E.g. are also expected to poses low height increment and short fronds (Escobar 2004). There have been various backcrossing programmes of hybrids to E.g. but the backcross generations show reversion back to E.g., and the desirable traits of unsaturated oil, low height increment and low vegetative growth are rapidly lost (Chin 1993; Oboh 1993 and Din and Rajanaidu 2000).

ASD Costa Rica has marketed compact palm cultivars and clones which were developed from an open pollinated hybrid between E.o. and E.g. (Alvarado et al. 2007), which had a short-trunked character but were low yielding. Sunilkumar et al. (2015) reported that hybrids of E.o. × E.g. had trunk heights of around 210 cm 12 years after planting and that the height trait was highly heritable. Some backcross progenies have been found with short stature and have comparable yields compared to conventional crossbreeding of E.g. (Escobar 2004; Alvarado et al. 2007) even though E.o. traits were lost through the backcrossing cycle.

8.6 Breeding Approaches

8.6.1 Recurrent Selection

The specific combining ability of a cross cannot be assessed without testing the progeny of that particular cross. In order to select for specific combining ability, a large number of cross comparisons must be made. Ideally a pair of lines that differs widely in gene frequencies at all loci that affect the character and show dominance is needed. It should be possible to build up these genetic differences by selection; thus, instead of the differences of gene frequencies being produced by the random process of inbreeding, they would be produced by the directed process of selection. This would be both more effective and more economical. Furthermore, both general and specific combining ability (GCA and SCA) would be selected simultaneously in a recurrent selection scheme.

Recurrent phenotypic selection tends to be more effective than mass selection. The basic outline of this process is illustrated in Fig. 8.5. A population is created by cross-pollination between two (or more) populations to create a base population. A large number of plants are grown from the base population, and a subsample of the most desirable phenotypes are identified and harvested as individual plants. These selected plants are randomly mated to produce a new improved population. This process is repeated a number of times.

The number of cycles performed will be determined by the desired level of improvement required over the base population, the initial gene frequency of the base population and the heritability of the traits of interest in the selection process. Recurrent phenotypic selection has been shown to be effective but mainly in cases where there is high heritability of the characters being selected for (e.g. some disease and pest resistance) and in species with short (annual) life cycles. The technique is less effective where traits have a lower heritability such as yield or quality traits (Brown and Caligari 2008).

Results of recurrent selection in oil palm by Socfindo at Aek Loba showed genetic progress of around 14% for oil production (selection of the best 8% of

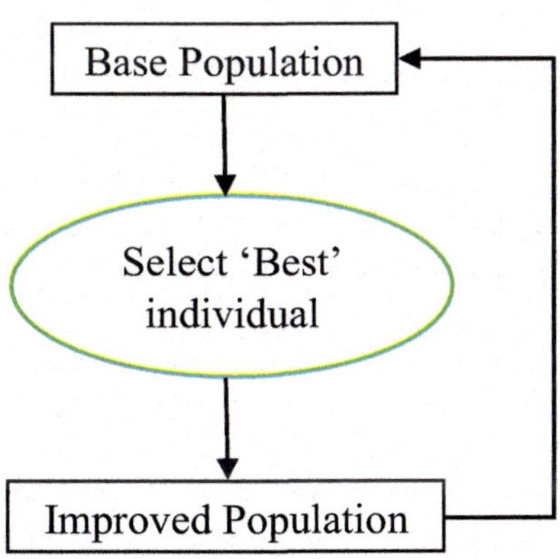

Fig. 8.5 Basic recurrent phenotypic selection scheme

crosses). However, given the highly additive nature of the transmission of production traits, by basing selection on the GCA of the parents, and even with a low selection rate (16%), it is possible to predict greater progress (17–18%, see also Durand et al. 2011).

Other trial results estimate 18% improvement in oil production after the first two cycles of recurrent selection. In the Ivory Coast, certain commercial plots planted between 1962 and 1972 were replanted from 1983 onwards with seed material obtained from the second recurrent cycle. Progress of 24% on average was seen for bunch production (Cochard et al. 1993).

Selection by IRHO in their recurrent selection programme involved a varied set of parameters in addition to yield traits. This included improvement of resistance to vascular wilt disease, reduction in height increment, ease of harvesting and an increase in unsaturated fatty acid content in the oil (Gascon et al., in Hartley 1988).

8.6.2 Early Mass Selection

There are great genetic differences in the material used for breeding in Africa compared to SE Asia since commercial establishment in the 1920s. The approach to improvement in the two regions has been different. In Africa, the poor quality of Dura fruit was apparent, and there was some hesitance in breeding for Tenera palms before the establishment of high-quality Tenera material. In SE Asia, the relatively high quality of Dura (emanating from the Deli ornamentals originally used as avenue trees) tended to confine very early work to mass selection in providing seed for further planting. In 1922 AVROS affirmed that the Deli type must remain as the standard oil palm for Sumatra until superior lines were bred (from newly imported materials). Under these circumstances, the general procedure was mass selection to provide progenies for extensive plantings, but that separate breeding programmes should be established to exploit outstanding individuals and exotic germplasm.

8.6.3 Dura Breeding

Seed gardens in SE Asia are now composed of Deli Dura female palms, but in Africa Deli Dura are not used automatically because of their susceptibility to vascular wilt disease. African Dura are being developed to match Deli Dura in terms of yield.

A number of plantation companies started oil palm breeding programmes independently of each other. The Marihat Research Station (RISPA) has an excellent history of oil palm breeding in Indonesia (see Sect. 8.2). In 1927 SOCFINDO established a Deli Dura breeding programme by planting legitimate progenies from selected palms at Mopoli, North Sumatra, Indonesia, and comprised 1859 oil palm trials. In 1920s in Malaysia, various Dura breeding programmes were carried out by the Department of Agriculture, Elmina selections, Guthrie Chemara, Banting and others (Rosenquist 1985).

Prior to its introduction in Costa Rica, the Deli Dura germplasm had undergone continuous selection and improvement since the early 1900s, first in Indonesia and later in Malaysia. Thus, the Deli Dura population introduced to Costa Rica was already improved compared to the original Deli Serdang population. This was also evident for the introductions to Indonesia that came from Dami, Papua New Guinea, as these had undergone two selection cycles from the original Banting and Ulu Remis breeding populations. The whole process guarantees the genetic stability and production potential of the female population and their descendants during the process of commercial seed production. The Dura breeding programme in ASD Costa Rica used various materials from Angola (from Bamenda, Kigoma and Yangambi regions). These materials were used in intra- and interpopulation crossings to develop superior Dura (Sterling and Alvarado 2002).

8.6.4 *Pisifera Breeding*

Because of their female sterility, Pisifera generally cannot be selected directly; instead, they are selected indirectly through progeny testing. In this scheme a number of provisionally selected Pisifera are crossed with several Dura. An additional reason for progeny testing is to confirm that the female-sterile palms are truly Pisifera as occasionally sterile Dura and Tenera can occur, and it is of course essential that these are recognised and eliminated from the seed production. After progeny testing, the best Pisifera are identified and selected as male parents for seed production. However, shortage of male inflorescences and hence of pollen may limit their use (Corley and Tinker 2003; Sterling and Alvarado 2002).

8.6.5 *Mutation Breeding*

Mutation breeding has proven to be a valuable tool in crop improvement with over 3000 officially released mutant cultivars in over 200 crop species (IAEA mutant database, http://mvgs.iaea.org). However, the list of mutant cultivars does not include any in oil palm, which is the only major crop (and the only oil crop) not to be improved by mutation breeding. Mutation breeding has been considered by Ishikawa et al. (2012). Mutation breeding generally aims to improve an elite line or superior cultivar for a given trait, e.g. yield, quality, disease resistance, pest resistance, tolerance to abiotic stress, etc., and offers a shortcut in the breeding process. Mutants may be induced by biological, chemical or physical mutagens though the latter, and especially gamma irradiation, has had the greatest impact in plant breeding with over 60% of mutant cultivars being developed from gamma ray treatments (Bado et al. 2015). Mutation breeding is a non-GM method that has been accepted and practiced as a conventional breeding method for many decades; the first mutant cultivar was a chlorina mutant in tobacco, released in 1936 in Indonesia (then the Dutch East Indies). Recent reviews and methods in plant mutation breeding can be found in Shu et al. (2012) and Bado et al. (2015).

In 1977 Dr. Wonky Appiah (Director of Oil Palm Research Institute, OPRI, Ghana) initiated work on mutation breeding in oil palm in collaboration with the Ghana Atomic Energy Commission (GAEC). The main purpose at the time was to use gamma treatments to promote germination of seed and pollen. However, a consequence was that OPRI produced the first induced mutant population (M1) in oil palm. Since most mutations are recessive and since M1 plants often suffer from chimeras and physiological disorders, it is normal practice to develop M2 and subsequent generations to screen for heritable mutant traits (Ukai and Nakagawa 2012; Prina et al. 2012). Oil palm takes 3–5 years to flower from planting a seed, i.e. at least 3–5 years to produce the M2 from the M1. The Ghana M1 population currently represents a unique resource for oil palm improvement. Pollen collection and storage of M1 palms re-established at OPRI in 2010. Subsequent work involved self-pollination to produce M2 generations and M2 seedling screening began in 2014. This work was supported by the International Atomic Energy Agency (IAEA) and Sumatra Bioscience. The irradiation treatments used in 1977 were relatively low for mutation induction purposes (dosage 1–5 Krad or 10–50 Gy), and in general the M1 palms appeared phenotypically normal. However, this may not necessarily be a setback as some desirable mutant traits, e.g. semi-dwarfism, are often present in relatively high frequencies (Forster et al. 2012), and mutant traits are expected to emerge in the near future.

Traits of interest, such as those required for mechanical harvesting (virescent fruit, long bunch stalks, fruit abscission, etc.), may be screened for at maturity in the M2 populations. Other traits of interest to Ghana are *Fusarium* wilt resistance and drought tolerance. In other parts of the world, target traits include *Ganoderma* resistance and improved oil quality (it should be noted here that oil palm is the only oil crop where oil quality has not been improved by mutation breeding (Vollmann and Raean 2010).

In addition to phenotypic screening, genotypic methods could be developed to screen mutants for desired traits. This is now more feasible in oil palm as the genome has been sequenced and genes of interest are being discovered, e.g. shell thickness, virescent and mantled fruits (see Sect. 8.11). In addition, analogues in other species, e.g. genes involved in oil biosynthesis, may be exploited to seek new variation for oil quality improvement and novel oils. Oil palm mutation breeding is in its infancy, but the enhanced capacity to detect and screen for mutations is likely to encourage more mutation breeding projects in oil palm, including gene editing technology. It is expected that mutation induction methods will be optimised and new mutant populations developed in oil palm (see Sect. 8.13).

8.7 Target Traits for Breeding

8.7.1 Fruit Type

Oil palm fruit types are mainly based on shell (endocarp) thickness and fruit colour. Shell thickness is controlled by a major gene, *Sh*. The gene for shell thickness was first reported in the Belgian Congo in the early 1940s (Beirnaert and Venerweyan 1941); depending on the genetic constitution, three phenotypes are produced:

- Dura: which has a thick shell, thin mesocarp and a relatively large kernel. The mesocarp has a rich oil content. Dura are generally used as female parents in oil palm hybridisations. The Dura genotype is *Sh*/*Sh* (homozygous dominant).
- Pisifera: this type has a thin shell or no shell, the mesocarp is relatively thick and the small kernel is surrounded by traces of a fibre ring. The mesocarp has a low oil content. Pisifera are generally used as male parents in crossing programmes. The Pisifera genotype is *sh*/*sh* (homozygous recessive).
- Tenera: a hybrid of Dura and Pisifera. The fruit is characterised by a thick mesocarp and a large kernel surrounded by a thin shell with a fibre ring. The mesocarp has a high oil content. The Tenera genotype is *Sh*/*sh* (heterozygous).

Abnormal fruits, mantled fruits, are a concern in oil palm. Mantled fruits are the result of a physiological disorder which has mainly been seen as a result of somaclonal variation This is a serious concern for the oil palm business and is discussed later (see Sect. 8.11).

Oil palm fruit colours include virescent, nigrescent and albescent types (Hartley 1988; Singh et al. 2014). The most common fruit type generally planted is nigrescent: deep violet to black when immature and red when ripe. Virescent types are relatively uncommon; here the fruit is green and turns orange when ripe. There is currently great interest in virescent fruit as the colour change at maturity is more distinct than nigrescent fruits; thus, the harvesting of ripe fruit bunches (and thereby high oil content) is simplified. The virescens gene (*Vir*) is dominant and is a natural mutant allele of nigrescens (Singh et al. 2014). Albescent fruit is white at maturity, the mesocarp lacks carotene and oil content is low; currently, albescence is of little interest to breeders (Corley and Tinker 2003).

8.7.2 Vegetative Growth/Flowering Time

Oil palm like many tree crops has a long juvenile stage, 3–5 years from germination to flowering. Modern cultivars start to produce flowers 24 months after planting seedlings. The ancestral Dura planted in Bogor were reported to take 4 years to flower from planting (Hartley 1988). Although there has been selection for precocity, the underlying genetics is not understood. Oil palm produces separate male and female inflorescences (monoecious); these are formed in each frond axil.

Sex ratio is of concern as harvestable products (fruits) arise from female bunches. It is known that long dry seasons favour the formation of male flowers (Corley and Tinker 2003). Other stresses, e.g. excessive pruning, can also diminish the number of female inflorescences and thereby yield (Adam et al. 2011; Hartley 1988). Currently, there is little information on the genetic controls of male/female flower production in oil palm, and as yet there is no variation identified for this trait and consequently no breeding work.

8.7.3 *Yield*

As discussed above, oil palm yield is linked to flowering, which in turn is influenced by the environment and by agronomic practices. Yield in oil palm is often discussed in terms of fresh fruit bunch (FFB, Cock et al. 2014). Bunch analysis is a major activity from which yield and yield components (CPO, PKO, etc.) are derived. CPO content of cultivars ranges from 22% to 30% and about 5% kernel oil (Berger 1983; Sambanthamurthi et al. 2000; Cadena et al. 2013). In order to obtain good data on yield and yield components, it is important that bunches are fully ripe when harvested and that harvesting techniques are standardised (Rao et al. 2001). Yield is a polygenic trait, controlled by many genes (Corley and Tinker 2003) and thus more challenging to breed for.

8.7.4 *Quality, Fruit and Kernel Oil*

In addition to fatty acids, palm oil contains many nutrients such as vitamin E, tocopherols, tocotrienols and carotenoids (Gibon et al. 2007; Dauqan et al. 2011) that function as antioxidants (Tang et al. 2015). Red (crude) and yellow (refined) palm oil contain about 1 g tocopherol and tocotrienol in 1 kg of oil; these protect the oil from oxidation at high temperatures such as deep-frying and other cooking processes, resulting in healthier food (Schroeder et al. 2006).

Palm oil quality is generally determined by the content of deleterious free fatty acids (FFAs); this is because FFAs are very easily oxidised (Maizura et al. 2008). FFA content increases shortly after the bunch is harvested (Corley and Tinker 2003). FFA increases with fruit damage caused by bruising, e.g. falling to the ground during harvest as well during transport to mills (Arifin and Ai 1989). FFA increases with time and therefore rapid processing is required to minimise FFA content. FFA content rises due to the activity of endogenous lipase enzymes (Morcillo et al. 2013). Low lipase activity is therefore of interest in breeding (Maizura et al. 2008).

8.7.5 Diseases

Prior to the World War II, oil palm was regarded as being free from any serious disease, but minor diseases included crown disease, bud rot and stem rot (Hartley 1988). Today, stem rot in SE Asia and wilt in Africa have become major disease concerns. Stem rot disease is caused by the soilborne fungus *Ganoderma boninense*; wilt disease is caused by the soilborne fungus *Fusarium oxysporum* (Turner 1981). In 2003 stem rot disease was reported to cause up to 30% yield losses; today, this can be up to 80% (Corley and Tinker 2003; Flood et al. 2000; Virdiana et al. 2011). At first, the disease was restricted to old palms and had little impact on yield (Turner 1981). But today, the disease is more common and attacks occur earlier in younger palms, resulting in significant yield losses (Flood et al. 2000). Disease symptoms are similar to drought, i.e. leaf spears fail to open, turn brown and die (Ariffin et al. 2000; Corley and Tinker 2003). This may be due to stem decay preventing water flow. In the end, *Ganoderma* fruiting brackets appear on the trunk, and the palm dies or is uprooted by the wind. Currently, the disease is managed by agronomic practices such as removal of remnants of the previous crop, treatment with antagonists (e.g. *Trichoderma* spp.) and in worst cases replanting (Virdiana et al. 2011). There is a search for genetic sources containing *Ganoderma* resistance/tolerance for breeding. Some partial resistance is available in contemporary breeding materials (Setiawati et al. 2010). Progenies derived from La Me material are reported to have low *Ganoderma* infection rates (Turnbull et al. 2014), and Idris et al. (2004) reported that DxP progenies derived from Congo × Cameroon material possessed partial resistance to *Ganoderma*. Screening methods rely on seedling responses to soil inoculated with the disease (usually *Ganoderma*-infected rubber wood blocks), followed up by field trials.

Wilt disease is mostly found in parts of Africa, particularly West Africa. This disease causes wilting, stunting and death (Turner 1981). *Fusarium oxysporum* is a soilborne fungus (Flood 2006) and will attack and enter plants through their roots and then enter and thrive in the xylem (Hubert and Sekou 2004). Flood and Mepsted (1990) described the symptoms of wilt disease. In a severe attack, plant death can occur rapidly, within 2–3 months after the first symptoms appear. Several factors, such as age, the crop cycle and the level of vulnerability of planting material, determine the severity of the symptoms (Hubert and Sekou 2004).

Various approaches have been taken to overcome wilt disease, e.g. various agronomic practices and breeding. Ntsomboh-Ntsefong et al. (2015) reported two *Fusarium*-tolerant progenies in material from Cameroon (with short stature). Ngado-Ebongue et al. (2013) found that progenies derived from crossing Ekona × Ekona and Ekona × Binga showed tolerance to wilt disease. As mentioned earlier interspecific crosses between *Elaeis guineensis* and *Elaeis oleifera* are also of interest in breeding disease-resistant planting materials.

8.7.6 Height

Oil palms can reach a height of about 40 m. Height is influenced by environmental factors, but has a strong genetic component. Palm height is closely related to the productivity of the palm and is a limiting factor in oil palm cultivation, with palms becoming too tall to harvest after 20 years.

Genetic variation exists for palm stature, e.g. short-trunked oil palms, Pobe dwarf and Dumpy types (Rosenquist 1985; Luyindula et al. 2005). However, these types are also associated with low yield, and there is a breeding challenge to produce a new type of oil palm which combines high yield with short stature to extend the economic life of oil palm plantations. Today, most plantations are felled 20–25 years after planting (Corley and Tinker 2003) when trees become too tall to harvest. Dwarfism and semi-dwarfism may be induced by mutation breeding (see Sect. 8.10).

8.7.7 Frond Length

Frond length, and hence canopy width, has an impact on palm spacing or plantation density and thus harvest potential. Frond length of most contemporary oil palm cultivars is between 5 and 9 m, for which the optimum number of plants is around 143 palm/ha. Higher densities, e.g. 186 palm/ha, can lead to increased abortion of female inflorescences (due to stress) and reduced yield (Breure et al. 1990).

Theoretically, a reduction of 1 m in frond length would allow an increase in palm density to 170/ha. A reduction of 2 m would allow 200 plants per hectare (Alvarado et al. 2007). In theory, an increase in density from the standard of 143 to 160 palm/ha could provide a 12% yield increase (Alvarado et al. 2007).

8.7.8 Mechanical Harvesting

Much of the work in oil palm plantations is manual. One of the jobs that needs a large work force is bunch harvesting; this is both inefficient and incurs large labour costs. Mechanised harvesting is therefore of interest (Nelson et al. 2006). There are a number of traits that could be exploited to aid the development of mechanised harvesting.

Bunch stalk length (Fig. 8.6): Currently most commercial palms have a short bunch stalk. A longer stalk would be more amenable to mechanical harvesting as it would be easier to cut the bunch off. Genetic variation exists for long bunch stalk and this being developed in many research stations and breeding programmes (Priwiratama et al. 2010). The colour change provided by virescent and late fruit abscission are other traits of interest in breeding for palms that may be harvested mechanically.

Fig. 8.6 Bunch stalk length variation: (**a**) long stalk; (**b**) medium stalk; (**c**) short stalk

8.7.9 Biofuel

World demand for energy is expected to increase and oil palm has potential as a biofuel crop (Nagi et al. 2008). Interest in new and more diverse oil palm products is also of interest, and a wide market for oil palm products provides greater stability for producers. One such emerging market for palm oil is as a biofuel (Mekhilef et al. 2011).

Additional sources of biofuel from oil palm other than that derived from CPO include bioethanol, bio-methanol, bio-briquettes, hydrogen and pyrolysis oil. Empty fruit bunches and palm press cake are oil palm by-products that contain cellulose that may be fermented to produce bioethanol (Shuit et al. 2009; Piarpuzan et al. 2011; Guitierrez et al. 2009). The oil palm trunk also has a high sugar content which may be tapped as a source for bioethanol production (Corley and Tinker 2003; Kosugi et al. 2010).

8.8 Field Trialling and Selection

8.8.1 Field Trials

In order to evaluate germplasm and material generated by breeding, breeders will normally carry out field trials. Field trials are designed to evaluate and screen/select for various traits including:

1. Yield
2. Parent selection (Dura and Pisifera)
3. Progeny testing
4. Suitability for production in different locations
5. Resistance/tolerance to pests and diseases
6. Specific characters (novel traits), e.g. ease harvesting, long stalk and late abscission

Oil palm breeding trials are usually planted using standard statistical designs (completely randomised or randomised complete block design) with various plot sizes and shapes; 8 or 16 palm plots are common in order to reduce bias caused by inter-palm competition.

Field trials take into account the crossing scheme. Once the crossing programme is completed, successful crosses can be evaluated. Field sites are selected which are homogenous in terms of soil fertility, drainage and topography, and for disease testing, e.g. *Ganoderma* trials, the area should be infected with the pathogen. Once the field site has been selected, the layout of field planting is set up based on the trial design. Individual palms of a progeny are planted into the right plot location and tagged to allow identification for subsequent scoring.

A palm census will normally be carried out 1 year after planting with the aim of replacing any dead/missing palms with equivalent material. Thus a reserve of seedlings is normally kept in the nursery. A number of standard traits are recorded from the field trial; these may be divided into:

(a) Yield: bunch number, bunch weight
(b) Growth: frond weight, frond length, leaf area, height, diameter
(c) Pest and disease: *Ganoderma*, *Fusarium* wilt
(d) Genetic disorders: crown disease
(e) Bunch characters: oil/bunch ratio, kernel/bunch ratio, kernel oil
(f) Oil quality: fatty acid profiles

Performance in different environments is achieved by replicating trials in a number of environments and thus also allowing progenies to be selected for more marginal environments and those that are adapted to a number of environments.

In oil palm breeding, there is a difficulty in knowing the genotype for shell thickness as this is only apparent at maturity (phenotypic screening) and will segregate differently depending on the cross. Thus TxT crosses give 1D:2T:1P; TxP will segregate 1T:1P; and TxD/DxT will segregate 1D:1T. It would be desirable to plant out plots containing only D, T or P genotypes, and since Ps are sterile, it would be preferable to remove these completely from yield trials. Today the genotype for shell thickness can be easily screened for using genetic markers taking DNA samples from seedling plants; thus D, T and P determinations can be made prior to field planting (see Sect. 8.9.4).

8.8.2 Bunch Analysis

Methods for analysing oil content in oil palm fruit bunches were set up at the West African Institute for Oil Palm Research (WAIFOR) in the early 1960s (see Rao et al. 1983 in Blaak et al. 1963). Over time, modifications have been made to improve the process (Blaak 1970; Rao et al. 1983; Junaidah et al. 2011).

Bunch and oil analysis activities begin with the harvesting of bunches according to ripe harvesting criteria. They are then weighed and fruit type determined (Dura, Tenera or Pisifera) by visual assessment. The bunch is then chopped up to separate the spikelets from the stalk. The stalk and a sample of spikelets are then weighed and the number of fruit per spikelet determined. The number of parthenocarpic and undeveloped fruits is noted and the total fruit set number determined.

A sample of 30 fruits is weighed and scraped to separate the mesocarp from the seed (nut). The mesocarps and seeds are weighed separately and dried in an oven at 105 °C for 24 h. Seeds are then dried in an oven for 8 h for Tenera seeds and 10 h for Dura seeds.

After drying according to standards, the dried mesocarp is cooled before grinding and sieving. After sieving, the mesocarp samples are dried in an oven for 30 min. A sample is then treated with an organic solvent such as hexane for 24 h to extract the oil.

The dried seeds are cracked open to separate the kernels and shells. There is normally one kernel per seed, but up to four can be encountered, but very rarely. The total kernel weight is obtained before and after drying.

Data are gathered for mesocarp/fruit (M/F) , oil/dry mesocarp (O/DM), shell/kernel (S/K), shell/fruit (S/F), oil/bunch (O/B) and kernel/bunch (K/B) ratios. These data are very useful to breeders in making selections.

8.8.3 Nursery Screening

Some traits can be screened for in the nursery. These include crown disease (a physiological disorder of young plants) and *Ganoderma* tolerance/resistance (see above). With respect to *Ganoderma*, germinated seeds are placed into plastic bags filled with soil in which a *Ganoderma*-infected rubber wood block is placed (Breton et al. 2009). Symptoms of the disease are then scored.

8.8.4 Genotypic Screening

Screening may be divided into two main classes: phenotypic and genotypic. Phenotypic indicators include morphological markers, e.g. flower colour, plant height, flowering time, etc., and also biochemical markers such as protein and isozyme variation (usually detected by protein banding patterns in gels). For plant breeding purposes, any marker system that is cheap and reliable is good, but there has been a massive development in DNA markers in recent years (Moose and Mumm 2008). DNA markers can be used to detect specific gene alleles either by linkage to the gene of interest or directly in detecting allelic (sequence) variation among alleles. DNA can also be used to fingerprint the genetic background. Thus, a breeder attempting to transfer a disease resistance gene into an elite genotype may wish to deploy markers for the disease resistance gene but also use genetic fingerprinting so that he/she protects the elite genetic make-up as much as possible.

Genetic markers have huge potential for oil palm not least because many commercially important traits are those of mature plants (shell thickness, fruit colour, oil content, etc.), and without genetic markers, it would take at least 5 years from sowing to assess and select the desired types. Using the marker technologies, breeders are able to make selections in the laboratory. Genotypic selection allows plants to be selected before they are planted in the field. For example, these tests can be done on nursery plants. Other traits of interest are listed in Sect. 8.8. Several marker systems are available, but a major bottleneck in genotyping is DNA extraction, and for breeding purposes, this usually involves several hundred individuals. DNA can be extracted using kits in automated or semiautomated systems or more tediously via manual preparations. Over 100 individual DNA extractions can be performed routinely in a day in designated laboratory. The major disadvantages of automated systems are that they are expensive to set up and run.

The release of the oil palm reference genome sequence in 2013 (Singh et al. 2013a) was a major step forward in the development of DNA markers for oil palm breeding. For some important traits, e.g. shell thickness, virescent fruit and mantled fruit commercial kits and services are becoming available (https://www.orionbiosains.com/about-orion).

8.9 Hybrid Breeding

8.9.1 Introduction

F_1 hybrids have been shown to produce spectacular yields compared to conventional cultivars. Success has been achieved in a number of outbreeding species, and F_1 hybrids are major goals for many plant breeding programmes. The prime example is maize, which like oil palm is a monoecious crop. F_1 hybrids in maize have produced a sixfold increase in yield. Although oil palm is an obvious candidate to benefit from F_1 planting materials, it has lagged far behind due to an inability to produce the homozygous parental lines required for F_1 hybrid breeding. However, it is estimated that F_1 hybrid oil palm would yield up to six times that of current average yields, i.e. from 3 tonnes crude palm oil/hectare/year (FAO statistics) to 18 tonnes/hectare/year, thus allowing a significantly reduced planted area to obtain the same yield. The parental homozygous lines may be produced by inbreeding, but as described above, this is a long process taking 20–30 years to produce the F5 generation (about 97% homozygous). A more attractive option is to produce homozygotes via doubled haploidy, the production of doubled haploids from haploids. Unfortunately, the standard in vitro methods of haploid production, such as pollen or microspore culture, used widely in other crops (Maluszynski et al. 2003) have proven recalcitrant in oil palm. A major breakthrough came with the discovery that haploid seedlings occurred naturally in oil palm albeit at extremely low frequencies and can be identified effectively (Nelson et al. 2009). Prescreening of 'off-type' seedlings (source material), the application of DNA markers (zygosity testing) and flow cytometry (ploidy detection) can be used to boost haploid production levels that rival that of other crops.

Haploids (Hs) are the precursors of doubled haploids (DHs), and their conversion to DHs may be spontaneous or induced. Tissue culture methods have been used to affect in vitro genome doubling in a range of crop plants (Maluszynski et al. 2003; Touraev et al. 2009). The doubling of haploid genomes creates completely homozygous lines in one step, from which parents may be selected for subsequent commercial F_1 hybrid seed production. Not all parental DH combinations will result in hybrid vigour (heterosis), and as in maize, desired parental combinations need to be identified. Genotypic analysis can aid the determination of heterotic groups. Since contemporary commercial oil palm has thin-shelled fruits (Tenera) and is produced by crossing thick-shelled (Dura) with shell-less (Pisifera) types (see Sect. 8.12), DHs need to be produced in both shell fruit types to produce F_1 Tenera cultivars, and again these types can be identified using DNA markers (see Table 6.1). The processes used to produce Hs, DHs and F_1 hybrids in oil palm are novel; they are based on the detection of naturally occurring haploids and do not involve GMO technology.

8.9.2 The Importance of F_1 Hybrids

F_1 hybrids are important in agriculture as they display hybrid vigour and heterosis (Jones 1997; Coors and Pandey 1999), which can be harnessed in raising yields. This has been a successful approach in many crops, notably maize, pepper, rice, rye, sorghum, sugar beet and sunflower (Coors and Pandey 1999). Heterosis is brought about by crossing two homozygous, but contrasting parental lines, which produce uniform, heterozygous F_1 progeny. Completely homozygous lines are ideal parents, and these can be obtained most conveniently through doubled haploidy. The starting materials for doubled haploids are haploids. Haploids are plants which carry the gametic rather than the normal sporophytic chromosome complement. Thus, normal oil palms are diploid with a chromosome complement of $2n = 32$; a haploid plant has the chromosome complement of $n = 16$ (which is the normal number for gametic cells, such as the sperm and egg). Haploid plants in themselves are of little interest in plant breeding as they are usually weak, non-viable and sterile. However, the doubling of the haploid chromosome complement to form DHs duplicates the genetic constitution (to restore the usual diploid ploidy level) and produces completely homozygous (genetically pure) lines. In inbreeding species, e.g. barley, rice and wheat, DH lines are true breeding, stable and uniform and can be developed directly and rapidly as cultivars for commercial exploitation. DHs are also used as parents to produce commercial F_1 seed. With the right parental pairing, the progeny of two DHs maximises heterozygosity and can result in hybrid vigour and a marked increase in yield. The F_1 progeny is also uniform, which allows the development of optimal agronomic practices, which in itself has led to remarkable yield increased in other outbreeding crops, notably maize (Crow 1998; Duvick 2001; Troyer 1991). Seed of F_1 cultivars is sold at a premium (which in part reflects its costlier production), and since it must be produced for each crop sowing, it creates a lucrative market for F_1 seed producers. F_1 hybrids are therefore highly prized to both the producers and the growers, and methods for converting crops to F_1 hybrids are of major interest. F_1 hybrids have been developed in many crops (Brown and Caligari 2008). Recent examples include rice (commercialisation started in 1996 in China, see Li et al. 2009), rye (hybrid rye now occupies about 70% of the rye hectarage in Germany, see Geiger and Miedaner 2009 for a review on rye breeding) and most recently wheat, barley and triticale (Longin et al. 2012). The possibility of converting oil palm into an F_1 hybrid crop became feasible only recently with the discovery of haploids and their detection using high-throughput biotechnologies (Nelson et al. 2009, Nasution et al. 2013).

8.9.3 Haploid/Doubled Haploid Techniques

Haploids occur naturally and have been reported in many species; however, the frequency is very low (about 1 in 100,000 seedlings) and has often been considered too low for practical purposes. The history of H/DHs in plant breeding and genetics began with the report of A. D. Bergner in 1921 (referred to in Blakeslee et al. 1922) that haploid plants could be found in Jimson weed, the first in a higher plant. Similar discoveries were made soon after in other angiosperms (reviewed by Kimber and Riley 1963). Chase (1952) is generally regarded as being the first to exploit Hs and DHs in plant breeding (of maize): the haploids were produced via parthenogenesis (Chase 1949). A major step forward came with the discovery that haploids could be generated from cultured anthers (Guha and Maheshwari 1964), and today in vitro methods are used routinely in breeding of a range of crop plants including maize (Touraev et al. 2009; Geiger and Gordillo 2010). Haploids can be artificially produced in large numbers using a range of methods that stimulate gametic cells to develop into embryos and/or plantlets. Today, the main methods of H/DH production are (1) wide crossing in which aberrant pollination/fertilisation results in egg cells developing into haploid embryos and (2) the culture of male or female gametic cells and tissues such as microspores, anthers and ovules in which normal development (into pollen and egg) is diverted into H/DH embryo production. These methods have been exploited in the development of protocols in over 200 plant species ranging from *Aconitum* to *Zingiber* (Maluszynski et al. 2003). The impact of doubled haploidy in plant breeding has been reviewed by Forster and Thomas (2005); the mechanisms involve are not fully understood but discussed in Touraev et al. (2009).

8.9.4 Early Attempts in F_1 Hybrid Production in Oil Palm

The commercial oil palm crop is Tenera (thin-shelled) and is produced as hybrids from Dura (thick-shelled) and Pisifera (no-shell) crosses (see Sect. 8.12). Seed production of Tenera types for commercial planting is a major business. The crop is therefore heterozygous and like maize is monoecious and open pollinated. Thus oil palm is well set up to be developed as an F_1 hybrid crop. As a consequence, there has been considerable interest in developing homozygous lines as parents for F_1 hybrid production in oil palm. F_1 hybrids can be achieved in maize by crossing inbred lines (produced by several rounds of artificially imposed self-pollination (see Sect. 8.7). This has been attempted in oil palm in several countries, such as Indonesia, DR Congo, Nigeria and Malaysia (Corley and Tinker 2003; Nelson et al. 2009; Wening et al. 2012a), but has largely been abandoned as oil palm suffers from inbreeding depression (many seedlings are weak and sterile), and the timescale in reaching homozygosity is daunting (15–40 years, due to the fact it takes 3–5 years from sowing a seed to getting the first seed of the next generation). In addition, this would need to be done for both Dura (female) and Pisifera (male) lines, if commercial Tenera F_1s are to be produced. Pooni et al. (1989) proposed a compromise, i.e. the production of superior lines and second cycle hybrids by inbreeding and selection. The superior parental lines are derived from F2 or F3 generations, i.e. only partially inbred, which would take 10–15 years to develop, and have up to 75% homozygosity. However, this idea would become redundant if doubled haploidy could be achieved in this timescale.

8.9.5 Haploid Screening in Oil Palm

There have been several attempts to apply haploid induction biotechnologies to oil palm, e.g. anther and microspore culture, but no haploid plants have yet been produced using these methods (Jones 1989; Dunwell et al. 2010; Kumar and Sparajanbabu 2013). Alternative methods have therefore been sought. Given the fact that haploids occur naturally, including in other palm species, e.g. coconut (Whitehead and Chapman 1962), it was surmised that haploid seedlings may be found in oil palm and that these may look morphologically different from normal diploid seedlings. Thus, 'off-type' seedlings that are normally discarded in commercial seed production were screened; these included twin seedlings, poor radicle and/or shoot emergence and low vigour. The frequency of haploids in 'off-type' seedlings is about 1 in 1000 (Nelson et al. 2009). Seed of commercial Tenera is sold pregerminated with a quality controls for normality and uniformity, and so discarded 'off-type' seed could be readily screened for haploids. Such screens began in 2004 and involved the selection of a range of 'off-types' for homozygosity using genetic markers and ploidy using flow cytometry and chromosome counts using microscope (Nelson et al. 2009; Dunwell et al. 2010). The discovery that spontaneous haploids did occur in oil palm was exciting and the screening process was scaled up, such that by 2009 over 95 million germinated seeds were screened with 1184 haploid seedlings being identified from 124,094 'off-types' (Nelson et al. 2009; Dunwell et al. 2010). The ability to screen for haploids among the abnormal seedlings greatly increased the efficiency of haploid detection. This is currently the only reported method of producing haploid plants in oil palm. Since haploids were screened in seedlings from commercial seed production and breeding crosses (Nelson et al. 2009), which covered a wide range of genotypes (Wening et al. 2009, 2011, 2012b), the haploid screen is thought to be relatively genotype independent, and haploids include both Dura and Pisifera types. Another major step forward was the deployment of high-throughput flow cytometry.

A high-throughput flow cytometry method for ploidy determination in oil palm has been developed (Nasution et al. 2013). This allows the analysis of about 1000 samples per day per flow cytometer and the detection of 1104 haploids from 386.787 'off-type' seedlings in a 4-year period from 2006 to 2010. This level of haploid production rivals other methods used in plant breeding in other crops. The method is practical in oil palm as there is an abundant supply of seeds and seedlings, but may also be applicable to other species that produce large amounts of seed.

Haploids of oil palm can be detected reliably in bulks made up of four to five individuals; root samples from four to five 'off-type' seedlings are bulked and chopped up in commercial Cystain®UV Ploidy solution with the addition of dithiothreitol (DTT) and polyvinylpyrrolidone (PVP), which are required to reduce the accumulation of secondary metabolites, such as phenols which interfere with the ploidy analysis. The preparations are then incubated in the dark at 5 °C and filtered. The suspension obtained is stirred vigorously and presented for flow cytometry. The analysis per sample takes about 1 min. In cases where a haploid peak is detected in the output histogram, the materials comprising that bulk are analysed individually to identify the haploid seedling.

8.9.6 Doubled Haploid Production in Oil Palm

Haploids in themselves are of little value in breeding as they tend to be weak and sterile. Their value is as a precursor to DHs, and therefore methods of doubling the chromosome complement are needed. This can be done in vivo or in vitro. Methods for doubling haploid plants of oil palm were investigated by Mienanti et al. (2009). These included submerging roots of haploid seedlings in a solution containing colchicine at various concentrations (2.5–10 mM) for 5 h. Various wetting agents and plant hormones were also investigated. Leaf materials from fronds emerging after treatment were sampled for ploidy determination. Most treatments resulted in $n/2n$ chimeras. The addition of the plant hormone, gibberellic acid (GA3), was found to be beneficial and resulted in up to 100% doubled cells in some cases.

Spontaneous doubling has been reported (Iswandar et al. 2010) during embryogenesis of callus cultures of oil palm using haploid plants as ortets and applying standard tissue culture methods for oil palm clonal production (see Sect. 8.11). Flow cytometry of embryogenic tissue cultures showed the presence of both haploid and doubled haploid cells. In some cases, completely doubled haploid ramets were regenerated suggesting that doubling occurred during the first rounds of mitoses in embryo development. However, fertile DHs have not yet been reported.

8.9.7 *Prospects for F_1 Hybrid Production in Oil Palm*

Breakthroughs in oil palm biotechnology over the past decade have delivered haploids and doubled haploids. The stage is now set for F_1 hybrid production. As yet there have been no reports of DHs flowering and thus no reports of inter-DH crossing to produce F_1 hybrids. Once fertile Dura and Pisifera DHs become available, crossing investigations can be initiated to determine the best parental combinations (possible existence of heterotic groups) and pave the way for the commercial production of F_1 hybrid oil palm cultivars with superior yields. This provides unrivalled opportunities of increasing yield per plantation with the real potential of relieving pressure on forest conversion.

Since parental stocks, manual pollination and seed nurseries are already part of Tenera commercial seed production (see Sect. 8.12), the conversion to F_1 hybrid commercial seed production is relatively straightforward as it only requires a change to Dura and Pisifera DH parental lines. The F_1 seed can be produced in large numbers and from a range of crosses. Growers will have access to a greater choice of seed, which is not subject to the potential of flowering abnormalities, e.g. mantled fruit, which are associated with clonal production (see Sect. 8.10).

8.10 Tissue Culture

Production of individual palms through conventional breeding is time consuming in perennial crops, including oil palm, as this can take more than 20 years. Clonal propagation through tissue culture was introduced as an option in speeding up and scaling up the production of specific selected palms. Tissue culture allows mass propagation of elite individual palms which express desirable traits; plants regenerated from tissue culture (ramets) are effectively clones of the donor plant (ortet). The first oil palm tissue culture started in 1960s and achieved initial success in the mid-1970s (Straritsky 1970; Rabéchault et al. 1970; Jones 1974). This led to more extensive research on oil palm tissue culture (Corley et al. 1979). However, the oil palm industry suffered a great disaster in the mid-1980s as many clones produced from tissue culture, when planted in the field, exhibited sterility due to malformation of flowers, commonly known as 'mantled', which was undetectable until the planted ramets began to flower and fruit about 3–5 years after planting (Durand-Gasselin et al. 1993, Fig. 8.7). The abnormality is a result of somaclonal variation (an epigenetic change caused by tissue culture) which leads to deficient stamen development and poor or incomplete fruit formation (Ho et al. 2009; Konan et al. 2010). Length of time in culture (particularly the callus phase), propagation methods, genotype or ortet type and concentration of plant growth regulators are some factors thought to generate variation in tissue culture (Pierik 1987). The oil palm industry is still extremely cautious about tissue cultured palms because of the risk of somaclonal variation. There have been great efforts to reduce/eliminate this in

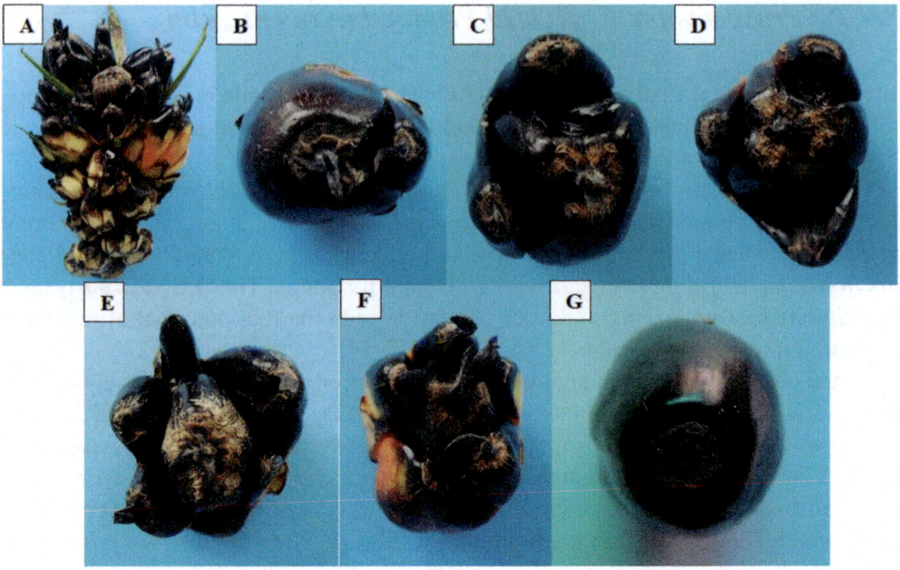

Fig. 8.7 Examples of abnormal fruits: (**a**) abnormal spikelet with abnormal fruits; (**b–f**) abnormal (mantled) fruit with supplementary carpels (1–5 carpels), compared to (**g**) normal

tissue culture-produced plants (ramets). A significant aid in this respect is the use of genetic markers to identify affected plants (see Section on Genetic Markers).

Since oil palm has only a single growing point (apical meristem), clonal propagation can only be done via tissue culture techniques in which ramets may be produced by regeneration or somatic embryogenesis. These techniques involve several stages: explant sampling, callus production, regeneration or embryogenesis, embryoid multiplication, shoot development, rooting and acclimatisation (Fig. 8.8) . This whole process takes 2–4 years depending on genotype and culture conditions.

8.10.1 Explant Sampling and Callus Induction

Various types of palm tissues can be used as explants, e.g. zygotic embryos, shoot meristems, young leaves from in vitro germinated seedlings or from nursery palms, immature leaves from adult palms, inflorescences, apical meristem and roots. The most commonly used material is young leaf spears from selected adult palms, which can provide thousands of explants (leaflet pieces) and give higher rates of ramet production (Rajanaidu et al. 1997). After the spears are harvested, the leaves are disinfected under a flow of sterile air and cut into small 1 cm^2 fragments, which are used to inoculated media plates containing specific nutrients and hormones. Callus induction occurs on MS (Murashige and Skoog 1962) medium supplemented with vitamins, casein hydrolysate and plant hormones (mostly

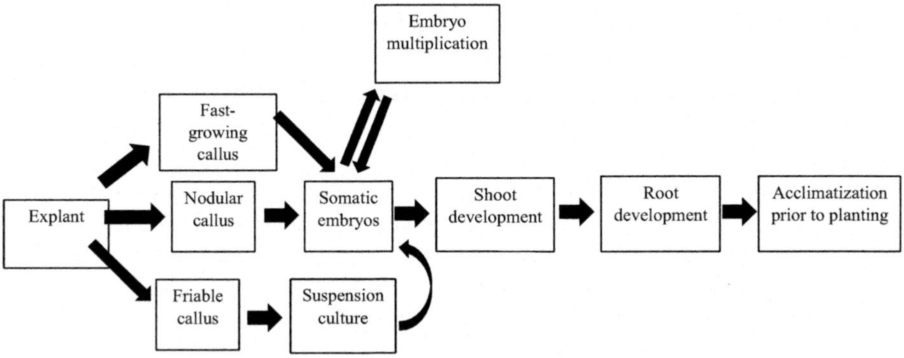

Fig. 8.8 Oil palm tissue culture scheme through callus, embryogenesis and plantlet development

auxin), e.g. 2,4-dichlorophenoxyacetic acid (2,4-D) and naphthalene acetic acid (NAA) and with or without activated charcoal, which controls hormone effects and removes inhibitors and toxins from the medium. The cultures are maintained in dark conditions at 27 ± 1 °C (90% relative humidity). The first calli emerge from leaf fragments after 6–8 weeks of culture. Three types of callus may be induced: nodular, friable and fast-growing callus; nodular calli are preferred for shoot/embryo initiation.

8.10.2 Clonal Production and Multiplication

Nodular calli induced from leaf segments are transferred onto solid media with lower concentrations of hormones to stimulate callus differentiation into embryo-like structures or somatic embryos (Duval et al. 1995). Additional cytokinin such as 6-benzylaminopurine (BAP), kinetin and 2-ip at relatively low concentrations may be needed in embryo production. Cultures are maintained in lit culture rooms with a 12/24 h photoperiod at 27 ± 1 °C and 50–60% relative humidity. Friable callus is preferred in setting up liquid embryogenic suspension cultures (de Touchet et al. 1991; Teixiera et al. 1995). The friable calli are maintained on liquid media containing low concentration of 2,4-D, under dark condition (27 ± 1 °C) on rotary shakers (140 rpm). Somatic embryos are induced after transfer to a hormone-free medium and transferred onto a solid medium (de Touchet et al. 1991). Fast-growing callus (FGC) can be induced when explants are cultured on media with high auxin concentrations (Smith and Thomas 1973; Hanower and Pannetier 1982). FGCs have a very high growth rate and composed of dispersed clumps of meristematic cells; once embryogenesis is initiated, it can be intense (Duval et al. 1995). However, FGCs have been identified as a source of 'mantled' somaclonal variation (Rival et al. 1997). Multiplication of embryoids is done by transferring embryogenic calli onto a proliferation medium and subcultured at 8-week intervals. In liquid suspension culture, the embryos with the desired size are sieved out every 4 weeks and transferred onto fresh solid media.

8.10.3 Shoot Development and Rooting

Embryogenic cultures may produce new adventitious somatic embryos depending on the genotype. The addition of BA (a cytokinin) into the medium can enhance shoot development of somatic embryos derived from cell suspension culture (Aberlenc-Bertossi et al. 1999). Histological observations show the development of the embryogenic axis (shoot apex and root pole) after BA treatment. Embryos may be maintained on shoot development medium for four to six cycles of 8 weeks each using the same medium composition. When plantlets are 6 cm, they may be transferred into tubes for root induction.

8.10.4 Acclimatisation

Rooted plantlets of 10–12 cm are washed with sterile water and treated with the auxin, indolebutyric acid (IBA), solution before transfer to planting trays containing an inert mixture of sand, coir dust, peat fibre and vermiculite in equal portion. The plantlets are placed in a shade house at a temperature of 27–30 °C. The plantlets are then covered with a plastic sheet to provide 75% shade and maintain a relatively high humidity, 90%. After 3–4 weeks, the shade and humidity are gradually reduced to 50% by removing the plastic cover, and the plantlets are kept for few months to harden off (Soh et al. 2009).

8.11 Seed Production

Following commercial breeding procedures described above, harvested bunches from seed gardens are sent to seed production facilities for processing. Seed is usually sold germinated for immediate nursery planting, and seed producers need to supply customers with thousands of germinated seed with a well differentiated plumule and radicle. Typically, plantation companies will buy seed lots of 5000–50,000 seed per consignment, while small holder farmers will often prefer to buy seedlings for field planting. Seed producers are therefore involved in high-throughput seed delivery. In marketing terms, the main qualities of commercial seed sales are:

1. High and early yield potential
2. Low Dura contamination
3. Disease resistance (*Fusarium oxysporum* in West and Central Africa)
4. Good nursery performance – low percentage of 'off-types'

Oil palm seeds do not germinate immediately because of an operculum which blocks the embryo from expanding and emerging from the germ pore. Usually, seeds are heat treated for 40–80 days (Alang 1982), and increase moisture content

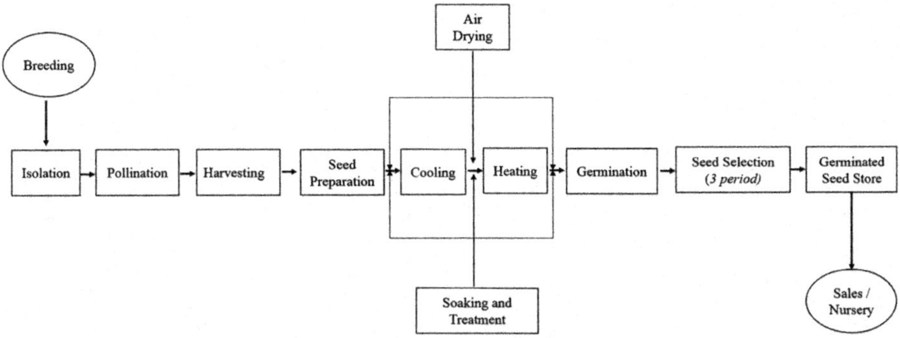

Fig. 8.9 Seed production scheme

results in the embryo being able to overcome the physical operculum barrier. Various seed producers of oil palm in the world have various methods to achieve high and uniform germination rates with a short a period as possible and to guard against fungal infection during storage. Variation in procedures probably reflects different germination requirements for different genetic materials. In seed production, the term 'seed' refers to a propagule for sale and nursery planting; botanically, this 'seed' is a 'nut' comprising of the true seed (embryo and endosperm) protected by a hard shell (endocarp).

The process depicted in Fig. 8.9 and described below encompasses the main features in producing germinated seed for commercial sale. It is also used by breeders to generate seedlings from crosses for progeny testing, trialling and further breeding.

Isolation

Selected female flowers from proven Dura mother palms grown in a seed garden are isolated as described above (see Section on Commercial crossing).

Pollination

The isolated female inflorescence is pollinated with pollen collected from proven Pisifera palms (details above; see Section on Commercial crossing).

Harvesting

Fruit bunches that have reached the age of approximately 150 days or have one loose fruit are harvested and transported to the processing site. The bunch label is retained until the bunch is received. A circular area around the mother palm must be kept clear of lose fruits before bunches are cut down to avoid contamination from other palms.

Chopping and fermentation

Bunch chopping is done to separate the spikelets from the stalk. Spikelets then fermented for approximately 4 days to loosen the fruits naturally for ease in subsequent de-pulping.

Detaching fruits
Fruits which remain attached to spikelets after the fermentation period are detached manually.

De-pulping
Fruits are placed in a de-pulping machine to separate the mesocarp from seed (nut). A vertical de-pulping machine that was first designed by Escobar (1980) is now widely used.

Scraping
Once seeds are separated from their mesocarps, the remaining fibres must be removed to reduce fungal infection. Seeds are then cleaned by washing.

Treatment and air-drying before storage in a cool room
The cleaned seeds are usually soaked in a disinfectant and then soaked in a fungicide and detergent/surfactant to remove surplus oil. Some seed production units may also use a bactericide. The seed lots are then dried for 24 h at ambient temperature in drying racks exposed to a good air flow.

Slow air-drying
Seeds undergo a slow and gentle air-drying process in drying racks for 24 h at a temperature of 19–23 °C; this equilibrates seed moisture content to 17–20% prior to storage.

Cooling and storage
Before entering the cold room, if storage is required, the seeds are placed into plastic trays or polythene bags and may be stored for 6 months with minimal loss in seed viability provided there is no significant infection. Longer periods of storage are also possible but for commercial seeds, usually it would not be extended beyond 6 months. Periodic inspections are carried out to ensure good sanitation is maintained, and any fungus-infected seeds are removed and destroyed. The remaining healthy seeds should be treated with fungicide and placed into a clean plastic tray or bag to continue storage. During storage, water condensation inside the trays or plastic bags must be dried out to prevent fungal infection.

First soaking
Before heating, seeds need to be soaked in clean water, usually for 5 days with daily water changes.

Treatment and air-drying before heating
After soaking, the seeds are washed with running water for about 2 min and then treated with detergent and fungicide by soaking. The seed is then dried in drying racks for 24 h. After 24 h, samples are taken to check seed and kernel moisture content with a target of 18–20%. Samples may also be taken to test embryo viability testing. Percentage viability is then calculated; under good systems, this should normally be greater than 90%.

Heating
Seeds with the target moisture content are placed in a hot room (at about 40 °C) or are resoaked or dried for a longer period. During the heating period, seeds are sprayed with distilled water. Seeds are also checked for fungal attack. Seed weights

Fig. 8.10 Germinated seeds ready for sale

may be taken every 20 days to ensure seed moisture remains in the required range. Fungal infected or rotten seeds are removed and the remaining healthy seeds are treated with a fungicide. After fungicide treatment, the seeds are dried and returned to the hot room to continue the heating process.

Second soaking

After 40–60 days, the seeds are removed from the hot room and soaked for a second time to increase the moisture content and aid the seed imbibition process.

Treatment and air-drying before germination

Seeds that have been soaked are washed and given a fungal and bacterial treatment; they are then dried and checked for moisture content (22%), before entering the germination room. Samples are also taken to check embryo viability before starting the germination process and to estimate germination percentage.

Germination

Treated seeds are moved to a germination room for up to 60 days, usually at ambient temperature. Seeds are checked regularly for germination with the emergence of a plumule and radicle. Generally, seeds start to germinate 5–7 days after entering the germination room and selection can start 2–3 days later. Seeds must be inspected continually for germination and fungal infection.

Germinated seed selection

Seed producers usually have a minimum target germination of 80% or more. Quality control measures are used for commercial production, and usually the maximum length of plumule and radicle are 1.5 cm and 2 cm, respectively. Abnormal and fungal infected seeds are removed and destroyed. Seeds that have not germinated may continue in the germination room. Generally, seeds are selected after 3–5 days with further selections up to 60 days. Seeds that do not germinate after 60 days are usually destroyed (Fig. 8.10).

8.12 Biotechnology

The interdisciplinary approach of breeding and biotechnology offers unique opportunities to improve the versatility and utility of oil palm. Biotechnological approaches applied to oil palm breeding have increased substantially in importance, not only in micropropagation (see Sect. 8.11) but also in marker-assisted selection (see Sect. 8.9.4), and have the potential to enhance physiological and molecular studies linking phenotype to genotype, e.g. the expression of genes controlling flowering, abscission, disease resistance, etc.

Molecular biology provides powerful tools for genetic investigation in disease (Tan et al. 2013; Al-Obaidi et al. 2014), inflorescences sex determination (Adam et al. 2007, 2011) and genetic mapping. Singh et al. (2013a) analysed genome sequences of *E. guineensis* and drafted a genome sequence of *E. oleifera* that provided insights into oil biosynthesis genes and their regulators. The same workers identified mutations in the *Sh* gene that specify the different fruit forms found in the oil palm (Singh et al. 2013b), which led to genetic characterisation of Dura, Tenera and Pisifera seedlings (Reyes et al. 2015). However, there are many more genes of interest that need to be discovered, for instance, those related to yield, quality, height, resistance to drought, wind or pest and disease, as well as genes controlling novel traits related to mechanical harvesting (see Sect. 8.8).

Major recent advances in the use of DNA markers in oil palm include those for shell thickness determination, fruit colour and epigenetic mechanisms underlying the mantled somaclonal variation in clonally micropropagated oil palms (see Table 6.1). Desired genotypes from segregating population can now be selected in oil palm, and this provides a new tool for plant breeding (Cros et al. 2015). There will be increased utility and application for these methods as more genes of interest are discovered and exploited in breeding.

Currently oil palm is a completely GM-free crop. Recent studies showed the possibility of improvement of oil quality (Tranberger et al. 2011; Morcillo et al. 2013) by modification of Elaeis fatty acid biosynthesis pathways (Montoya et al. 2013, 2014). Transgenic oil palms with improved fatty acid profiles may be produced for specific food and industrial use, but this has yet to be taken up as it means confronting sensitive issues faced by producers, processors and end users in accepting genetically modified (GM) oil palm. The USA considers the technology to be safe, while in Europe there is a strong movement that considers genetically modified organisms (GMOs) to be potentially harmful and unacceptable as food crops. GM crops are considered unsafe by some groups as they claim there is a lack of knowledge in biological consequences of the technique; thus, GMO crops are subject to stringent regulations. Some parties may never accept GMOs, despite good safety and environmental credentials, but the views of the general population and environmentalists are less clear and may gradually change through time and as methods become more sophisticated and benign. In this respect, new techniques in gene editing are becoming interesting as several countries do not classify this as a GM technique.

Potential new products from oil palm include polyhydroxyalkanoate (PHA, polymers with diverse plastic-like properties) and polyhydroxybutyrate (PHB, a biodegradable plastic). At present, bacterial fermentation is used for production (Urtuvia et al. 2014), but production of PHA from plants such as oil palm (Tilbrook et al. 2011) is feasible (Masani and Parveez 2008; Zhang et al. 2013; Parveez et al. 2015).

8.13 Future Prospects and Outlook

Oil palm is a major world crop and will remain the number one oil crop for the foreseeable future. The prospects for oil palm improvement through plant breeding are extremely bright as there is immense scope for both conventional and new breeding practices.

Conventional breeding relies on access to genetic variation, and although this has been a bottleneck in the past, rich germplasm resources have been assembled through cooperation between major oil palm-producing countries (SE Asia) and countries with primary, wild-type germplasm (West and Central Africa). In other crops, such as wheat and rice, access to new genetic variation, especially genes for semidwarf stature, ushered in massive increases in yield which stabilised food security, the 'Green Revolution'. The germplasm available to breeders has never been so great. And in addition to tapping natural variation, mutation induction programmes have been initiated in oil palm; these promise to produce desired traits directly into elite lines, thus speeding up the breeding process and providing new plant forms and products. With the development of cryopreservation in oil palm, these genetic stocks no longer need to be maintained as living palms, which take up vast areas of land, but may be stored in the laboratory.

Traditional breeding has been carried out using phenotypic selection, typically from observations and data collection from field trials. With the advent of genetic markers, breeding can now be carried out via genotypic selection whereby DNA of progeny are screened for markers associated with genes of interest; thus, shell thickness determinations can be made prior to field planting saving both time (it takes 3–5 years to first bunch production for phenotyping from seed germination) and space (sterile Pisifera can be eliminated from yield trials). A significant breakthrough in respect to genomic selection is the development of whole genome sequencing of oil palm. This now facilitates the identification of genes controlling traits of interest and at the same time provides genomic tags for detection. In addition to using DNA diagnostics for target traits (aggressive breeding), widespread DNA markers can be used to monitor the genetic background, thus allowing the selection of genetic backgrounds known to be important for successful cultivar performance (defensive breeding).

There are many traits that can now be targeted for future oil palm improvement, and it is impossible to predict all that will emerge as important, but they will most certainly include traits for yield (including hybrid vigour) in terms of bunch production but more importantly increases in % oil, disease (*Fusarium* and *Ganoderma*) resistance, mechanical harvesting (smaller bunches, fruit colour, long bunch stalk, late and low individual fruit abscission), material for more marginal environments and oil quality (product diversification, oils for specific end users). Input use efficiency is clearly a complex of physiological systems but all with a genetic base that can now be delved into and exploited. The higher yield will probably be partially achieved by increasing harvest index, and increasing yields per hectare may be achieved by planting smaller palms at higher density and via F_1 Hybrid cultivation.

If weather patterns continue to change, e.g. resulting in dry spells being normal in environments where traditionally there has been no water deficit, then selecting material for such environments will become a priority, and increasing yields without also increasing the monthly peaks (and troughs) of bunch production will become an issue.

There are currently principally two serious diseases that threaten the oil palm industry, *Fusarium oxysporum* and *Ganoderma boninense*. There are also a few insect pests that often threaten oil palm plantations but with informed approaches can be kept under control. The current oil palm planting practices are an issue: the planting of large continuous areas of oil palm with successive replanting with no rotation, fallow or break crop has to be changed, but as part of an integrated management approach, the planting of commercial material with levels of resistance/tolerance will become increasingly important. It is becoming accepted throughout the industry that it is pointless fighting against the biology but a need to work with it. The use of integrated management is thus becoming more and more one that is seen as being essential.

A general activity in breeding is the capture of both phenotypic and genotypic data from individuals of progeny from crosses made. These datapoints are now numbered in the hundreds of thousands and will continue to grow. Data storage therefore requires large computing capacity, and data evaluation will require skills in bioinformatics, which is becoming established as a central new tool in plant breeding.

Future challenges for oil palm breeders are to produce new cultivars that give higher yield per unit land area, per worker and per unit area of water which meet new market demands and are more sustainable and environmentally friendly, and so doing do not just continue the economic viability of this crop but also create a truly more sustainable and environmentally friendly oil palm industry. Meeting standards for this 'Evergreen Revolution' can only be realised by investment in research and development.

Acknowledgements This chapter was composed by members of Verdant Bioscience's Breeding and Biotechnology teams; the authors acknowledge inputs and comments from all colleagues in agronomy, crop protection, land surveys and marketing.

References

Aberlenc-Bertossi F, Noirot M, Duval Y (1999) BA enhances the germination of oil palm somatic embryos derived from embryogenic suspension cultures. Plant Cell Tissue Organ Cult 56(1):53–57

Adam H, Jouannic S, Escoute J et al (2005) Reproductive developmental complexity in the African oil palm (*Elaeis guineensis*). Am J Bot 92:1836–1852

Adam H, Jouannic S, Orieux Y et al (2007) Functional characterization of MADS box genes involved in the determination of oil palm flower structure. J Exp Bot 58(6):1254–1259

Adam H, Collin M, Richaud F et al (2011) Environmental regulation of sex determination in oil palm: current knowledge and insights from other species. Ann Bot 108(8):1529–1537

Alang ZC (1982) Some physiological effects of the commercial heat treatment of oil palm (*Elaeis guineensis*) seeds. In: Pushparajah E, Chew PS (eds) The oil palm in agriculture in the eighties, vol 1. Incorp. Soc. Planters, Kuala Lumpur, pp 207–222

Al-Obaidi JR, Mohd-Yusuf Y, Razali N et al (2014) Identification of proteins of altered abundance in oil palm infected with *Ganoderma boninense*. Int J Mol Sci 15(3):5175–5192

Alvarado A, Escobar R, Peralta F et al (2007) Compact seeds and clones and their potential for high density planting. ASD Oil Palm Paper 31:1–8

Ariffin D, Idris AS, Singh G (2000) Status of *Ganoderma* in oil palm. In: Flood J, Bridge PD, Holderness M (eds) Ganoderma: disease of perennial crops. CABI Publishing, Oxfordshire, pp 49–68

Arifin AA, Ai TY (1989) The effect of handling subsequent quality of crude palm oil. In: Proceeding 1989 PORIM international palm oil development conference, Kuala Lumpur 5-9 September 1989. Palm oil research institute of Malaysia, pp 482–485

Bado S, Forster BP, Nielen S et al (2015) Plant mutation breeding: current progress and future assessment. In: Janick J (ed) Plant breeding reviews, vol 39. Wiley-Blackwell, New Jersey, pp 23–88

Barcelos E, de Almeida RS, Cunha RNV et al (2015) Oil palm natural diversity and the potential for yield improvement. Front Plant Sci 6:190

Basiron Y (2007) Palm oil production through sustainable plantations. Eur J Lipid Sci Tech 109:289–295

Beirnaert A, Vanerweyen R (1941) Contribution à l'étude génétique et biométrique des variétés d'*Elaeis guineensis* Jacquin. INEAC Sér Sci 27:1–101

Berger KG (1983) Production of palm oil from fruit. J Am Oil Chem Soc 62(2):206–210

Billotte N, Marseillac N, Risterucci AM et al (2005) Microsatellite-based high density linkage map in oil palm (*Elaeis guineensis* Jacq.) Theor App Genet 110:754–765

Orion Biosains (2016) Shell, vir, mantled, Selangor https://www.orionbiosains.com/contact. Accessed 7 June 2016

Blaak G, Sparnaij LD, Menendez T (1963) Breeding and inheritance in the oil palm (*Elaeis guineensis* jacq): part II methods of bunch quality analysis. J W Afr Inst Oil Palm Res:146–155

Blakeslee AF, Belling J, Farnham ME et al (1922) A haploid mutant in the jimson weed, *Datura stramonium*. Science 55:646–647

Breton F, Rahmaningsih M, Lubis Z et al (2009) Early screening test: a routine work to evaluate resistance/susceptibility level of oil palm progenies to basal stem rot disease. In: MPOB International Palm Oil Congress (PIPOC 2009), Kuala Lumpur, 9-12 November 2009. MPOB, p 1–11.

Breure CJ (1982) Factors affecting yield and growth of oil palm tenera in West New Britain. Oleagineux 37(5):213–223

Breure CJ, Menendez T, Powel MS (1990) The effect of planting density on the yield components of oil palm (*Elaeis guineensis*). Exp Agric 26(1):117–124

Brown J, Caligari PDS (2008) An introduction to plant breeding. Wiley-Blackwell, New Jersey

Cadena T, Prada F, Perea A et al (2013) Lipase activity, mesocarp oil content, and iodine value in oil palm fruits of *Elaeis guineensis*, *Elaeis oleifera*, and the interspecific hybrid O×G (*E. oleifera* × *E. guineensis*). J Sci Food Agric 93:674–680. doi:10.1002/jsfa.5940

Chase SS (1949) Monoploid frequencies in a commercial double cross hybrid maize, and its component single cross hybrids and inbred lines. Genetics 34:328–322

Chase SS (1952) Production of homozygous diploids of maize from monoploids. Agron J 44:263–267

Chin CW (1982) Segregation of fertility, fruit forms and early yield performance of selected sibbed and outcrossed fertiliser pisifera progenies. In: Pushparajah E, Chew PS (eds) The oil palm in agriculture in the eighties, vol 1. Incorporated Society Planters, Kuala Lumpur, pp 63–68

Chin CW (1988) Outlook on fertile pisifera breeding. In: Halim Hassan A et al (eds) Proceedings 1987 international oil palm conference "*Progress and prospects*". Palm Oil Research Institute of Malaysia, Kuala Lumpur, pp 107–111

Chin CW (1993) Progress and prospects of *oleifera* hybrids and backcrosses in breeding. In: Basiron T et al (eds) Proc. 1991 PORIM international palm oil conference – agriculture. Palm Oil Research Institute, Malaysia, pp 557–563

Chin CW (1995) Felda's experience with fertile pisifera breeding. In: Jalani BS et al (eds) Proceedings 1993 PORIM International Palm Oil congress – Agriculture. Palm Oil Research Institute Malaysia, Kuala Lumpur, pp 489–501

Cochard B, Noiret JM, Boudouin L (1993) Second cycle reciprocal recurrent selection (rss) in oil palm, *Elaeis Guineensis Jacq.* Result of Deli x La mé hybrid test. In: Proceeding 1993 PORIM international Congress "Update and Vision", Kuala Lumpur, 20–25 September 1993

Cochard R, Durand-Gasselin T, Adon B (2000) Oil palm genetic resource in the cote d'ivore – composition, assessment and use. Paper presented at International Symposium Oil Palm Genetic Resources and Utilization. Malaysia Palm Oil Board, Kuala Lumpur, 8–10 June 2000

Cochard B, Adon B, Rekima S et al (2009) Geographic and genetic structure of African oil palm diversity suggests new approaches to breeding. Tree Genet Genomes. 2009 5(3):493–504

Cock J, Donough CR, Oberthur T et al (2014) Increasing palm oil yields by measuring oil recovery efficiency from the field to the mill. Available via researchgate. https://www.researchgate.net/profile/Christopher_Donough/publications. Accessed 15 Mar 2016

Coors JG, Pandey S (1999) The genetics and exploitation of heterosis in crops. American Society of Agronomy, Crop Science Society of America, Wisconsin

Corley RHV (1983) Potential productivity of tropical perennial crops. Exp Agric 19:217–237. doi:10.1017/S0014479700022742

Corley RHV (1998) What is the upper limit to oil extraction ratio? In: Henson RIE, Jalani BS (Eds). Proceedings 1996 International conference 'Oil and kernel production in oil palm – a global perspective' – Palm Oil Research Institute Malaysia, Kuala Lumpur, pp 256–269

Corley RHV, Gray BS (1976) Growth and morphology. In: Corley RHV, Hardon JJ, Wood BJ (eds) Oil palm research: developments in crop science 1. Elsevier, Amsterdam, pp 7–21

Corley RHV, Lee CH (1992) The physiological basis for genetic improvement of oil palm in Malaysia. Euphytica 60(3):179–184

Corley RHV, Tinker PB (2003) The oil palm, 4th edn. Wiley-Blackwell, New Jersey

Corley RHV, Wooi KC, Wong CY (1979) Progress with vegetative propagation of oil palm. Planter 55(641):377–380

Crone GR (1937) The voyages of Cadamosto and other documents on Western Africa in the second half of the fifteenth century. Hakluyt Society, London, p 80

Cros D, Denis M, Bouvet JM et al (2015) Long-term genomic selection for heterosis without dominance in multiplicative traits: case study of bunch production in oil palm. BMC Genomics 16:651

Crow JH (1998) 90 years ago: the beginning of hybrid maize. Genetics 148:923–928

Dauqan E, Sani HA, Abdullah A et al (2011) Vitamin E and beta carotene composition in four different vegetable oils. Am J App Sci 8(5):407–412

De Poerck RA (1950) Contributions a l'étude du palmier a huile African *Elaeis guineensis* Jacq. Oleagineux 5:623–662

De Touchet B, Duval Y, Pannetier C (1991) Plant regeneration from embryogenic suspension cultures of oil palm (*Elaeis guineensis* Jacq.) Plant Cell Rep 10(10):529–532

Din MA & Rajanaidu N (2000) Evaluation of *Elaeis oleifera*, interspecific hybrids and backcrosses. Paper presented at international Symposium 'Oil palm genetic resources and utilisation' 8–10 June, Malaysian Palm Oil board, Kuala Lumpur

Dransfield J, Uhl NW (1998) Flowering plant-monocotyledons: alismatanae and commelinanae (except graminae). In: Kubitski K (ed) Palmae. Springer-Verlag, Berlin, pp 306–389

Dransfield J, Uhl NW, Asmussen CB et al (2005) A new phylogenetic classification of the palm family, Arecaceae. Kew Bull 60(4):559–569

Dumortier F, Konimor J (1999) Selection and breeding progress at Dami OPRS, Papua New Guinea. In: Rajanaidu N, Jalani BS (eds) Proceedings 1996 seminar 'sourcing of oil palm planting materials for local and overseas joint ventures'. Palm Oil Research Institute, Malaysia, pp 142–170

Dumortier F, van Amstel H, Corley RHV (1992) Oil palm breeding at Binga, Zaire, 1970–1990. Unilever Plantations, London

Dunwell JM, Wilkinson MJ, Nelson SPC et al (2010) Production of haploids and doubled haploids in oil palm. BMC Plant Bio 10:218–243

Durand-Gasselin T, Duval Y, Baudouin L, Maheran AB, Konan K, Noiret JM (1993) Description and degree of the mantled flowering abnormality in oil palm (*Elaeis guineensis* Jacq) clones produced using the orstom-CIRAD procedure. In: Rao V, Henson IE, Rajanaidu N (ed). Proceeding 1993 ISOPB International Symposium on Recent Developments in Oil Palm Tissue Culture and Biotechnology. Kuala Lumpur, 24–25 Sept 1993, pp 48–63

Durand-Gasselin T, Hayun S, Jacquemard JC et al (2011) Palm oil yield potential of oil palm (*Elaeis guineensis*) seeds developed in a network by CIRAD and its partners. TDG ISOPB Veleur des Semences Cirad UK

Duval Y, Engelmann F, Durand-Gasselin T (1995) Somatic embryogenesis in oil palm (*Elaeis guineensis* Jacq.) In: Bajaj YPS (ed) Biotechnology in agriculture and forestry: somatic embryogenesis and synthetic seed I. Springer, Berlin, pp 335–352

Duvick DN (2001) Biotechnology in the 1930s: the development of hybrid maize. Nature Rev Genet 2:69–73

Endress PK (2011) Angiosperm ovules: diversity, development, evolution. Ann Bot 107:1465–1489

Escobar CR (1980) An improved oil palm depulper for single bunch lots. Planter 56(657):540–548

Escobar R (2004) Strategies in production of oil palm compact seeds and clones. ASD Oil Palm Paper 27:1–12

Flood J (2006) A review of *Fusarium* wilt of oil palm caused by *Fusarium oxysporum* f. sp elaeidis. Am Phytopathol Soc 96(6):660–662

Flood J, Mepsted R (1990) Vascular wilt of oil palm – a potential problem for Malaysia? Planter 66(776):581–585

Flood J, Hasan Y, Turner PD et al (2000) The spread of *Ganoderma* from infective sources in the field and its implications for management of the disease in oil palm. In: Flood J, Bridge PD, Holderness M (eds) *Ganoderma* diseases of perennial crops. CABI Publishing, Surrey, pp 101–112

Foreign Agriculture Service/USDA (2015) Palm oil: world supply and distribution. In: Oilseeds: world market and trade. http://apps.fas.usda.gov/psdonline/circulars/oilseeds.pdf. Accessed 22 Feb 2016

Forster BP, Thomas WTB (2005) Doubled haploids in genetics and plant breeding. In: Janick J (ed) Plant breeding reviews, vol 25. Wiley, Oxford, pp 57–88

Forster BP, Franckowiak JD, Lundqvist U et al (2012) Mutant phenotyping and pre-breeding in barley. In: Shu QY, Nakagawa H, Forster BP (eds) Plant mutation breeding and biotechnology. CAB International and FAO, Dundee, pp 327–346

Geiger HH, Gordillo GA (2010) Doubled haploids in hybrid maize breeding. Maydica 54:485–499

Geiger HH, Miedaner T (2009) Rye breeding. In: Carena MJ (ed) Handbook of plant breeding: cereals. Springer, Stuttgart, pp 157–181. 10/1007/978-0-387-72297-9

Gibon V, Greyt WD, Kellens M (2007) Palm oil refining. Eur J Lipid Sci Tech 109:315–335

Guha S, Maheshwari SC (1964) In vitro production of embryos from anthers of Datura. Nature 204:497

Guitierrez LF, Sanchez OJ, Cardona CA (2009) Process integration possibilities for biodiesel production from palm oil using ethanol obtained from lignocellulosic residues of oil palm industry. Bioresour Technol 100(3):1227–1237

Hanower J, Pannetier C (1982) In vitro vegetative propagation of the oil palm, *Elaeis guineensis* Jacq. In: Plant tissue culture 1982: proceedings, 5th International Congress of Plant Tissue and Cell Culture, Tokyo, 11–16 July 1982

Hardon JJ (1970) Inbreeding in populations of the oil palm (*Elaeis guineensis* Jacq.) and its effect on selection. Oleagineux 25:449–456

Hardon JJ, Turner PD (1967) Observations on natural pollination in commercial plantings of oil palm (*Elaeis guineensis*) in Malaya. Exp Agric 3(2):105–116

Hardon JJ, Corley RHV, Lee CH (1987) Breeding and selecting the oil palm. In: Abbott AJ, Atkin RK (eds) Improving vegetatively propagated crops. Academic, London, pp 63–81
Hartley CWS (1988) The oil palm, 3rd edn. Longman, London
Ho YW, Tan CC, Soh AC et al (2009) Biotechnological approaches in producing oil palm planting material—a success story. Int J Oil Palm 6:86–93
Hubert DF, Sekou D (2004) Status on oil palm vascular wilt. In: Proceeding of the International Conference on Pest and Diseases of Importance to the Oil Palm Industry, Kuala Lumpur, 18–19 May 2004, pp 30–36
Idris AS, Kushairi A, Ismail S et al (2004) Selection for partial resistance in oil palm progenies to *Ganoderma* basal stem rot. J Oil Palm Res 16(2):12–18
Ishikawa S, Ishimaru Y, Igura M et al (2012) Ion-beam irradiation, gene identification, and marker-assisted breeding in the development of low-cadmium rice. Proc Natl Acad Sci 109:19166–19171
Iswandar HE, Dunwell JM, Forster BP et al (2010) Doubled haploid ramets via embryogenesis of haploid tissue cultures. In: Proceeding on Advances in Oil Palm Tissue Culture, Indonesia 29 May 2010, pp 100–104
Jones LH (1974) Propagation of clonal oil palms by tissue culture. Oil Palm News 17:1–8
Jones LH (1989) Prospects for biotechnology in oil palm (*Elaeis guineensis*) and coconut (*Cocos nucifera*) improvement. Biotech Gen Eng Rev 7:281–296
Jones LH (1997) The effect of leaf pruning and other stresses on sex determination in oil palm and their representation by a computer simulation. J Theo Bio 187:241–260
Junaidah J, Kushairi, Jones B et al (2011) Innovation for oil extraction method using nmr in bunch analysis. Paper presented at international seminar on Breeding for Sustainability Oil Palm. Joint organised by International Society for Oil Palm Breeder (ISOPB) and Malaysian Palm Oil Board (MPOB), Kuala Lumpur, 18 Nov 2011, pp 1–18
Kajale LB, Ranade SG (1953) The embryo sac of *Elaeis guineensis* Jacq: a reinvestigation. J Ind Bot Soc 32:110–107
Kimber G, Riley R (1963) Haploid angiosperms. Bot Rev 29:480–531
Konan KE, Durand-Gasselin T, Kouadio YJ et al (2010) In vitro conservation of oil palm somatic embryos for 20 years on a hormone-free culture medium: characteristics of the embryogenic cultures, derived plantlets and adult palms. Plant Cell Rep 29(1):1–13
Kosugi A, Tanaka R, Magara K et al (2010) Ethanol and lactic acid production using sap squeezed from old oil palm trunk felled for replanting. J Biosci Bioeng 110(3):322–325
Kumar SK, Sparjanbabu DS (2013) Haploid breeding in palms – a brief review. Adv Crop Sci Tech 1:113. doi:10.4172/2329-8863.1000113
Li J, Xin Y, Yuan L (2009) Hybrid rice technology development: ensuring China's food security. International Food Policy Research Institute, Washington, DC
Longin CFH, Mühleisen J, Maurer HP et al (2012) Hybrid breeding in autogamous cereals. Theor Appl Genet 124:1087–1096
Luyindula N, Corley RHV, Mantantu N (2005) A comparison of the Deli dumpy and pobe dwarf short-stemmed oil palms and their outcrossed progenies. J Oil Palm Res 17:152–159
Maizura I, Kushairi A, Mohd Din A et al (2008) PS13: breeding populations selected for low lipase. MPOB Information Series 425
Maluszynski M, Kasha KJ, Forster BP et al (2003) Doubled haploid production in crop plants: a manual. Kluwer Academic Publishers, Dordrecht
Masani MY, Parveez GK (2008) Development of transformation vectors for the production of potentially high oleate transgenic oil palm. Electron J Biotechnol 11(3):1–9
Mekhilef S, Singh S, Saidur R (2011) A review on palm oil biodiesel as a source of renewable fuel. Renew Sust Energ Rev 15(4):1937–1949
Menendez T, Blaak G (1964). Plant breeding division. In: 12th Annual Report West African Institute Oil Palm Research 49–75, Benin City
Mienanti D, Sitorus AC, Forster BP et al (2009) Chromosome doubling of oil palm (*Elaeis guineensis* Jacq.) haploids. In: Proceedings of PIPOC 2009, Kuala Lumpur, 9–12 Nov 2009, AP46, pp 948–953

Mohd Din A, Rajanaidu N, Jalani B (2000) Performance of *Elaeis oleifera* from Panama, Costarica, Columbia and Honduras in Malaysia. J Palm Oil Res 12(1):71–80

Montoya C, Lopes R, Flori A et al (2013) Quantitative trait loci (QTLs) analysis of palm oil fatty acid composition in an interspecific pseudo-backcross from *Elaeis oleifera* (H.B.K.) Cortés and oil palm (*Elaeis guineensis* Jacq.) Tree Genet Genomes 9:1207–1225

Montoya C, Cochard B, Flori A et al (2014) Genetic architecture of palm oil fatty acid composition in cultivated oil palm (*Elaeis guineensis* Jacq.) compared to its wild relative *E. oleifera* (H.B.K) Cortes. PLoS ONE 9(5):e95412

Moose SP, Mumm RH (2008) Molecular plant breeding as the foundation for 21st century crop improvement. Plant Physiol 147:969–977

Morcillo F, Cros D, Billotte N (2013) Improving palm oil quality through identification and mapping of the lipase gene causing oil deterioration. Nat Commun 4:2160. doi:10.1038/ncomms3160

Murashige T, Skoog F (1962) A revised medium for rapid growth and bio assays with tobacco tissue cultures. Physiol Plant 15(3):473–497

Nagi J, Ahmed SK, Nagi F (2008) Palm biodiesel an alternative green renewable energy demands of the future. ICCBT 7:79–94

Nasution O, Rusfiandi R, Sitorus AC et al (2009) Cytological studies of pollen development in oil palm (*Elaeis guineensis* Jacq.). In: Proceedings of PIPOC 2009, Kuala Lumpur, 9–12 Nov 2009, AP47, pp 954–961

Nasution O, Sitorus AC, Nelson SPC et al (2013) A high-throughput flow cytometry method for ploidy determination in oil palm. J Oil Palm Res 25:265–271

Nelson SPC, Juhyana Y, Sitepu B et al (2006) Future mechanisation – the role of the breeder. Int Oil Palm Conf Bali 2006:1–19

Nelson SPC, Wilkinson MJ, Dunwell JM (2009) Breeding for high productivity lines via haploid technology. In: Proceedings of the 2009 International Palm Oil Congress – Agriculture, Biotechnology & Sustainability Conference. MPOB, Bangi, 9–12 Nov 2009, pp 203–225

Ngado-Ebongue GF, Eha CE, Ntsomboh-Ntsefong G et al (2013) Breeding oil palm (*Elaeis guineensis*) for *Fusarium* wilt disease tolerance: an overview research program and seed production potentialities in Cameroon. Inter J Agric Sci 3(5):513–520

Ntsomboh-Ntsefong GN, Goldima M, Aurelie NN et al (2015) Vascular wilt disease tolerance status of some oil palm progenies in relation to local strain of *Fusarium*. Inter J Curr Res Biosci Plant Biol 2(8):111–122

Oboh BO (1993) The development of short stemmed oil palm 1 – the *Elaeis oleifera* approach. In: Basiron Y et al (eds) Proceedings 1991 PORIM International Palm Oil Conference – Agriculture. Palm Oil Research Institute Malaysia, Kuala Lumpur, pp 492–496

Okwuagwu CO (2011) The genetic base of the NIFOR oil palm breeding programme. Nigerian Institute for Oil Palm Research

Okyere-Boateng G, Dwarko DA, Kaledzi PD, Nuertey BN (2008) Collection, conservation and evaluation of the disappearing oil palm (*Elaeis guineensis* J) landraces in Ghana. Int J Pure Appl Sci 1(3):18–31

Opsomer JK (1956) Les premières descriptions de palmier à huile *Elaeis guineensis* Jacq. Bull Séances Acad R Soc Colon Outre Mer 2:253–272

Pamin K (1988) A hundred and fifty years of oil palm development in Indonesia: from the Bogor Botanical Garden to the industry. In: Jatmika A et al (eds) Proceedings of 1998 International Oil Palm Conference 'Commodity of the past, today and the future'. Indonesian Oil Palm Research Institute, Medan, pp 3–23

Pamin K, Lubis RA, Kusnadi TT et al (1990) Yield potential of planting material and reassessment of oil palm breeding program in Indonesia. In: Proceeding of International Society for Oil Palm Breeders (ISOPB), Phuket, 29–30 Oct 1990, pp 11–23

Parveez GK, Bahariah B, Ayub NH et al (2015) Production of polyhydroxybutyrate in oil palm (*Elaeis guineensis* Jacq.) mediated by microprojectile bombardment of PHB biosynthesis genes into embryogenic calli. Front Plant Sci 6:598

Piarpuzan D, Quinten JA, Cardona A (2011) Empty fruit bunch from oil palm as a potential raw material for fuel ethanol production. Biomass Bioenergy 35(3):1130–1137

Pierik RLM (1987) In vitro culture of higher plants. Kluwer Academic Publishers, Dordrecht

Pooni HS, Cornish MA, Kearsey MJ et al (1989) The production of superior lines and second cycle hybrids by inbreeding and selection. Elaeis 1:17–30

Prina AR, Landau AM, Pacheco MG (2012) Chimeras and mutant gene transmission. In: Shu QY, Forster BP, Nakagawa H (eds) Plant mutation breeding and biotechnology. CABI International and FAO, Dundee, pp 181–189

Priwiratama H, Djuhyana J, Nelson SPC et al (2010) Progress of oil palm breeding for novel traits: late abscission, virescens and long bunch stalk. Paper presented at international oil palm conference, Yogyakarta, 1–3 June 2010

Pye O, Bhattacharya J (2013) The palm oil controversy in Southeast Asia: a transnational perspective. ISEAS Publishing, Singapore, p 283

Rabechault H, Guénin G, Ahée J (1970) Studies on the in vitro culture of oil palm embryos: VDL A comparison of different nutrient media. Oleagineux 25:519–524

Rajanaidu N, Jalani BS (1994) Oil palm genetic resources collection, evaluation, utilization and conservation. Presented at PORIM Colloquium on Oil Palm Genetic Resources, Kuala Lumpur, 13 Sept 1994

Rajanaidu N, Rohani O, Jalani BS (1997) Oil palm clones: current status and prospects for commercial production. Planter 73(853):163–184

Rankine I, Fairhust TH (1999) Management of phosphorus, potassium and magnesium in mature oil palm. Better Crops Inter 13(1):10–15

Rao V, Soh AC, Corley RHV et al (1983) A critical reexamination of the method of bunch quality analysis in oil palm breeding, PORIM Occasional Paper. Palm Oil Research Institute of Malaysia, Kuala Lumpur, 28p

Rao V, Gomez M, Chayawat N et al (2001) Some influence oil/bunch, OER and KER. In: Proceeding of The 2001 PIPOC International Palm Oil Congress (Agriculture), Selangor. 20–22 Aug 2001, pp 167–191

Rees AR (1965) Evidence of the African origin of the oil palm. Principles 9:30–36

Reyes PA, Ochoa JC, Montoya C et al (2015) Development and validation of a bi-directional allele-specific PCR tool for differentiation in nurseries of Dura, Tenera, and Pisifera oil palm. Agron Colomb 33:5–10

Richardson DL, Alvarado A (2003) ASD oil palm germplasm from Nigeria. ASD Oil Palm Pap 26:1–32

Rival A, Beule T, Barre P et al (1997) Comparative flow cytometric estimation of nuclear DNA content in oil palm (*Elaeis guineensis* Jacq.) tissue cultures and seed-derived plants. Plant Cell Rep 16:884–887

Rosenquist EA (1984) Notes on inbreeding in the oil palm. Oil palm breeders meeting, Unifield T.C. Ltd. Mimeograph

Rosenquist EA (1985) The genetic base of oil palm breeding populations. International Workshop on Oil Palm Germplasm and Utilisation, Kuala Lumpur, 26–27 Mar 1985

Rutgers AAL, Blommendaal HN, Van Heurn FC, Heusser C, Mass JGJA, Tampolsky C (1922) Investigation of oil palm. Ruygrok & Co, Batavia

Sambanthamurthi R, Kalyana S, Tan Y (2000) Chemistry and biochemistry of palm oil. Prog Lipid Res 39:507–558

Sapey E, Adusei-Fosu K, Agyei-Dwarko D et al (2012) Collection of oil palm (*Elaeis guineensis* Jacq) germplasm in the northern region of Ghana. Asian J Agric Sci 4(25):325–328

Sayer J, Ghazoul J, Nelson P et al (2012) Oil palm expansion transforms tropical landscapes and livelihoods. Glob Food Secur 1:114–119

Schroeder MT, Becker EM, Skibsted LH (2006) Molecular mechanism of anti-oxidant synergism of tocotrienol and carotenoids in palm oil. J Agric Food Chem 54(9):3445–3453

Setiawati U, Rahmaningsih M, Breton F et al (2010) Screening of sumatra bioscience progenies for ganoderma partial resistance in field and nursery trial. Paper presented at international oil palm conference, Yogyakarta, 1–3 June 2010

Sheil D, Casson A, Meijaard E et al (2009) The impacts and opportunities of oil palm in Southeast Asia: what do we know and what do we need to know? CIFOR, Bogor

Shu QY, Forster BP, Nakagawa H (2012) Plant mutation breeding and biotechnology. CABI International and FAO, Dundee, p 608

Shuit SH, Tan KT, Lee KT (2009) Oil palm biomass as a sustainable energy source: Malaysian case study. Energy 34:1225–1235

Siew WL (2002) Palm oil. In: Gunstone FD (ed) Vegetable oil in food technology: composition, properties and uses. Wiley-Blackwell, New Jersey, pp 25–58

Singh R, Ong-Abdullah M, Low ETL et al (2013a) Oil palm genome sequence reveals divergence of interfertile species in old and new worlds. Nature 500(7462):335–339

Singh R, Low ET, Ooi LC et al (2013b) The oil palm SHELL gene controls oil yield and encodes a homologue of SEEDSTICK. Nature 500(7462):340–344

Singh R, Low ET, Ooi LC et al (2014) The oil palm *virescens* gene controls fruit colour and encodes a R2R3-MYB. Nat Commun 5:4106

Smith WK, Thomas JA (1973) The isolation and in vitro cultivation of cells of *Elaeis guineensis*. Oleagineux 28(3):123–127

Soh AC (1986) Expected yield increase with selected oil palm clones from current D x P seedling materials and its implications on clonal propagation, breeding and ortet selection. Oléagineux 41:51–56

Soh AC, Kee KK, Goh KJ (2006) Research and innovation towards sustainable palm oil production. J Sci Technol Trop 2:77–95

Soh AC, Choo KW, Yuk WH et al (2009) Oil palm. In: Vollmann J, Rajean I (eds) Oil crops, handbook of plant B SHU QY, FORSTER BP and NAKAGAWA. 2012. Plant mutation breeding and biotechnology. CABI International and FAO, Dundee, p 608

Soon RBF, Hang HW (2001) Oil palm response to N, P, K and Mg fertilizer on two major soil type in Sabah. In: Proceeding of the PIPOC Palm Oil Congress (Agriculture), Kuala Lumpur, 20–22 Aug 2001, pp 244–259

Sparnaaij LD, Menendez T, Blaak G (1963) Breeding and inheritance in the oil palm (*Elaeis guineensis*). Part I: the design of a breeding program. J W Afr Inst Oil Palm Res 4:126–155

Srisawat T, Kanchanapoom K, Pattanapanyasat K et al (2005) Flow cytometric analysis of oil palm: a preliminary analysis for cultivars and genomic DNA alteration. Songklanakarin J Sci Technol 27:645–652

Staritsky G (1970) Tissue culture of the oil palm (*Elaeis guineensis* Jacq.) as a tool for its vegetative propagation. Euphytica 19(3):288–292

Sterling F, Alvarado A (2002) Historical account of ASD's oil palm germplasm collection. ASD Oil Palm Pap 24:1–16

Sunilkumar K, Mathur RK, Sparjanbabu DS et al (2015) Evaluation of interspecific oil palm hybrids for dwarfness. J Plant Crops 43(1):29–34

Syed RA, Law IH, Corley RHV (1982) Insect pollination of oil palm: introduction, establishment and pollinating efficiency of *Elaeidobius kamerunicus* in Malaysia. Planter 58:547–561

Tampubolon FH, Daniel C, Ochs R (1989) Oil palm response to nitrogen and phosphate fertilizer in Sumatra. In: Proceeding of PORIM International Palm Oil Conference, Kula Lumpur, 5–9 Sept, pp 419–428

Tan YC, Yeoh KA, Wong MY et al (2013) Expression profiles of putative defence-related proteins in oil palm (*Elaeis guineensis*) colonized by *Ganoderma boninense*. J Plant Physiol 170:1455–1460

Tang Y, Li X, Chen PX et al (2015) Characterization of fatty acid, carotenoid, tocopherol/tocotrienol compositions and anti-oxidant activities in seed of three *Chenopodium quinoa* Willd. genotype. J Agric Food Chem 171:502–508

Taniputra B, Lubis AU, Pamin K et al (1987) Progress of oil palm industry in Indonesia in the last fifteen years (1971–1985). In: Hassan H et al (eds) Proceeding of international oil palm conference – progress and prospects, Kuala Lumpur, 29 June-1 July 1987. pp. 27–35

Teixeira JB, Söndahl MR, Nakamura T et al (1995) Establishment of oil palm cell suspensions and plant regeneration. Plant Cell Tissue Organ Cult 40(2):105–111

Tilbrook K, Gebbie L, Schenk PM et al (2011) Peroxisomal polyhydroxyalkanoate biosynthesis is a promising strategy for bioplastic production in high biomass crops. Plant Biotechnol J 9:958–969

Touraev A, Forster BP, Mohan Jain S (2009) Advances in haploid production in higher plants. Springer, Dordrecht

Tranbarger TJ, Dussert S, Joët T et al (2011) Regulatory mechanisms underlying oil palm fruit mesocarp maturation, ripening, and functional specialization in lipid and carotenoid metabolism. Plant Physiol 156(2):564–584

Troyer AF (1991) Breeding corn for the export market. Proceedings of the 26th Annual Corn and Sorghum Research Conference. Am Seed Trade Assoc 46:165–176

Tuo Y, Koua HK, Hala N (2011) Biology of *Elaeidobius kamerunicus* and *Elaeidobius plagiatus* (Coleoptera: Curculionidae) main pollinators of oil palm in West Africa. Eur J Sci Res 3:426–432

Turnbull N, de Franqueville H, Breton F et al (2014) Breeding methodology to select oil palm planting material partially resistant to Ganoderma boninense. Paper presented at the 5th quadrennial International Oil Palm Conference, Bali, 21 June 2014

Turner PD (1981) Oil palm disease and disorders. Oxford University Press, Kuala Lumpur

Ukai Y, Nakagawa H (2012) Strategies and approaches in mutant population development for mutant selection in seed propagated crops. In: Shu QY, Forster BP, Nakagawa H (eds) Plant mutation breeding and biotechnology. CABI International and FAO, Dundee, pp 209–221

Urtuvia V, Villegas P, Gonzales M et al (2014) Bacterial production of biodegradable plastics polyhydroxyalkanoates. Int J Biol Macromol 70:208–213

Virdiana I, Flood J, Sitepu B et al (2011) Integrated disease management to reduce future ganoderma infection during oil palm replanting. In: Proceeding of the PIPOC 2011 International Palm Oil Congress (Agriculture, Biotechnology and Sustainability), Kuala Lumpur 15–17 Nov 2011, pp 130–134

Vollmann J, Rajean I (eds) (2010) Oil crops. Handbook of plant breeding. Springer, New York

Wening S, Wilkinson MJ, Djuhjana J et al (2009) Genetic relationship studies of Sumatra bioscience oil palm germplasm using ISSR and AFLP profiles. In: Proceedings of PIPOC 2009, Kuala Lumpur, 9–12 Nov 2009, pp 903–912

Wening S, Wilkinson MJ Djuhjana J et al (2011) Genetic diversity of commercial oil palm seed production parents. Pin: proceedings of PIPOC 2011, Kuala Lumpur, 15–17 Nov 2011, pp 193–197

Wening S, Croxford AE, Ford CS et al (2012a) Ranking the value of germplasm: new oil palm (*Elaeis guineensis*) breeding stocks as a case study. Ann Appl Biol 160:145–156

Wening S, Fillianti H, Prasetyo JHH et al (2012b) Homozygosity of parental palm used in single seed descent programmes. Oil Palm Bull 65:1–5

Whitehead ADRA, Chapman GP (1962) Twinning and haploidy in *Cocos nucifera*. Nature 195:1228–1229

Zeven AC (1965) Oil palm groves in southern Nigeria. J W Afric Inst 1965

Zhang Y, Sun W, Wang H et al (2013) Polyhydroxybutyrate production from oil palm empty fruit bunch using *Bacillus megaterium* R11. Bioresour Technol 147:307–314

Chapter 9
Sugarcane (*S. officinarum* x *S. spontaneum*)

Marvellous Zhou

9.1 History of Cultivation

Sugarcane has been cultivated for thousands of years, beginning in the prehistoric era (Ming et al. 2006). Sugar was mainly produced from three *Saccharum* species. *S. officinarum* was the main species grown in New Guinea, *S. barberi* in India and *S. sinense* in China. The species *S. barberi* was distributed from India to various countries in the Middle East, the Mediterranean and the New World in the late fourteenth century. *S. officinarum*, also known as noble cane, was the main sugar-producing species and is characterised by thick stalks with high-sucrose content and only found under cultivation in the tropical regions of the world. The rediscovery of the *S. officinarum* clones in the eighteenth century provided the basis of commercial production of sugarcane in many countries (Berding et al. 2004) and led to rapid spread of sugar industries in the tropical and subtropical regions (Ming et al. 2006).

Sugarcane is a plant that accumulates high sucrose content in its stalk. Sugarcane is a perennial grass that does not tolerate severe frosts. Sugarcane centre of origin is in Southeast Asia around New Guinea where farmers chewed sugarcane plant for its sweet juice (Barnes 1964). The earliest known sugar production began in Northern India. Sugarcane cultivation spread to tropical and subtropical regions of the world. Currently, sugarcane is grown in south-western Europe, Africa, Asia, Australia, the USA, Mexico and Southern America (FAOSTAT 2014). Most of the world's sugarcane is grown between 22°N and 22°S and some up to 33°N and 33°S of the equator extending from tropical to subtropical zones.

Brazil is the largest producer of sugar with countries like China, India and Pakistan within the top five producers. Sugarcane cultivation plays a significant role in the economy of many countries. Brazil, India and China are prominent producers

M. Zhou (✉)
Department of Plant Breeding, South African Sugarcane Research Institute,
Mount Edgecombe, South Africa
e-mail: Marvellous.Zhou@sugar.org.za

H. Campos, P.D.S. Caligari, *Genetic Improvement of Tropical Crops*,
DOI 10.1007/978-3-319-59819-2_9

of sugar, while India and China are major consumers of sugar, and these three countries control the world markets (Gopinathan 2010). Approximately 70% of the world sugar is from sugarcane and 30% from sugar beet. Sugarcane is also becoming a major biofuel crop with ethanol being produced from sugarcane juice, while the residual fibre can be burned to generate electricity (Leal et al. 2013).

9.2 Biology

The ability of the sugarcane plant to produce and accumulate high sucrose content in the stems has made it a major crop for sugar production (Alexander 1985). Several studies on sugarcane morphology (Artschwager 1925, 1930, 1940, 1951; van Dillewijn 1952) provide comprehensive description of the plant. Sugarcane has the C4 pathway of photosynthesis (Hatch 2002). The C4 pathway provides a highly efficient mechanism of converting carbon dioxide to carbohydrates.

The only time sugarcane is propagated from true seed is during plant breeding. During propagation in commercial planting and testing stages in plant breeding, the crop is planted from vegetative material. The vegetative propagation provides a stable genetic makeup with no recombination after crossing.

The sugarcane plant is made up of the roots, stalks, leaves and flowers (van Dillewijn 1952). The roots form the underground portion of the plant and are used to anchor the whole plant. The roots are divided into anchor roots that provide support for the plant and superficial roots that are largely used for extracting nutrients and water from the soil (Barnes 1964).

After germination, the plant forms shoots and tillers. The tillers develop into stalks where sugar is stored during photosynthesis. Tillering and number of stalks per plant vary across genotypes and form the basic yield component for cane yield. The stalk is made up of the node and internode. The node is where the leaves attach to the stalk and also the location of the buds. The internode is located between the nodes and is made up of large storage tissue for sucrose. The buds are located at the base of each internode and are the growing point from where the plant shoot emerges at planting.

The major components of the sugarcane leaf are the sheath, midrib and leaf blade (Nelson 2011). The leaf sheath wraps around the stalk and is attached at the node and acts as an anchor for the leaf blade. The blade is attached to the leaf sheath and grows on either side of the midrib. The leaf midrib is made up of thick tissues and offers support or anchorage to the leaf blade. The leaf sheath and blade sizes vary across genotypes. The leaf sheath is sometimes loosely attached in some genotypes that are referred as self-thrashing.

Flower induction in sugarcane is controlled by photoperiod. During flowering, the sugarcane plant apical meristem changes from vegetative to reproductive (van Dillewijn 1952; Moore 1971; Clements 1975). After the transition, the formation of leaves ceases and the sugarcane flower develops, signalling the end of vegetative growth. The flower will eventually emerge in May to August in the Southern

Hemisphere. The biology of flowering in sugarcane represents a significant constraint to its breeding efforts, and in order to address its needs of a narrow photoperiod and temperature range, most sugarcane breeding programmes have developed either very specialised crossing stations, sometimes far away from breeding activities and commercial planting, or sophisticated chambers or greenhouse-like infrastructure to control temperature and photoperiod. Both approaches add significant expenses and logistic complexity to sugarcane breeding programmes.

9.3 Genetics

Saccharum species have a complex genome and are characterised by high levels of polyploidy. Several species make up the *Saccharum* complex. These include *Saccharum officinarum*, *Saccharum spontaneum*, *Saccharum robustum*, *Saccharum barberi*, *Saccharum sinense* and *Saccharum edule*. The chromosome numbers vary across species. The chromosome number of *S. officinarum* is $2n = 80$ with a basic chromosome number of 10. *Saccharum spontaneum* has a chromosome number of $2n = 40$–128 (Sreenivasan et al. 1987), with a basic chromosome number of eight or ten. *Saccharum robustum* is a diverse sugarcane species that grows in the wet tropics and is known to have $2n = 60$ and $2n = 80$ chromosome numbers. *Saccharum barberi* and *S. sinense* are thought to be intergeneric hybrids produced by interbreeding of other species. *Saccharum barberi* has a chromosome number of $2n = 111$–120 and *S. sinense* have $2n = 80$–124 chromosomes (Daniels and Roach 1987), and both are speculated to be hybrids of *Saccharum* spp. complex. *Saccharum edule* is a cultivated species thought to be a product from introgression of *S. officinarum* or *S. robustum* with other species. The chromosome number of S. *edule* is $2n = 60$–80 with prevalent aneuploidy. Modern cultivars are made up of 70–80% of chromosomes derived entirely from *S. officinarum*, and 10% to 20% are from *S. spontaneum*, with 10% to 20% recombinations (Grivet et al. 1996; D'Hont et al. 1996; Piperidis and D'Hont 2001). The modern cultivar range in chromosome numbers from 90 to 130.

Modern sugarcane hybrids are polyploids made up of two genomes. The two genomes that have contributed to modern hybrids are *S. officinarum* and *S. spontaneum*. *S. officinarum* is the source of sugar, while *S. spontaneum* is the source of genes for wide adaptability and disease and pest resistance. *S. spontaneum* is found in almost all continents and is widely adapted to several environments including deserts, swamps and hot and cold environments. As a result, *S. spontaneum* is a source of genes for wide adaptability to modern sugarcane commercial hybrids. *S. officinarum* is found only in Papua New Guinea, its centre of origin. In its centre of origin, *S. officinarum* is found in gardens where it has been cultivated for centuries. Additionally, very few clones of *S. officinarum* exist, and all are found under domestication. In contrast, *S. spontaneum* is found in all continents and in the wild. Several clones of *S. spontaneum* exist in the wild and in wild species collections of several breeding programmes.

Sugarcane polyploidy ranges from 8 to 14 copies of chromosomes, with individual chromosomes and alleles in varying numbers. The genome of modern sugarcane cultivars is highly polyploid (~12×), characterised by frequently unbalanced number of chromosomes which is also known as "aneuploidy" (D'Hont 2005). Its size and complexity represent a major challenge for the isolation of agronomical important genes (Sreenivasan et al. 1987). The nature of polyploidy varies with sugarcane species. For example, *S. edule* is a form of aneuploidy. *S. officinarum* is not a simple polyploid, and it is both allopolyploid and autopolyploid which behaves like a diploid (Stevenson 1965).

9.4 Germplasm Intra- and Interspecific

Sugarcane germplasm is made up of *Saccharum* complex and related species. The *Saccharum* complex is made up of *S. officinarum*, *S. spontaneum*, *S. robustum*, *S. sinense*, *S. barberi* and *S. edule*. Other related species include *Erianthus*, *Miscanthus*, *Narenga* and *Sclerostachya* (Daniels and Roach 1987). Two species, *S. spontaneum* and *S. robustum*, are found growing in the wild. The other four species, *S. officinarum*, *S. barberi*, *S. sinense* and *S. edule*, are only found under cultivation. Therefore, the truly wild *Saccharum* are *S. spontaneum* and *S. robustum*, while the rest are largely cultivated species. *S. officinarum* was cultivated for centuries in New Guinea, while *S. barberi* was cultivated in India and *S. sinense* was cultivated in China. *S. officinarum* was used widely for sugar production and originated from New Guinea and Melanesian or Polynesian islands (Brandes 1958; Mukherjee 1957). Some scientists believe *S. officinarum* evolved from *S. robustum* (Grivet et al. 2006). The low fibre and low impurities in juice of *S. officinarum* make it a suitable source of genetic material for developing cultivars for commercial sugar processing.

Interspecific hybridisation between *S. officinarum* and *S. spontaneum* and further backcrossing to *S. officinarum* in a process known as nobilisation produced most of the modern sugarcane cultivars (Price 1963). Two to three backcross cycles to *S. officinarum* produced genotypes with high sucrose content. Recent introgression breeding has continued to utilise *S. officinarum* as a donor parent. *Erianthus* genus has attracted a lot of interest among sugarcane breeders because of its ability to impart resistance to pests and diseases. Resistance to borers and white grubs (Mukunthan and Nirmala 2002) has been reported in research. *Miscanthus* has also attracted interest in recent years as a source of genes for biomass breeding for second-generation ethanol production.

9.5 Crossing Approaches

Crossing approaches used are aimed at increasing genetic diversity of commercial traits. The crossing approaches aim to maximise potential of additive genetic effects for the several traits of commercial importance. Therefore, parent selection identifies those genotypes that possess high breeding values. The breeding value of parents is determined using progeny data obtained from family plots in the early stages of clonal selection plots (Aitkin et al. 2009). Only parents known to have high breeding values are selected for planting in crossing facilities. The parental genotypes are classified in populations or groups, and each group is made up of genotypes that possess high breeding values of the required traits. Parent selection then selects individual genotypes from these groups which are then subjected to photoperiod treatment in matching treatments. Because of variable flowering and days to flowering in subtropical countries such as South Africa, crossing between groups with known desirable trait combinations helps achieve better cross combinations than aiming for predesigned biparental crosses.

At SASRI, the selected parents are planted in six photoperiod treatments, three in the photoperiod house and three in the glasshouse. The photoperiod treatments are designed to have early, mid- and late flowering each for the photoperiod house and glasshouse. The early flowering treatment in the photoperiod house is matched with an early flowering treatment in the glasshouse. Each genotype is planted to at least six plants in each target photoperiod treatment to ensure the probability of recovering a flower for crossing. The photoperiod house is used for generating male parents, while the glasshouse is used for generating female parents. Male parents are expected to excel in the production of pollen and should possess at least 30% pollen viability measured using iodine pollen stain. The photoperiod house is used for producing male parents because there is better control of temperature, humidity and light to induce flowering as well as maintain pollen viability. The female parents are expected to produce very little pollen, less than 30%. Further, female parents should have very low levels of selfing. Selfing is tested by collecting seed from a flower that has not been cross-pollinated and determining the amount of viable seed compared to another flower that has been cross-pollinated. High level of selfing is undesirable in female parents. The photoperiod treatment allocation is based on knowledge of data collected from previous genotype flowering time. Early flowering genotypes in the glasshouse are designed to produce female flowers to be crossed with early flowering male flowers from the photoperiod house.

The glasshouse also acts as a crossing facility at SASRI. The glasshouse is subdivided into compartments where crosses are made. The compartment acts as a barrier to prevent pollen contamination between adjacent crosses. During crossing, the male plant is placed above the female plant in the compartment. Every morning between 6 and 8 AM, the male flower is gently shaken to allow the pollen to be released and drift downwards to the female anthers below to facilitate crossing. The temperature in the glasshouse is maintained between 18 and 25 °C, and humidity is maintained above 70%, the optimum temperature and humidity for pollen development and viability. After 2 weeks from crossing, pollen shedding reduces significantly, and the female plant is then transferred to a ripening area under the same temperature and humidity control to facilitate seed development, maturity and ripening. The matured seed is harvested, dried and stored in vacuum-sealed bags at −20 °C until required for sowing. The male plant is discarded.

Three mating designs are commonly used at time of crossing in sugarcane. These are biparental crosses, melting pot and male only. With biparental crosses, one female genotype is crossed with one male genotype. This mating design provides greatest genetic control as the source of pollen is known. This design also provides better evaluation of parental genotypes. The second mating design used is the melting pot where several male plants from different genotypes are used to pollinate a single female plant. The source of pollen remains unknown. This type of design can be used with, for example, at least two male genotypes that possess several desirable traits required from the progenies but not possessed by the female parent. The design has also been used during introgression breeding where several male genotypes are crossed to a single female with desirable background genetic makeup. The seed is collected from the female parent and the male plants are discarded. The third crossing design is the male only. With male only, at least two male genotypes that both produce high pollen are allowed to outcross and exchange pollen. Seed is harvested from both plants. The pollen source is unknown particularly when more than two parents are involved. This design is used to sample different cross combinations. The melting pot and male only are sometimes combined during crossing. In this case, several male plants donate pollen to a female plant as well exchange pollen among themselves. Seed is collected from the female plant as well as the male plants.

9.6 Breeding Approaches

Sugarcane being a clonal plant limits the breeding schemes that can be applied. However, since the beginning of the modern error of sugarcane breeding, a few breeding schemes have been used successfully, and new schemes continue to evolve in future. Backcross breeding has been applied in early days of nobilisation (Price 1963). During backcross breeding, *S. officinarum* clones (female parent) were crossed with *S. spontaneum* (male parent) to produce F1 progenies. The progenies were evaluated, and selected progenies backcrossed to *S. officinarum* as the female parent to produce BC1 population. The objective of backcrossing was to eliminate undesirable *S. spontaneum* genes and increase desirable genes from *S. officinarum* such as high sucrose content and low fibre. The progenies were further evaluated and selected. The selected progenies were then again backcrossed to *S. officinarum* to produce BC2 populations. The process is repeated until the progenies have high sucrose content with good traits from *S. spontaneum* such as wide adaptability, high ratooning ability and pest and disease resistance.

The second breeding scheme used has been the proven cross and proven parent system. This breeding scheme was used to evaluate crosses and parents (Skinner et al. 1987). The proven cross was implemented by evaluating the number of advanced genotypes across the different stages of a breeding programme. Crosses that produced the highest numbers of advanced genotypes across stages were classified as proven crosses and were repeated several times, and more seedlings were raised from these crosses for further selections. In the same system, parents that produced proven crosses were also considered as proven parents and were used more frequently in future crosses. The advantage was that very little statistics other than advancement numbers were used to determine proven cross and proven parents. Major advances in several breeding programmes were made using the proven cross system (Heinz and Tew 1987; Skinner 1982; Sukarso 1986). However, the disadvantage was that it took years to have advancement data and quantify the value of proven crosses and parents. Further, because of the method of collecting data, bias towards few families also occurred because the so-called proven cross continued to be planted in large numbers at the expense of testing new crosses and parents creating a potential to limit introduction of new genetic diversity in breeding programmes (Walker 1963).

In recent years, family evaluation and selection has created a new breeding scheme that evaluates families or crosses using objective data (Jackson and McRae 1998; Kimbeng and Cox 2003; Pedrozo et al. 2011; Zhou 2014). With family evaluation, progenies from a family are planted in replicated family plots. Data collected are analysed to determine elite families from where selection for elite progenies is focused. The same data used for family evaluation are also used to evaluate the parents that produced the families. Breeding value which is the ability of a parent to produce elite families is estimated from the family data. The advantage of family evaluation is that new families can be objectively compared to known elite families using objective data and a statistic. With respect to parent evaluation, breeding values of new parents are also compared to those of known parents to determine their value in breeding programmes. The bias associated with proven cross and parent is significantly reduced, and new crosses and parents get evaluated objectively.

Another breeding scheme adopted is introgression breeding. A lot of interest in sugarcane introgression breeding has increased to broaden genetic diversity of breeding and selection populations (Berding and Roach 1987; Jackson et al. 2014). Greater interest to introgression with *S. spontaneum*, *Erianthus* and *Miscanthus* has increased. *S. spontaneum*, because of its wide genetic diversity emanating from the wide diversity of clones from diverse centres of origin, has been exploited to diversify the genetics of commercial breeding populations (Berding and Roach 1987; Hale et al. 2010, 2014). The diversity being exploited is expected to increase the adaptability range of future commercial cultivars, create high biomass cultivars for bioenergy (Wang et al. 2008) and increase pest (Jackson and Dunkelman 1974; White et al. 2001) and disease resistance (Koike 1980; Hale et al. 2010; Costet et al. 2012). *Erianthus* genus introgression is being used to tap into its high levels of pest and disease resistance and high biomass production (Jackson and Henry 2011). Several breeding programmes have incorporated these introgression breeding schemes.

9.7 Traits

In sugarcane, traits of economic importance include cane yield, sucrose content, yield stability and ratooning ability, pest and disease resistance, early maturity and agronomic characteristics. The importance of the different traits varies widely across different environments. Cane yield is the primary trait that is important to the sugarcane grower and determines the quantity of raw material supplied for sugar processing. Another commercially relevant trait is the tonnage of sugar per hectare. Generally, growers prefer high cane yield. Associated with cane yield is the yield stability, which is the ability of a cultivar to produce high yield across different environments such as soil types, management, seasons and weather-determined growing conditions. High yield stability provided sustainability for the sugarcane grower (Zhou and Shoko 2012a). Ratooning ability refers to the practice used in sugarcane growing where several crops are harvested from the same planting (Zhou and Shoko 2012b). With ratooning, after harvesting, the underground stems produce shoots and roots and regrow to produce the next crop. In extreme cases, some cultivars such as NCo310 (Donovan 1996) and NCo376 (Nuss 2001; Zhou 2004) are known to have produced high cane yield after more than 35 years of sugarcane ratooning. Ratooning significantly increases sugarcane production economics because it costs significantly less to maintain a crop than to plant and establish. The expensive land preparation required and sourcing vegetative planting material that is bulky and costly to deliver to the planting field all add to the value of ratooning in sugarcane production. Therefore evaluation and selection for ratooning require that field trials be harvested for more than two ratoon crops in variety development trials in South Africa.

Sugar is the product that is marketed in sugarcane production. Sugar yield is a product of cane yield and sucrose content. Therefore breeding for high sucrose content is another important objective in sugarcane (Lingle et al. 2010). High sucrose content cultivars that produce high cane yield increase economics of sugarcane production. Sucrose content is also associated with crop maturity. Cultivars that mature early and those that accumulate high sucrose content during unfavourable growing conditions are more desirable. During cultivar evaluation, sucrose content is also tested from 4 months before the targeted month of harvesting to determine maturity trends and curves that are in turn used to guide time of harvest of the cultivar when recommended for release. When sugarcane is grown further away from the processing factories, it is more efficient to transport per unit of sugar produced to high-sucrose cultivars for processing. High-sucrose cultivars also increase factory processing efficiency because it is more efficient to extract sugar than in low-sucrose cultivars. High-sucrose cultivars also possess low fibre, further reducing the loss of sugar in waste fibre during processing.

Pest and disease reduce yield and quality by damaging the plant in the field. In South Africa, the major pest is *Eldana saccharina* (*eldana*) stem borer (Atkinson and Nuss 1989; Rutherford 2015). The stem borer causes significant losses in yield and quality by tunnelling the internodes and feeding of the nutrients in the plant tissues. It is the larvae stage of the borer that causes economic damage. Chemical control of the larval stage is impossible because the larvae stay and feed inside the stalk. Therefore chemical control is only possible for the adult moth which emerges from the stalk. As a result, repeated chemical sprays are required to target repeated adults emerging from the stalk.

In South Africa, the major diseases of sugarcane are mosaic (Sugarcane mosaic virus), smut (*Sporisorium scitamineum.*), rusts (*Puccinia melanocephala*; *Puccinia fulvous*; *Puccinia kehnii*), ratoon stunting disease (*Leifsonia xyli*) and pokkah boeng (*Fusarium spp.*). Smut causes the plant to produce spindly stalks that are not millable resulting in yield losses. The infected plants produce smut whips that carry and spread billions of spores that further infect adjacent plants increasing the infection and losses to smut. While smut can be controlled by chemical dipping at planting and rouging smut-infected stalks before they spread disease, resistant varieties form an important part of integrated disease management. Mosaic infects the plant through aphids and infects the leaves resulting in reduced growth of plants and loss of yield. Leaf rusts include brown and tawny rust in South Africa. Orange rust, although not present in South Africa, is spreading to most sugarcane-growing regions. Rusts that infect leaves cause death of leaf tissue, reducing the photosynthetic area which leads to loss of yield. Breeding for these diseases provides the most effective control measure. The outbreak of orange rust in Australia and Brazil resulted in a major shift in cultivars and significant efforts to breed for resistance. Recent outbreak of smut in Australia resulted in intensive breeding for smut resistance. These outbreaks caused major panic and yield losses but also highlighted the impact of breeding for resistance in sugarcane.

9.8 Field Trialling and Selection Approaches

Sugarcane breeding starts with crossing to generate variability and is followed by several field selection and testing stages. Several breeding programmes across the world have adopted different field selection and testing stages. However, the overall aim of these stages is to identify clones with superior combinations of commercial traits and to test their adaptability in the target environments for commercial production. In general, all breeding programmes start with seedlings germinated from true seed. The populations of seedlings undergo selection through several stages until fewer and most adapted genotypes with desirable commercial potential are identified and recommended for release. In this review, the field trialling and selection approaches used at the South African Sugarcane Research Institute (SASRI) breeding programmes will be described. Where possible, references will be made to similarities and differences with other breeding programmes.

The SASRI breeding programmes have five stages starting with seedlings (Table 9.1). True seeds for all breeding programmes are germinated in a glasshouse located at Mount Edgecombe research station in Durban. Crosses made from elite parents are selected for sowing. Crosses tested in previous trials that have shown potential are selected for planting in large numbers to generate populations for selecting commercial genotypes. Additionally, crosses made from test parents are planted in small numbers to test their commercial potential.

Before sowing, trays are prepared with three layers made up of the bottom layer of concrete stones, followed by a layer of sieved river sand and then a top layer of

Table 9.1 Stages of the SASRI breeding programmes

Stage	Description	Genotypes	Plot size	Replication
1	Seedlings	250,000	1 plant	Families replicated
2	Mini-lines	175,000	1 row × 1 m	Families replicated
3	Single lines	20,000	1 row × 8 m	Families replicated
4	Observation	2000	2 rows × 8 m	Three replication per genotype
5	Variety trial	125	5 rows × 8 m	At least 3 reps and 5 locations

peat moss. Peat moss is a decomposed plant material with very high organic matter. The fuzz seed of the selected crosses is sown on the top layer of peat moss. The fuzz is manually spread by hand on the layer of peat moss. The spread fuzz is then covered with a layer of peat moss. The sown seen is taken into a glasshouse and laid on rails and then watered regularly. The temperature in the glasshouse is maintained between 30 and 35 °C for optimum germination. The seed starts germinating after 36 h. After 1 week from sowing, seedlings are counted to determine the germination. The seedlings are then taken outside the glasshouse for further growth and hardening. At 5 weeks after germination, the seedlings are shipped to the respective breeding stations for transplanting into bricks.

The germination estimates are used to generate experimental designs for the seedlings to be transplanted into bricks. The seedling numbers in a tray (made up of one cross) are divided by three. Each set of seedlings in a tray is randomly assigned to one of three replicates to generate a randomised block design. This design is used to plant seedling in bricks and is carried across the next stages (1, 2, 3) where individual genotypes are not replicated.

Each brick is made up of six holes, each hole being 5 cm wide by 5 cm long and 5 cm deep. The bricks are laid on a concrete slab. The plant material is made up of a mixture of sand, topsoil and compost in a 1:1:1 ratio. Each seedling in a cross is planted in a hole. Seedlings from a plot within a replicate will be from an individual cross. After transplanting, the seedlings are watered three times a day using an automated irrigation system. NPK fertiliser is applied in a ratio of 5:1:5 every week to replenish the nutrients leached by frequent irrigation. The uniform soil media provide for uniform growth of seedlings. At this stage, the crosses of families are evaluated for growth vigour and diseases. The most frequently observed disease is brown rust. Seedlings showing symptoms of brown rust are discarded from the seedling nursery, a form of selection. The seedlings are left in the nursery for 9 months until they form internodes and stalks that are at least 1 m long. The 1 m stalks per seedlings are harvested and bundled together for transplanting in the field (Table 9.1).

In stage 2 (mini-lines), the 1 m stalks are planted in 1 m long plots spaced 1 m apart within the row and 1.1–1.5 m apart between two adjacent rows. The rows are set up in a tramline fashion where two adjacent rows are followed by an unplanted row. The unplanted row provides access to the plots for data collection and assessment during visual selection. The seedlings in the field are managed like commercial crops. Therefore the first field stage in the breeding programme is planted like clonal crops. The mini-lines are left to grow until crop maturity for each of the breeding programmes. During growth, regular inspections are done, and incidences of diseases such as smut, mosaic, rust, pokkah boeng are recorded. At crop maturity, the first 20 mini-lines in each plot are used to collect data. The total number of stalks for each mini-line is counted; the height of the mini-line and the diameter of three representative stalks are measured. The stalk parameters are used to estimate the cane yield of each mini-line using an empirical formula (Miller and James 1974; Gravois et al. 1991; Chang and Milligan 1992). One stalk each is randomly sampled from each of the first 20 mini-lines in a plot and bundled together. Each stalk is assessed for *eldana* entry and exit holes, and the number of stalks with *eldana* entry and exit holes is recorded. The bundles are analysed for quality parameters such as Brix and sucrose and fibre content at the sucrose laboratory. The data for yield, quality, *eldana* and disease scores is analysed for family effects. Families that possess the best combinations of yield, quality, *eldana* and disease scores are identified from where visual selection of genotypes with desirable commercial potential is focused. More genotypes are selected from families with higher potential.

Stage 3 (single lines) is planted from genotypes selected from stage 2. Each genotype is planted to 1 row by 8 m long spaced 1.1–1.5 m between rows, depending on the agro-ecological region. After every 20 genotypes, a control cultivar is planted. The crop is managed in a similar way like a commercial sugarcane crop. During crop growth, disease inspections and scores are recorded regularly for all the important diseases. At crop maturity, the number of stalks is counted, plant heights are measured at three positions in the plot and stalk diameter is measured from three random stalks. These measurements are used to estimate cane yield using empirical formula (Miller and James 1974; Gravois et al. 1991; Chang and Milligan 1992). Twelve random stalks are randomly chosen from each plot and used to estimate quality parameters from the sucrose laboratory. These 12 stalks are also examined for *eldana* entry and exit holes, and the number of bored stalks is recorded. The data is analysed using logistic regression procedures in SAS to estimate selection prediction (Zhou 2013). Genotypes with optimum combinations of the commercial traits are selected using analysis output and visual field assessment for further testing.

Stage 4 (observation variety trials) is planted from elite genotypes selected from stage 3. Each genotype is planted to three replications. The designs used are incomplete block designs because of the large number of genotypes as well as the desire to control field variability. Alongside the trial, a propagation plot of each genotype included in the trial is planted. The trials are managed in a similar way to commercial crops. During crop growth, the trials are regularly inspected for diseases. At crop maturity, harvesting involves cutting and weighing each plot separately. The cane is weighed using a hydraulic operated and tractor-mounted scale. Twelve stalks are randomly picked from each plot and sent to the laboratory to determine quality parameters. Twenty stalks from each plot are randomly chosen and split. The number of internodes on each stalk is counted. The number of internodes with *eldana* entry and exit holes is counted and recorded and used to estimate percentage of *eldana* bored internodes. The data is collected in the plant and ratoon crop. The data collected is subjected to combined analysis of plant and first ratoon and used to determine the genotypes that possess the best combination of commercial traits. The genotypes that possess the best combination of desirable traits must also produce high yields in both plant and first ratoon crops. The selected genotypes are planted in the final stage of testing.

Stage 5 (variety trials) is planted to the elite genotypes with the greatest chance of achieving commercial potential. Each trial is designed using incomplete block designs to increase efficiency of accounting for field variability. Each trial is planted at three replications. Each trial series is planted at five locations to sample the variations in the agro-ecological regions for differences in soil type and other production conditions. The genotypes are tested for genotype by environment interaction (GxE) and wide adaptability. The trial data are collected in the plant, first and second ratoon crops to measure ratooning ability. The same set of genotypes is planted in *eldana*, smut and mosaic screening trials to determine reaction to pest and diseases for commercial recommendation. Data for cane yield, quality, *eldana* and diseases is collected as for observation variety trials. The combined analysis across trials and ratoons is used to identify the most adaptable genotypes that are recommended for release as cultivars.

9.9 Tissue Culture and Clonal Production

The only stage in sugarcane breeding planted from true seed is the seedling stage. All other stages are propagated from vegetative cuttings from the stalks. Therefore stages 2, 3, 4 and 5 are planted from vegetative material. As a result, no genetic recombination occurs after crossing. In each breeding cycle, only one opportunity for genetic recombination exists at crossing.

Tissue culture has also been used to accelerate propagation of cultivars. Tissue culture has been adopted in several countries including South Africa because it facilities propagation of large numbers of plants in a very short time. Tissue culture also provides the ability to produce disease-free plantlets for commercial seed cane production and allows the movement of disease-free planting materials over large distances at lowest cost as plantlets. However, the capital equipment and expertise to have a fully functioning laboratory are high for small breeding programmes.

9.10 Seed Production and Marketing

In sugarcane, there is very limited controlled seed production and marketing. The most critical seed cane is the initial seed cane source after a variety has been released. After release, because of the vegetative nature of the plant propagation, most sugarcane growers propagate their own seed cane material for their commercial plantings. However, there are some fundamental procedures that are recommended to ensure good quality seed cane and limit the spread and movement of pests and diseases as well as ensure good establishment required to guarantee good and sustainable commercial yields.

Firstly, two major stages of commercial seed cane production are recognised. These are foundation and commercial seed cane source. The foundation seed cane is propagated from seed cane material that has been hot water treated. Hot water treatment is used to control and eliminate ratoon stunting disease, a bacterial disease of sugarcane that causes yield reduction in ratoon crops. Hot water treatment also kills *eldana* larvae, further ensuring *eldana*-free plant material. The hot water-treated vegetative material is planted into a nursery. This nursery will provide seed cane for planting the commercial nursery. The commercial nursery acts as a source of seed cane for commercial plantings. Seed cane marketing is localised as growers exchange seed cane among themselves.

9.11 Biotechnology

Advances in biotechnology have produced several new technology and techniques that will increasingly be applied in plant breeding. Molecular breeding is increasingly being used to supplement and complement conventional sugarcane breeding. Plant breeders and molecular biologists are increasing exploring mechanism and methods of integrating molecular techniques with conventional breeding. In sugarcane breeding, the applications of molecular technology particularly DNA markers have been in fingerprinting clones for precise identification (Pan 2010). DNA markers have also been used for determining hybrids after crossing and the parents of progenies (D'Hont et al. 1995). The application of DNA markers in genetic diversity studies (Alwala et al. 2005; Arro et al. 2006) has aided in importation of new clones as well its utilisation.

Progress in marker trait association studies (Zhou et al. 2012) and QTL studies (Alwala et al. 2009) has shed light on potential identification of markers for future applications in marker-assisted selection. Despite the wide range of research on genetic linkage maps (Alwala et al. 2008; Aitken et al. 2007), most of the markers identified have proved not useful for application in marker-assisted selection. One of the reasons is the limited genetic coverage of markers derived from linkage maps constructed from biparental crosses. Recently, genome-wide association studies have offered potential for identifying potential markers that have potential for application in marker-assisted selection. Markers associated with *eldana* damage are being explored and have offered potential. However, these markers are still at the validation stage and require further testing in larger and independent populations.

Genetic markers such as Bru1 gene (Daugrois et al. 1996) have been applied in several breeding programmes (McCord and Migneault 2016) where they have helped quantified the level of brown rust resistance genetic background in breeding populations. But more markers are required that cover several important economic traits. Genomic selection offers exciting potential in the future and is expected to increase estimation of breeding values of parents as well as predict genetic gains for traits under selection.

9.12 Future Prospects and Outlook

The greatest challenge facing sugarcane breeding is achieving sustainable increase in cultivar genetic gains (Edme et al. 2005; Zhou and Gwata 2016; Burnquist 2013). Currently, sugarcane achieves some of the lowest genetic gains compared to other crop species (Burnquist 2013). The long breeding cycles in sugarcane limit the ability to accelerate genetic gains. Because sugarcane is a clonally propagated crop, the only opportunity to create genetic recombinations occurs at crossing. All subsequent stages of cultivar development are propagated from vegetative material with no further opportunity for genetic recombination.

Family evaluation has been proved to increase efficiency of breeding programmes in several studies (Jackson and McRae 1998; Kimbeng and Cox 2003; Pedrozo et al. 2011; Zhou 2014). However, the greatest challenge facing several sugarcane breeding programmes has been the ability to collect data required to evaluate families. Further, collecting this data for several traits of commercial importance has created challenges. The cost of data collection and the equipment required such as automatic weighing machines are expensive and in certain parts of the world are not locally manufactured and available. Coupled with data collection for family evaluation, data analysis techniques are limited particularly for resource-poor breeding programmes with limited access to statisticians and biometricians as well as the software required to efficiently perform the analysis. Future developments of data collection methods and equipment that allows measurements from large populations would enhance family evaluations.

Parent evaluation using family data has faced similar challenges to family evaluation. Future outlook should focus on data analysis to estimate breeding values of genotypes. Incorporating genealogy data (Aitkin et al. 2009) in estimates of breeding values is known to increase the accuracy of estimates. However, software and training for the analysis is limited and largely unavailable to some breeding programmes where expertise is limited or unavailable. Developing these analytical skills will increase efficiency of several sugarcane breeding programmes.

Introgression breeding to tap the genetic diversity that exists in wild species of sugarcane is expected to increase the genetic diversity of breeding and selection populations and ultimately produce cultivars with wider genetic diversity. However, introgression efforts are limited and sometimes impeded by the inability to synchronise flowering between wild species and commercial-type clones. This phenomenon is more complex in subtropical countries where sugarcane does not flower naturally every year. Further, introgression is also limited by the limited availability of wild germplasm collections. Future outlook of introgression can be improved through subtropical breeding programmes performing their crossing in tropical countries on contract. Further, increase in sharing of wild germplasm will benefit diversifying breeding germplasm across several sugarcane breeding programmes.

Genetic diversity of breeding, selection and cultivar populations is required to provide genes required to cope with increase in pests and diseases of sugarcane. In recent years, pests such as yellow sugarcane aphid, thrips, mites, *chilo* (*Chilo sacchariphagus*), *eldana*, sugarcane borer (*Diatraea saccharalis*) and Mexican rice borer (*Eoreuma loftini*) have spread into new environments and spread within environments to cause more economic damage to commercial crop. Diseases such as tawny rust and orange rust are spreading across the sugarcane areas increasing yield losses in commercial crops. It is through preserving the diversity of populations that allow breeding programmes to develop future varieties that possess resistance to sustain commercial production.

Marker identification and marker-assisted breeding will provide valuable tools to increase efficiency of sugarcane breeding. Firstly, identifying robust markers that can predict trait performance is required. Genome-wide association studies offer potential to increase genome coverage in search of markers and genes associated with important commercial traits. Efficient markers will increase genetic gains of breeding for traits such as pest and disease resistance. However, the challenge is collecting accurate and sufficient phenotype data particularly for cryptic pests such as *eldana* that are manpower intensive to quantify damage. Mechanisms to make such data collection efficient and fast will increase accuracy and availability of data required for marker identification, validation and testing in large breeding populations.

Determining and tracking population and cultivar genetic gains is required in the future to ascertain the efficiency of breeding programmes. Quantifying these genetic gains also provides the breeders with mechanism to validate the appropriateness and effectiveness of breeding programmes. Genetic gains also provide a measure of the value or return on investment in breeding programmes. They also guide investors in breeding programmes to invest more resources with a reasonable projection of the economic future value that can be derived. Estimates of genetic gains can also be used to solicit funding for breeding programmes because they can help put monetary future value associated with genetic resources.

Niche breeding is expected to further provide opportunity for increasing genetic gains in breeding programmes. Niche breeding means genetic resources are directed and deployed in specific environments where maximum genetic gains can be achieved. Further, niche breeding means smaller breeding programmes per environment that will be adequate to achieve the needs. Further, because of small populations, shortening breeding programmes will be possible. Shortening breeding programmes is also known to accelerate genetic gains and increase the transmission of benefits from breeding programmes to growers and communities. Further, such breeding programmes because of expected increased efficiencies will also lead to reduced costs of breeding programmes and therefore a higher return on investment.

References

Aitken KS, Jackson PA, McIntyre CL (2007) Construction of genetic linkage map of *Saccharum officinarum* incorporating both simplex and duplex markers to increase genome coverage. Genome 50:742–756

Alexander AG (1985) The energy cane alternative. Elsevier Science Publishers, Amsterdam

Alwala S, Kimbeng CA, Veremis JC (2008) Linkage mapping and genome analysis in a *Saccharum* interspecific cross using AFLP, SRAP and TRAP markers. Euphytica 164:37–51

Alwala S, Kimbeng CA, Veremis JC, Gravois KA (2009) Identification of molecular markers associated with sugar related traits in a Saccharum interspecific cross. Euphytica 167:127–142

Alwala S, Suman A, Arro JA, Veremis JC, Kimbeng CA (2005) Target region amplification polymorphism (TRAP) for assessing genetic diversity in sugarcane germplasm collections. Crop Sci 46(1):448–455

Arro J, Veremis JC, Kimbeng CA, Botanga C (2006) Genetic diversity and relationships revealed by AFLP markers among Saccharum spontaneum and related species and genera. JASSCT 26:101–115

Artschwager E (1925) Anatomy of the vegetative organs of sugar cane. J Agric Res 30:197–221

Artschwager E (1930) A comparative study of the stem epidermis of certain sugarcane varieties. J Agric Res 41:853–865

Artschwager E (1940) Morphology of the vegetative organs of sugarcane. J Agric Res 60:503–509

Artschwager E (1951) The role of the ligule in sugarcane taxonomy. Am J Bot 38:144–146

Atkin FC, Dieters MJ, Stringer JK (2009) Impact of depth of pedigree and inclusion of historical data on the estimation of additive variance and breeding values in a sugarcane breeding program. Theor Appl Genet 119:555–565

Atkinson PR, Nuss KJ (1989) Association between host-plant nitrogen and infestation of the sugarcane borer, *Eldana saccharina* Walker (*Lepidoptera: Pyralidae).* Bull Entomol Res 79:489–506

Barnes AC (1964) The sugar cane Botany, cultivation and utilization. World Crops Book. Nescience Publishers, New York, p 456pages

Berding N, Hogarth DM, Cox M (2004) Plant improvement of sugarcane. In: Glyn J (ed) Sugarcane. Blackwell Science, Oxford, pp 20–53

Berding N, Roach BT (1987) Germplasm collection, maintenance and use. Sugarcane improvement through breeding. Elsevier, Amsterdam, pp 143–210

Brandes E (1958) Origin, clasification and characteristics. In: Artschwager E, Brandes EW (eds) Sugarcane (Saccharum officinarum L), vol 1-35. USDA handbook, Washington, pp 260–262

Burnquist WL (2013) Sugarcane research and development: a view from the private sector. Proc Int Soc Sugar Cane Technol 28:1–7

Chang YS, Milligan SB (1992) Estimating the potential of sugarcane families to produce elite genotypes using univariate cross prediction methods. Theor Appl Genet 84:662–671

Clements HF (1975) Flowering of sugarcane: mechanics and control, Hawaii Agricultural Experiment Station Technical Bulletin No 92. University of Hawaii, Honolulu

Costet L, Raboin LM, Payet M, D'Hont A, Nibouche S (2012) A major quantitative trait allele for resistance to the sugarcane yellow leaf virus (Luteoviridae). Plant Breed 131:637–640

Daniels J, Roach BT (1987) A review of the origin and improvement of sugarcane. Copersucare International Sugarcane Breeding Workshop, Brazil, pp 1–32

Daugrois JH, Grivet L, Roques D, Hoarau JY, Lombard H, Glaszmann JC, D'Hont A (1996) A putative major gene for rust resistance linked with an RFLP marker in sugarcane cultivarR570. Theor Appl Genet 92:1059–1064

D'Hont A (2005) Un-ravelling the genome structure of polyploids using FISH and GISH: examples of sugarcane and banana. Cytogenet Genome Res 109:27–33

D'Hont A, Rao P, Feldmann P, Grivet L, Islam-Farridi N, Taylor P, Glaszmann JC (1995) Identification and characterisation of intergeneric hybrids, *S. officinarum* × *Erianthus arundinaceus* with molecular markers and in situ hybridization. Theoretical and Applied Genetics 91:320–326

D'Hont A, Grivet L, Feldmann P, Rao S, Berding N, Glaszmann JC (1996) Characterisation of the double genome structure of modern sugarcane cultivars Saccharum spp by molecular cytogenetics. Mol Gen Genet 250:405–413

Donovan PA (1996) An empirical evaluation of the sugarcane variety NCo310. Proc South Afr Sugarcane Technol Assoc 70:93–102

Edme SJ, Miller JD, Glaz B, Tai PYP, Comstock JC (2005) Genetic contribution to yield gains in the Florida sugarcane industry across 33 years. Crop Sci 45:92–97
FAOSTAT (2014) Crop production data, food and agriculture organization of the United Nations, available online at http://faostat.fao.org/site/567/default.aspx#ancor. Accessed 19 Oct 2016
Gopinathan MC (2010) Research and technology transfer strategies for the next decade an Indian example. Proc Int Soc Sugar Cane Technol 27:1–15
Gravois KA, Milligan SB, Martin FA (1991) Indirect selection for increased sucrose yield in early sugarcane testing stages. Field Crop Res 26:67–73
Grivet L, D'Hont A, Roques D, Feldmann P, Lanaud C, Glaszmann JC (1996) RFLP mapping in cultivated sugarcane *(Saccharum* spp.): genome organization in a highly polyploid and aneuploid interspecific hybrid. Genetics 142:987–988
Grivet L, Glaszmann JC, D'Hont A (2006) Molecular evidences for sugarcane evolution and domestication. In: Motely T, Zerega N, Cross H (eds) Darwin's harvest. New approaches to the origins, evolution and conservation of crops. Columbia University Press, New York, pp 49–66
Hale A, Hoy J, Veremis J (2014) Identification of freeze tolerant Saccharum spontaneum accessions through pot-based study in sugarcane germplasm enhancement for adaptation to temperate climates. Biomass Bioenergy 61:53–57
Hale AL, Viator RP, Veremis J (2010) Identification of sources of resistance to sugarcane red rot. Proc Int Soc Sugar cane Technol 27:1–8
Hatch MD (2002) C4 photosynthesis: discovery and resolution. Photosynth Res 73:251–256
Heinz DJ, Tew TL (1987) Hybridization procedures. In: Heinz DJ (ed) Developments in crop science 11: sugarcane improvement through breeding. Elsevier, New York, pp 313–342
Jackson R, Dunkelman P (1974) Relative resistance of Saccharum spontaneum clones to the sugarcane borer. Proc ISSCT 14:513–515
Jackson P, Hale A, Bennett G, Lakshmanan P (2014) Sugarcane. Alien Gene transfer in crop plants, vol 2. Springer, New York, pp 317–345
Jackson PA, Henry R (2011) Erianthus. Wild crop relatives: genomic and breeding resources. Springer, Heidelberg, pp 97–107
Jackson PA, McRae TA (1998) Gains from selection of broadly adapted and specifically adapted sugarcane families. Field Crop Res 59(3):151–162
Kimbeng CA, Cox MC (2003) Early generation selection of sugarcane families and clones in Australia. A review. J Am Soc Sugar Cane Technol 23:23–29
Koike H (1980) Evidence of resistance in Saccharum spontaneum and Saccharum related genera to sugarcane mosaic virus strains H and I. Proc ISSCT 17:1523–1527
Leal MV, Walter A, Seabra JA (2013) Sugarcane as an energy source. Biomass Convers Biorefinery 3:17–26
Lingle SE, Johnson RM, Tew TL, Viator RP (2010) Changes in juice quality and sugarcane yield with recurrent selection for sucrose. Field Crop Res 118:152–157
McCord PH, Migneault AJ (2016) Genotyping sugarcane for the Brown rust resistance locus *Bru*1 using unlabeled probe melting. SugarTech 18:401–406
Miller JD, James NI (1974) The influence of stalk density on cane yield. Proc ISSCT 15:177–184
Ming R, Moore P, Wu K (2006) Sugarcane improvement through breeding and biotechnology. Plant Breed Rev 27:15–118
Moore PH (1971) Investigation on the flowering of Saccharum I. Ontogeny of the inflorescence. Can J Bot 49:677–682
Mukherjee SK (1957) Origin and distribution of Saccharum. Bot Gaz 199:55–61
Mukunthan N, Nirmala R (2002) Tolerance to white grub, Holotrichia serrata in Saccharum and Erianthus spp. Sugar Cane Int 22:26–28
Nelson T (2011) The grass leaf developmental gradient as a platform for a system understanding of the anatomical specialisation of C4 leaves. J Exp Bot 62:3039–3048
Nuss KJ (2001) The contribution of variety NCo376 to sugar production in South Africa from 1955 to 2000 and its value as a parent in the breeding programme. Proc South Afr Sugarcane Technol Assoc 75:154–158

Pan YB (2010) Databasing molecular identities of sugarcane (Saccharum spp) clones constructed with microsatellite (SSR) DNA markers. AJPS 1:87
Pedrozo CA, Barbosa MHP, da Silva FL, de Resende MDV, Peternelli LA (2011) Repeatability of full-sib sugarcane families across harvests and the efficiency of early selection. Euphytica 182:423–430
Piperidis G, D'Hont A (2001) Chromosome composition analysis of various Saccharum interspecific hybrids by genomic in situ hybridization GISH. Proc Int Soc Sugar Cane Technol 24:565–566-47
Price S (1963) Cytogenetics of modern sugarcane. Econ Bot 17:97–105
Rutherford RS (2015) IPM for Eldana control: an Integrated Pest Management (IPM) approach for the control of the stalk borer *Eldana saccharina* Walker (Lepidoptera: Pyralidae). ISBN 1-874903-41-7. South African Sugarcane Research Institute, Durban, South Africa
Skinner JC (1982) Efficiency of bunch-planted and single-planted seedlings for selecting superior families in sugarcane. Euphytica 31(2):523–537
Skinner JC, Hogarth DM, Wu KK (1987) Selection methods, criteria and indices. In: Heinz DJ (ed) Sugarcane improvement through breeding. Elsevier, Amsterdam, pp 409–453
Sreenivasan TV, Ahloowalia BS, Heinz DJ (1987) Cytogenetics. In: Heinz DJ (ed) Sugarcane improvement through breeding. Elsevier, Amsterdam, pp 211–251
Stevenson GC (1965) Genetics and the breeding of sugarcane. Longman, London, p 284
Sukarso G (1986) Assessment of family selection in original seedlings of sugarcane at Pasuruan. Proc Int Soc Sugar Cane Technol 19:440–446
Van Dillewijn C (1952) Botany of sugarcane. HH Veenaman and Zonen, Wageningen, p 371
Walker DIT (1963) Family performance at early selection stages as a guide to the breeding programme. Proceedings of the International Society of Sugarcane Technologists 11, pp 469–482
Wang LP, Jackson PA, Li FS (2008) Evaluation of sugarcane x progeny for biomass composition and yield components. Crop Sci 48:951–961
White W, Miller J, Milligan S, Burner D, Legendre B (2001) Inheritance of sugarcane borer resistance in sugarcane derived from two measures of insect damage. Crop Sci 41:1706–1710
Zhou MM (2004) Performance of varieties N14 and NCo376 in the South East Lowveld of Zimbabwe. Proc South Afr Sugarcane Technol Assoc 78:153–160
Zhou MM (2013) Using logistic regression model for selection in non-replicated sugarcane breeding populations. Euphytica 191:415–428
Zhou MM (2014) Family evaluation for sugarcane yield, using data estimated from stalk number, height and diameter. J Crop Improv 28(3):406–417
Zhou MM, Gwata ECT (2016) Quantifying sugarcane cultivar genetic gains in the midlands region of South Africa. Agron J 108:342–348
Zhou MM, Shoko MD (2012a) Simultaneous selection for yield and ratooning ability in sugarcane genotypes using analysis of covariance. South Afr J Plant Soil 29(2):93–100
Zhou MM, Shoko MD (2012b) Simultaneous selection for yield and stability in sugarcane using parametric statistics. J Agric Sci Technol 2(4):400–410
Zhou MM, Suman A, Kimbeng CA (2012) Molecular markers associated with starch content and implications for sugarcane introgression breeding using *Saccharum spontaneum*. J Agric Sci Technol 2(11):1127–1137

Index

H. Campos, P.D.S. Caligari, *Genetic Improvement of Tropical Crops*,
DOI 10.1007/978-3-319-59819-2

C

H

Printed by Printforce, the Netherlands